BRITISH CERAMIC PROCEEDINGS
NO. 59

ENGINEERING WITH CERAMICS

These proceedings are dedicated to the memory of Prof. Roger Davidge (1936–1997).

ENGINEERING WITH CERAMICS

Edited by

W. E. Lee
Dept of Engineering Materials,
University of Sheffield

and

B. Derby
Dept. of Materials,
University of Oxford

British Ceramic Proceedings
No. 59

Book 713
First published in 1999 by
IOM Communications Ltd
1 Carlton House Terrace
London SW1Y 5DB

ISBN 1-86125-087-8

IOM Communications Ltd
Is a wholly-owned subsidiary of
The Institute of Materials

Typeset by IOM Communications Ltd

Printed and bound in the United Kingdom
at the Univesity Press, Cambridge

Contents

FRACTURE, EROSION AND WEAR

NANOCOMPOSITES

PROCESSING

REFRACTORIES

Recollections of Roger Davidge

It is an honour to be asked to provide remarks and personal recollections about Roger. Both of us had the same thesis advisor (Peter Pratt), but did not overlap. So even as a PhD student I was aware of this person who wrote a superb thesis on deformation phenomena in alkali halides. Subsequently, I was flattered to be offered my first job in his group at Harwell. I worked for Roger for four years, which for me comprised the career-making experience. He introduced me to many things; especially ceramics and fracture mechanics. I will always be indebted. The qualities that Roger showed through his career and personal life set standards of quality, integrity and generosity for me and others taken under his wing. He encouraged free thinking, welcomed new ideas and nurtured the progress of those fortunate enough to be in his group. His critiques were incisive, but made with a lightness that encouraged more dialogue. This style brought the best out of his colleagues and disciples and provided a pattern to emulate.

The quality and breadth of his professional contributions are captured by his book as well as several seminal articles. I particularly recall his work on stress/time/probability (STP) diagrams, which has profoundly affected the thinking about the design and implementation of ceramics. That effort started while I was still at Harwell, so I witnessed the beginning. Then Roger put together an impressive collection of ideas into a concise visualization of the failure envelope. We miss his technical contributions and his enthusiasm for the field, as well as his personal warmth and encouragement.

Professor A. G. Evans
Harvard University, USA
September 1998

Foreword

Roger Davidge was one of the UK's most distinguished and influential ceramic scientists and a kind and considerate man. His death at the early age of 61 has been a source of sadness to his many friends in the ceramics community, both nationally and internationally. To acknowledge the great significance of his contributions in the field the Ceramic Science Committee of the Institute of Materials dedicated the 1998 Annual Ceramic Convention Symposium on Engineering With Ceramics, held in Cirencester on April 21st and 22nd, to his memory. Roger conducted a wide range of studies in fracture, toughness, fatigue, thermal shock and the statistics of mechanical failure of structural ceramics making an outstanding contribution to our knowledge of these subjects. The broad theme of the symposium was naturally mechanical behaviour and these proceedings contain many papers covering this aspect. Developments in techniques for measuring mechanical properties, as well as results from ceramics ranging from nanocomposites to multilayer capacitors, are included. Further papers dealing with fracture, erosion, wear, processing and refractories would all have been of interest to Roger. We hope these proceedings are a fitting tribute.

W. E. Lee and B. Derby, August 1998

KEYNOTE ADDRESSES

Characterisation of Sub-Critical Crack Growth in Ceramics Using Indentation Cracks

DAVID J. GREEN and PRIYA J. DWIVEDI*

The Pennsylvania State University, University Park, PA 16802, USA

VINCENZO M. SGLAVO

Università di Trento, I-38050 Trento, ITALY

ABSTRACT

The introduction of indentation cracks into brittle materials has proved to be a useful tool in characterising sub-critical crack growth, notably in efficiently measuring the kinetic growth parameters and in defining whether the material exhibits a fatigue threshold. The accuracy of the sub-critical crack growth parameters obtained using indentation mechanics can be excellent, provided the stress intensity factor associated with the indentation cracks is well characterised. Indentation cracks can also be used to measure crack velocity as a function of the applied stress intensity factor by direct observation. In such testing, it is critical that the changes in crack shape as the crack extends are known or accurately predicted. Numerical simulations suggest that the shape changes can be influenced not only by the testing geometry but also the growth kinetics. Finally, it is shown that the fatigue threshold can be determined by allowing median cracks to extend sub-critically during indentation.

1 INTRODUCTION

In order to utilise ceramics in technological applications, it is critical to characterise any sub-critical crack growth that can occur in the chosen material as this can define the lifetime of the material. Over the last 20 years a design methodology has evolved in which the initial strength distribution of a material is coupled with the parameters that define the sub-critical growth behaviour.[1-4] This approach allows one to define the lifetime of a component that is subjected to any stress history in terms of a failure probability. The aim of the current paper is to review the ways in which indentation cracks can aid in characterising sub-critical crack growth, notably in (1) measuring the kinetic parameters associated with the growth and (2) defining whether the material exhibits a threshold for this growth. It will be shown that indentation techniques represent a powerful tool for the characterisation of the fatigue behaviour of brittle materials. Surface cracks can be easily introduced in brittle materials, such as glass, using a sharp hardness indenter, e.g., Vickers. Although these cracks are intentionally introduced, they act in a similar fashion to 'natural' flaws with the advantage of being more reproducible and yet, large enough to be observed by an optical microscope.

* Now with Corning Inc., Corning, NY 14831, USA

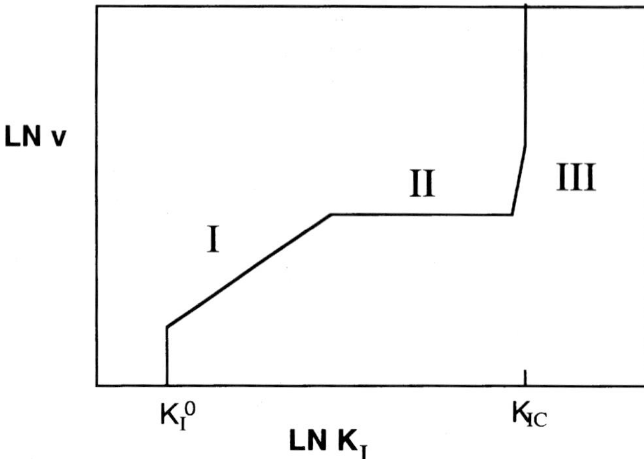

Figure 1 Schematic dependence of sub-critical crack growth velocity on applied stress intensity factor (log–log plot).

2. BACKGROUND

Figure 1 shows schematically the form of the sub-critical crack growth behaviour that is often depicted in ceramics.[1–3] The figure describes the crack velocity v as a function of the applied stress intensity factor K. At low values of K, there often appears to be a threshold value K_I^0 below which crack growth does not occur. At higher values of K, there is a strong sensitivity of v on the value of K. This region (Region I) is usually associated with chemically assisted crack growth and is found to depend on the concentration of the environmental species that is 'aiding' the growth process. A plateau region (Region II) often follows and this is often associated with the inability of the reacting species to keep pace with the crack tip motion (transport limited growth). The final region (Region III) relates to the crack growth behaviour in vacuum and is usually associated with thermal activation of bond rupture. At high temperatures, it is also found that localised creep damage can give rise to sub-critical crack growth.

For the engineering use of ceramics, the primary emphasis has been on the threshold and Region I crack growth behaviour. An engineering component will usually spend most of its 'lifetime' in these regions. It has been become popular to use an empirical power law to describe this behaviour for $K \geq K_I^0$, in which the crack velocity is expressed as

$$v = v_0 \left(\frac{K}{K_{IC}} \right)^n \tag{1}$$

where n and v_0 are termed the sub-critical crack growth parameters. These parameters will be dependent on the choice of material, environment and temperature. To analyse the strength degradation as a result of sub-critical crack growth, it is often assumed that only Region I sub-critical crack growth occurs with no threshold. This may seem rather restrictive but threshold values often occur at such low fractions of K_{IC}, that ignoring its presence has a

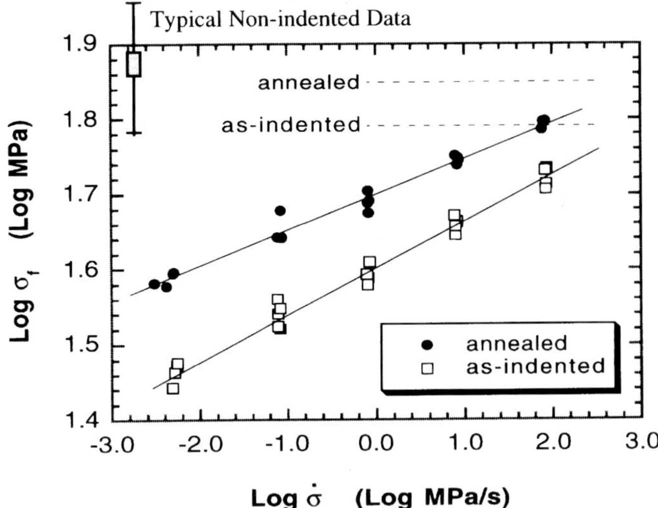

Figure 2 Dynamic fatigue plot for as-indented and annealed specimens. Dashed lines represent the strength measured in the inert environment. Included is a typical example (with standard deviation) of data obtained from normal strength (i.e., non-indented) samples.

negligible (conservative) effect on the fatigue behaviour and lifetime predictions. Clearly, it is still critical within this approach to show that assuming the threshold to be absent is justified and hence, it is useful to at least estimate the value of K_I^0 for a given situation. The presence of a threshold can also be of interest to designers, since for applied stresses below this value, strength becomes time invariant, i.e., the strength does not degrade. In the lifetime analysis for ceramic components, eqn (1) is often coupled with a statistical (e.g., Weibull) description of the initial strength distribution.[1-4] This approach has now been developed by NASA into a sophisticated software package, CARES/Life.[5,6]

3. DETERMINATION OF SUB-CRITICAL CRACK GROWTH PARAMETERS

The sub-critical crack growth parameters are usually determined using dynamic or static fatigue tests. The former approach is often preferred as it involves shorter testing times. An attractive approach is to determine the dynamic strength with specimens containing Vickers indentation cracks and the theoretical approach for analysing such data has been performed previously.[7,8] Figure 2 shows dynamic fatigue data for a commercial soda–lime–silica glass[9] and it is shown the use of indentation cracks significantly reduces the scatter associated with the strength values, thus allowing data to be gathered more efficiently. This use of indentation cracks does complicate the data analysis as the stress intensity factor must include a residual stress component. For some materials this component can be removed and the analysis simplified, e.g., by annealing, as was performed for some of the data in Fig. 2.[9]

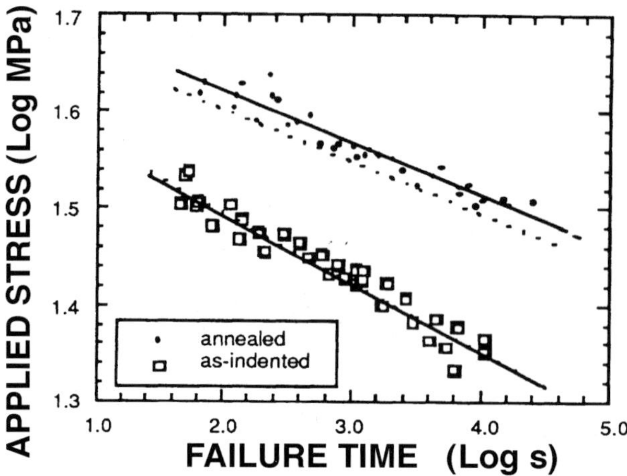

Figure 3. Static fatigue plot for as-indented and annealed specimens. Dashed and solid lines represent the theoretical lifetime predictions calculated by using constant and variable ψ and χ values, respectively.

The stress intensity for a Vickers indentation crack is usually written

$$K = \frac{\chi P}{c^{3/2}} + \sigma \psi c^{1/2} \tag{2}$$

where P is the indentation load, c the crack depth, σ the applied stress, χ the residual stress factor and ψ the crack shape factor. Table 1 shows a comparison of the v_0 and n values obtained from various types of dynamic fatigue tests for a commercial soda-lime-silica (SLS) glass.[9] The values of χ and ψ in eqn 2 are often assumed to be constant but as shown by Sglavo and Green this can lead to significant discrepancies in lifetime predictions.[9–10] For example, Fig. 3 shows static fatigue data for the same SLS glass that was used to obtain the dynamic fatigue data in Fig. 2. It is found that the prediction of the static fatigue results from the values in Table 1 can lead to some discrepancies, particularly for the annealed data. Sglavo and Green[11] have shown that during sub-critical growth, the radial indentation cracks can undergo a complex interaction with the lateral cracks, which will influence the values of χ and ψ in eqn (2). Sglavo and Green[10] have, therefore, suggested an approach to determine the variations in χ and ψ and this methodology leads to an improved prediction of both the static (see Fig. 3) and cyclic fatigue behaviour of SLS glass.[9–10]

An alternative approach to determining the sub-critical crack growth parameters is to use pre-cracked (fracture mechanics) specimens and directly measure crack velocity as a function of the applied stress intensity factor. Dwivedi and Green[12] have shown that Vickers indentation cracks can be used in this approach but have suggested the cracks be extended beyond the region influenced by residual stresses and lateral cracks. This allows the residual stress term to be omitted from eqn 2 and hence simplifies the analysis. Using a special experimental set-up, Dwivedi and Green[12] measured the crack velocity associated with the surface trace of the median-radial cracks under a constant applied load in bending. In this work, they noted that significant changes can occur in crack shape as the crack

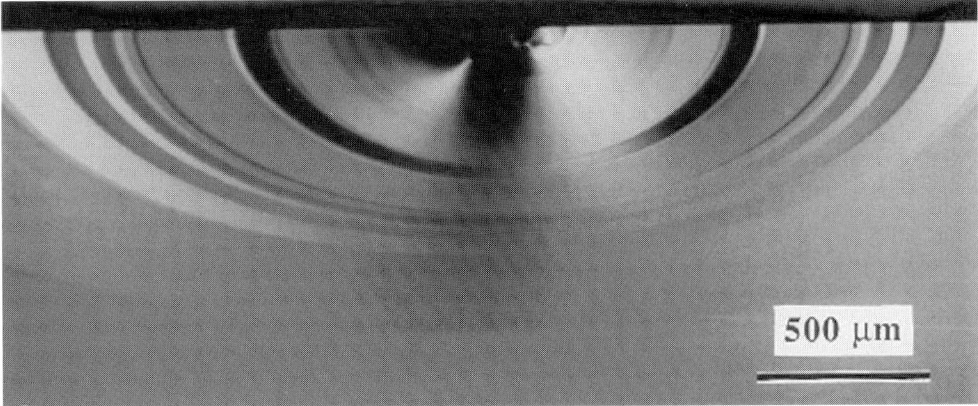

Figure 4 The change in the shape of an indentation crack as it extends in a sub-critical fashion. The test was performed in bending on a soda–lime silica glass.

Table 1 Fatigue parameters for as-indented and annealed SLS glass specimens calculated from dynamic fatigue tests in deionised water (Sgalvo and Green).[9]

Specimen Condition	n	v_0 (mm/s)
As-indented	20.1 ± 0.7	28.8 ± 6.4
Indented + annealed	19.9 ± 0.7	6.4 ± 1.4

extends (Fig. 4). The shape changes were measured experimentally and this allowed the correct value of ψ to be used in eqn (2) using the analysis of Newman and Raju.[13] Figure 5 shows the stress intensity factor–crack velocity relationships for a commercial SLS glass and the results are summarised in Table 2.

As an extension of this work, Dwivedi and Green[14] considered whether the changes in crack shape could be predicted *a priori*, given the sub-critical crack growth parameters for the SLS glass (i.e., the inverse problem). The simulation was initiated with a semi-circular crack of a size similar to that produced by indentation. The crack velocity was then determined from the stress intensity factor values at the surface and the point furthest away from the surface. After a small time interval, the process was re-iterated so the evolution in crack shape could be predicted. The results are shown in Fig. 6 and excellent agreement was found. Similar computations were also performed using different starting crack sizes and shapes. In all cases, the cracks were found to initially change shape rapidly but then followed an 'equilibrium' path. These simulations were extended to show that the change in crack shape depended on the mode of loading and on the sub-critical crack growth exponent.[14] For example, the crack shape would not be expected to change as rapidly in uniaxial tension as bending, because there is not a stress gradient acting along the crack path as the crack moves away from the surface.

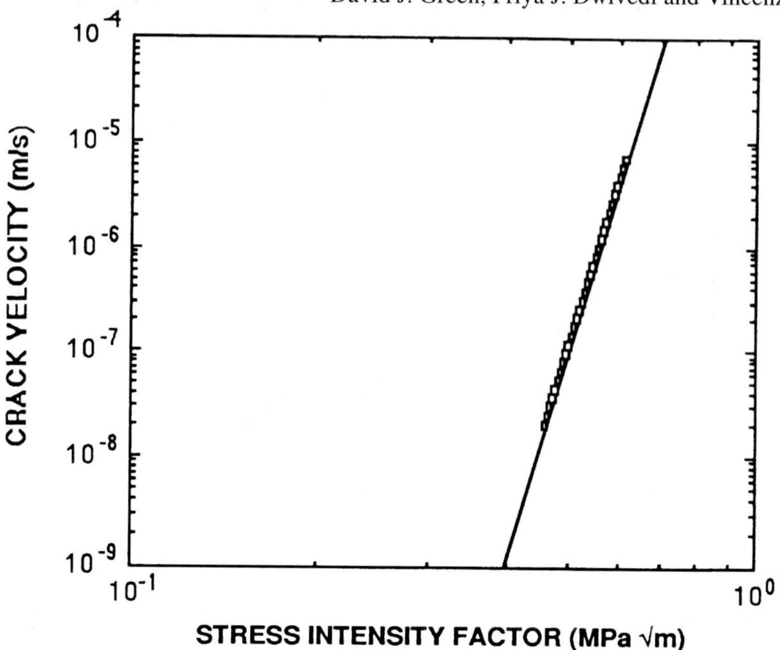

Figure 5 Stress intensity factor/crack velocity (schematic) plot obtained from direct observations of indentation crack growth. The data were obtained in ambient conditions (e.g., 26.7 °C, 65% relative humidity) on a soda–lime silica glass.

Table 2 Fatigue parameters for a commercial SLS glass specimens calculated from direct observation of indentation crack growth and dynamic fatigue, ambient conditions (Dwivedi and Green).[12]

Specimen Condition	n	v_0 (mm/s)
Direct	20.4	0.3
Dynamic Fatigue		
– natural flaws	22	2.6
Dynamic Fatigue		
– indentation flaws	21	2.4

4. THRESHOLD STRESS INTENSITY FACTOR MEASUREMENTS

As pointed out earlier, it is important in characterising sub-critical crack growth that some knowledge be gained about whether there is a threshold for this growth. Initial evidence for the presence of a threshold in SLS glass was found in static fatigue testing of strength specimens. This 'endurance limit' was typically 0.15 to 0.40 times the strength measured in liquid nitrogen.[15-21] The measurement of this threshold often involves long testing times and is complicated by the scatter in strength. In an attempt to reduce testing times, a test

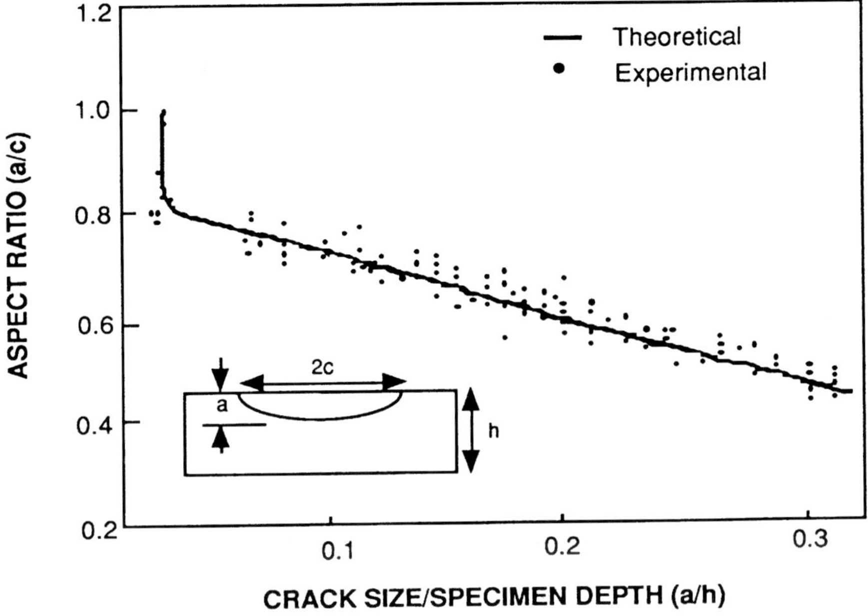

Figure 6 The change in the aspect of an indentation crack as it extends in a sub-critical fashion from observations such as shown in Fig. 4. The data are compared to a numerical simulation that predicted the shape changes from the stress intensity factors at the surface intersection and the deepest point on the crack front.

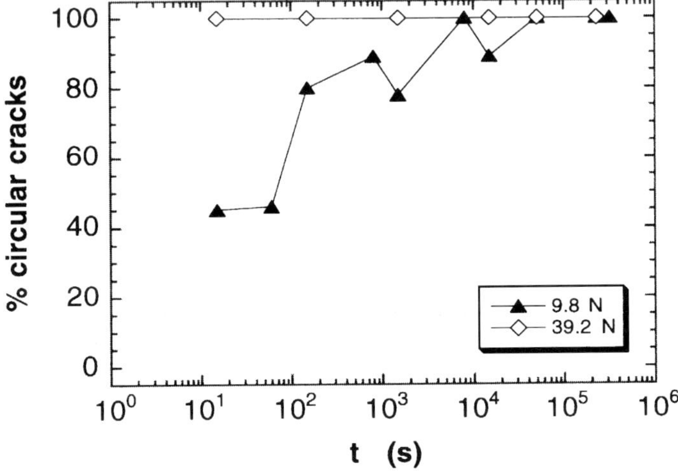

Figure 7 The percentage of median cracks formed as a function of indentation load and time for soda–lime–silica glass.

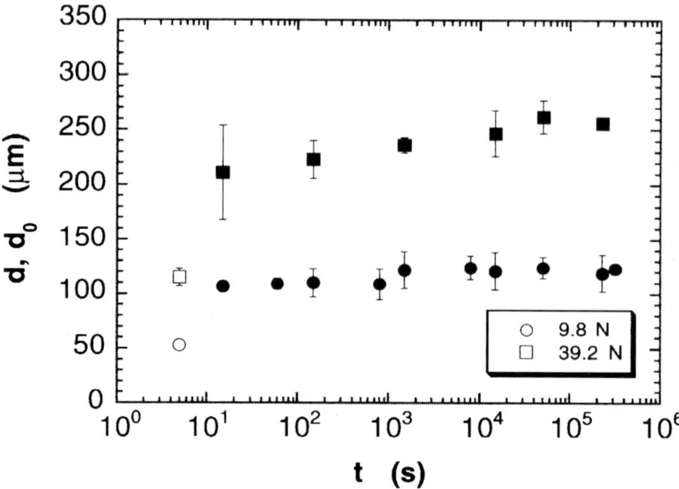

Figure 8 Crack depth measurements for soda–lime–silica glass as a function of indentation time at maximum load. The empty symbols represent the depth measured on indentation cracks obtained in silicone oil. (The latter data are plotted at t = 5 s for improved clarity).

Table 3 Geometric stress factors and threshold stress intensity factors for soda-lime-silica (SLS) and soda–alumina–silica (SAS) glasses.

	χ (9.8 N)	χ (3.92 N)	K_{th} (MPa √m) (9.8 N)	K_{th} (MPa √m) (39.2 N)
SLS	0.031 ± 0.004	0.027 ± 0.006	0.23 ± 0.03	0.25 ± 0.02
SAS	0.029 ± 0.006	0.029 ± 0.009	0.24 ± 0.04	0.27 ± 0.05

known as Interrupted Static Fatigue (ISF) is sometimes used. In this test, specimens are held at a constant load for a given time period and specimens (not broken during the stress hold) are broken by a subsequent increasing stress. The basic premise is that if the specimens that survive the stress are not weakened, the applied stress was below the fatigue limit. The difficulty with this approach is that under some conditions, the strength loss even above the threshold can be too small to measure. The ISF test was recently analyzed theoretically by Sglavo and Green[22] and two testing procedures were proposed. These procedures both involved establishing that a proposed threshold value shows invariance as the hold time increases. These ideas were subsequently applied to a commercial SLS glass tested in water[23] but even with testing times as long as 20 days, no definitive evidence of the threshold was established.

Evidence for a fatigue limit in silicate glasses has also been found in tests using artificially-cracked specimens in which v is measured directly as a function of K. These tests are often performed using the Double Cantilever Beam geometry and values of the threshold

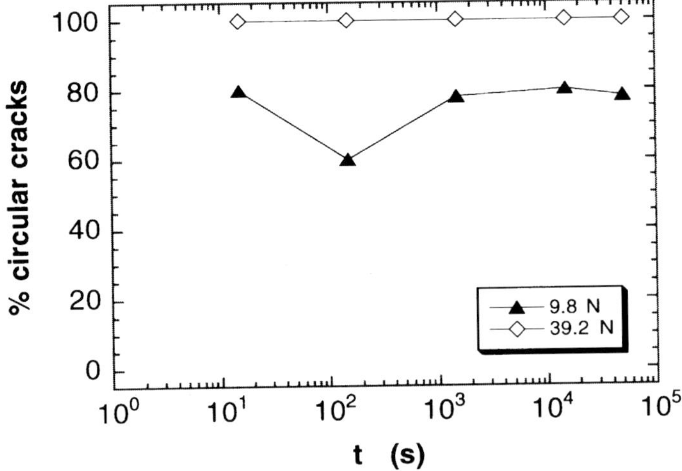

Figure 9 The percentage of median cracks formed as a function of indentation load and time for soda–lime–silica glass in the acidic solution (HCl 0.3N)

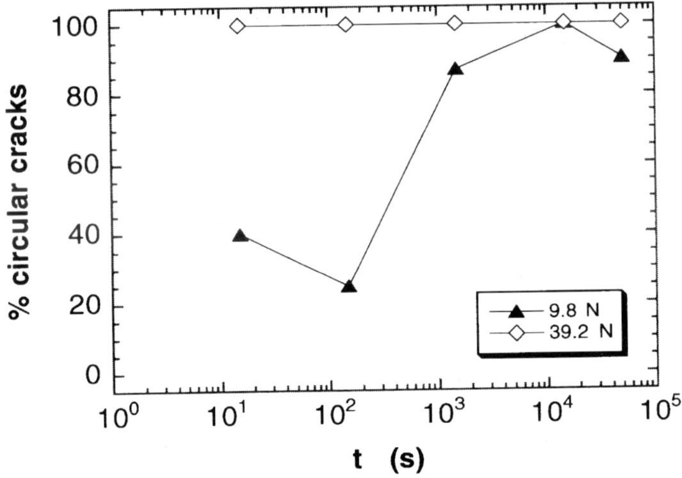

Figure 10 The percentage of median cracks formed as a function of indentation load and time for soda–lime–silica glass in the basic solution (NaOH 0.3N).

for commercial SLS glasses in liquid water are usually in the range 0.25 to 0.29 MPa √m.[24–27] These values are higher than those found by Sglavo and Green[23] using the ISF test. An alternative approach for tests in which crack growth is directly observed is to introduce an indentation crack. In an active environment, the crack will continue to grow (sub-critically) after indentation as it is driven by the residual stress component of K, i.e., the first term in eqn (2). As the crack extends, K drops because of the inverse dependence on crack size and

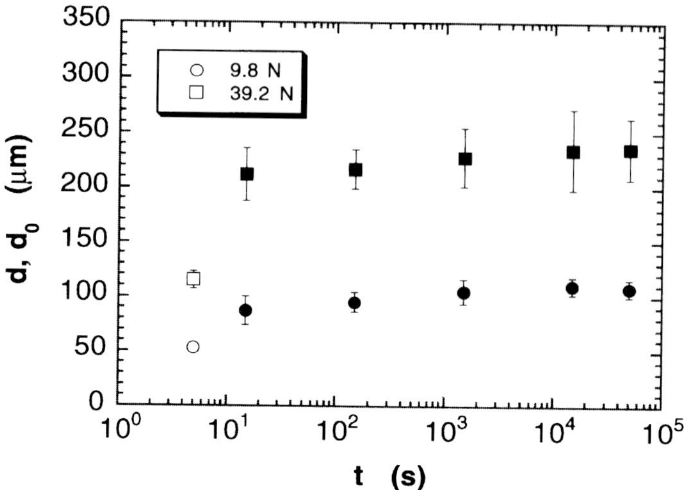

Figure 11 Crack depth measurements for soda–lime–silica glass in the acidic solution (HCl 0.3 N). The empty symbols represent the depth (d_0) measured on indentation cracks obtained in silicone oil. (The latter data are plotted at t = 5 s for improved clarity).

thus, in theory the crack will reach a limiting length once K reaches its threshold value.[27] This approach has recently refined by Salomonson *et al.*,[28] in which they have introduced an experimental procedure to calibrate the χ value, which decreases as the crack extends. One concern with this approach is that median/radial Vickers indentation cracks can sometimes interact with the lateral cracks in a complex way and can lead to some uncertainty in the form of the K relationship.[11]

An alternative approach that still uses indentation cracks has recently been put forward by Sglavo and Green.[29] In earlier work,[30] they had shown that median indentation cracks can be grown sub-critically and in a controlled way during the *application* of the indentation load, primarily at the maximum load. The stress intensity factor for this situation still has the form $K = \chi P/c^{3/2}$ but for this *applied* stress field, one would expect that χ can be considered to be approximately constant. The median cracks are, therefore, expected to reach an asymptotic crack length as they approach the threshold value of K. The value of χ can be obtained from the inert strength of the indented specimens provided the fracture toughness of the material is known.[29]

Figure 7 shows the percentage of times that median cracks are formed for SLS glass in water at two different indentation loads and Fig. 8 shows the growth behaviour of these cracks during indentation. The presence of an asymptotic crack length is particularly clear in the data at the lower indentation load and the calculated threshold K values are given in Table 3. Included in the table are data for a sodium aluminosilicate (SAS) glass.[29] The threshold values obtained using the indentations cracks are in good agreement with those found by DCB testing. An attractive feature of using the indentation technique to measure K_I^0 is that tests can be easily performed in various liquids. For example Sglavo and Green[29] extended their study to include the effect of pH on the threshold value in SLS glass. Figures 9 and 10 show the ease of median crack formation in acidic and alkaline environments.

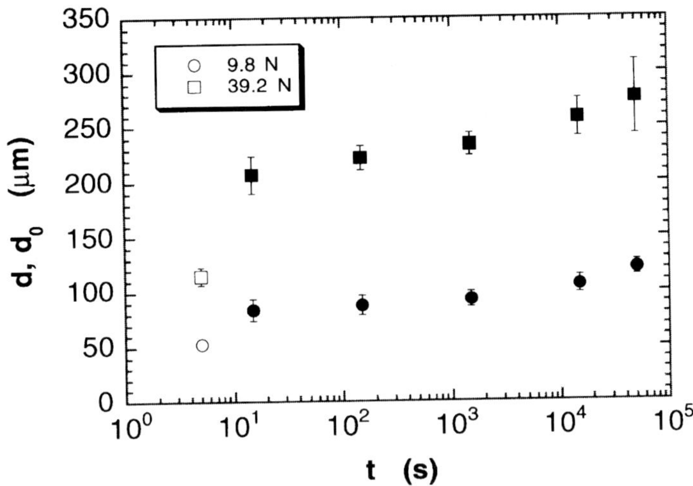

Figure 12 Crack depth measurements for soda–lime–silica glass in the basic solution (NaOH 0.3 N). The empty symbols represent the depth (d_0) measured on indentation cracks obtained in silicone oil. (The latter data are plotted at $t = 5$ s for improved clarity).

Interestingly, it appears that cracks are more difficult to nucleate in the alkaline environment but once nucleated they appear to grow faster. This is confirmed by the crack extension data obtained in acidic (0.3N, HCl) and alkaline (0.3N, NaOH) environments, as shown in Figs 11 and 12. For the study of pH effects, Sgalvo and Green [29] did not find any evidence of a threshold in the alkaline solution. There was, however, evidence for a threshold value in the acidic solution (0.27 MPa √m), which is slightly higher than that obtained in neutral water. These observations are qualitatively similar to those obtained by Gehrke *et al.* on a variety of silicate glasses. [31] It is usually considered that sub-critical crack growth in silicate glasses is primarily caused by hydoxyl ions attacking the siloxane bond in the silicate network. [32–34] In acidic solutions, however, sub-critical crack can still occur because of an ion exchange reaction between hydronium ions and the alkali ions in the glass. [32–34] It is considered that this reaction gives rise to a tensile layer on the glass fracture surface that continues to aid crack growth. One possible interpretation of the threshold K is that it represents a cessation of the (stress-induced) ion exchange reaction, which in turn stops the hydrolysis reaction. In contrast, for the basic solutions, the ion exchange reaction is not needed and the crack depth shows a continuous increase at the higher loading times and no threshold is apparent.

5. CONCLUSIONS

It is shown that the introduction of indentation cracks using a sharp indenter is useful in characterising sub-critical crack growth in ceramics and glasses in several ways. For example, in dynamic, static and cyclic fatigue testing, strength is less scattered if indented samples are used in the strength measurements. This approach does complicate the data analysis but once obtained, accurate lifetime predictions are possible, provided the stress intensity factor associated with the indentation cracks is well characterised. The sub-critical growth of indentation cracks under load can also be observed directly, allowing an alternative measurement of the sub-critical crack growth parameters. In these tests, the crack growth can be complex in the initial stages, so it is recommended that measurements be made after these initial interactions, i.e., after the crack has undergone a significant extension. In the direct observation approach and in subsequent lifetime predictions, it is critical to know the changes in crack shape that occurs as the crack extends. Numerical simulations suggest that the shape changes depends not only on the loading and specimen geometry but also on the nature of the sub-critical crack growth process. Finally, it is shown that the sub-critical growth of median cracks during indentation can be used to estimate the fatigue limit. This approach is also attractive in that changes in the environment can be easily studied. For example, it was shown that the threshold values for soda–lime silica glass was increased in an acidic environment, whereas in an alkaline environment no evidence for a threshold was found.

6. REFERENCES

1. R. W. Davidge, *Mechanical Behaviour of Ceramics,* Cambridge University Press, 1979.
2. D. J. Green, *Introduction to Mechanical Properties of Ceramics,* Cambridge, UK. Cambridge University Press, 1998.
3. J. E. Ritter, Jr., in *Introduction to Mechanical Behaviour of Ceramics,* G. de Portu ed., CNR/IRTEC, 1992, 105–21.
4. B. R. Lawn, *Fracture of Brittle Solids,* 2nd edn, Cambridge University Press, 1993.
5. N. N. Nemeth, L. A. Janosik, J. P Gyekenyesi and L. M. Powers, *Am. Ceram. Soc. Bull.*, 1993, **72** (12), 59–66, 69.
6. N. N. Nemeth, L. M. Powers, L. A. Janosik, J. P. Gyekenyesi, *NASA Tech. Memo.* 106475, Jan. 1994.
7. E. R. Fuller, B. R. Lawn and R. F. Cook, *J. Am. Ceram. Soc.,* 1983, **66** (5), 314–21.
8. B. R. Lawn, D. B. Marshall, G. R. Anstis and T. P. Dabbs, *J. Mater. Sci.,* 1981, **16**, 2846–54.
9. V. M. Sglavo and D. J. Green, submitted to *J. Mater. Sci.*, 1998.
10. V. M. Sglavo and D. J. Green, *Engg. Fract. Mech.,* 1996, **55** (1), 35–46.
11. V. M. Sglavo and D. J. Green, *Acta Metall. Mater.,* 1995, **43** (3), 965–72.
12. P. Dwivedi and D. J. Green, *J. Am. Ceram. Soc.,* 1995, **78** (8), 2122–28.
13. J. C. Newman and I. S. Raju, *Engg. Fract. Mech.*, 1981, **15**, 185–92.
14. P. Dwivedi and D. J. Green, *J. Am. Ceram. Soc.,* 1995, **78** (5), 1240–46.
15. D. A. Stuart and O. L. Anderson, *J. Am. Ceram. Soc.*, 1953, **36** (12), 416–24
16. E. B. Shand, *J. Am. Ceram. Soc.*, 1954, **37** (2), 52–60.
17. E. B. Shand, *J. Am. Ceram. Soc.*, 1954, **37** (12), 559–72.
18. R. E. Mould and R. D. Southwick, *J. Am. Ceram. Soc.*, 1959, **42** (12), 582–92.
19. R. E. Mould, *J. Am. Ceram. Soc.*, 1961, **4** (10), 481–91.
20. M. Watanabe, R. V. Caporali and R. E. Mould, *Phys. Chem. Glasses*, 1961, **2** (1), 12–23.
21. R. J. Charles and W. B. Hillig, in *Symposium on Mechanical Strength of Glass and Ways of Improving It,* Union Scientifique Continentale du Verre, Charleroi, Belgium, 1962, 511–27.
22. V. M. Sglavo and D. J. Green, *J. Eur. Ceram. Soc.*, 1995, **15** (8), 777–85.
23. V. M. Sglavo and D. J. Green, *J. Eur. Ceram. Soc.,* 1996, **16** (6), 645–51.
24. S. M. Wiederhorn and L. H. Bolz, *J. Am. Ceram. Soc.,* 1970, **53** (10), 543–48.
25. T. A. Michalske in *Fracture Mechanics of Ceramics,* Vol. 5, R. C. Bradt *et al.*eds, Plenum Press, New York, 1983, pp. 277-89.

26. C. J. Simmons and S. W. Freiman, *J. Am. Ceram. Soc.,* 1981, **64** (11), 683–86.
27. K. Wan, S. Lathabai and B. R. Lawn, *J. Eur. Ceram. Soc.*, 1990, **6** (8), 777–86.
28. J. Salomonson, K. Zeng and D. J. Rowcliffe, *Acta Metall. et Mater.*, 1996, **44**, 259–68.
29. V. M. Sglavo, L. Burndi and D. J. Green, Paper No. SXIV-09-96, Annual Meeting of The American Ceramic Society 1996, submitted to *J. Am. Ceram. Soc.*
30. V. M. Sglavo and D. J. Green, *J. Am. Ceram. Soc.,* 1995, **78** (3), 650–56.
31. E. Gehrke, C. Ullner and M. Hahnert, *J. Mater. Sci.,* 1991, **26**, 5445–55.
32. S. M. Wiederhorn, *J. Am. Ceram. Soc.,* 1972, **55** (2) 81-85.
33. S. M. Wiederhorn and H. Johnson, *J. Am. Ceram. Soc.,* 1973,. **56** (4), 192-97.
34. B. C. Bunker and T. A. Michalske, in *Fracture Mechanics of Ceramics,* Vol. 8, R. C. Bradt *et al.* eds, Plenum Press, 1986, 391–411 .

Freeform Fabrication of Ceramics

JOHN W. HALLORAN

University of Michigan, Ann Arbor, MI 48109-2136, USA

ABSTRACT

Fabrication of ceramic objects by a solid freeform fabrication (SFF) techniques is reviewed, with emphasis on ceramic versions of commercial Rapid Prototyping (RP) methods. A wide range of ceramics have been demonstrated, including aluminium oxide, silica, hydroxyapatite, silicon nitride, and lead zirconate titanate, with techniques based on extrusion, ink-jet deposition, green tape lamination, and photopolymerisation. The quality of SFF ceramics compares well with conventionally processed ceramics.

1. INTRODUCTION

Sophisticated ceramic objects can be produced directly from computer aided design (CAD) files using the methods commonly called referred to as 'solid freeform fabrication' (SFF)[1] or 'rapid prototyping' (RP).[2] These methods create three-dimensional physical objects representing Computer Aided Design (CAD), medical Computed Tomography (CT) or Magnetic Resonance Imaging (MRI) files. Over the past five years, a wide variety of ceramic objects have been freeform fabricated from structural, functional, and bioceramics materials, using diverse SFF techniques. This paper is a brief review of a small selection of these developments, with a brief background on SFF and CAD, and a description of several of the major techniques and applications.

2. BACKGROUND ON SFF AND CAD

An SFF process builds a shape without tooling. It begins with a CAD model – usually a solid model or a surface model. The software slices the solid model to define a large number of layers. A computer-controlled device then forms each layer (by cutting, curing, depositing, or otherwise) in a solid material, creating the solid object from hundreds of individual layers. Since the object grows layer-by-layer, it is possible to realise shapes and material combinations which are impossible to achieve with conventional methods.

The present SEF field has emerged in only the past decade, although it has antecedents. Techniques for building solid objects from many layers stretches back to the late 19th century, and there has been a small amount of related activity particularly in the 1960s to early 1980s. But the present commercial SFF industry emerged rather suddenly between 1984-1986. The companies which currently dominate the SFF industry were founded during this two year period, including 3D Systems (Valencia, CA), Stratasys (Eden Prairie, MN), Helisys (Torrance, CA), and DTM (Austin, TX). These were entrepreneurial 'start-up' companies, founded to exploit particular SFF techniques: stereolithography (SLA) by 3D Systems, fused deposition modelling (FDM) by Stratasys, laminated object manufacturing (LOM) by Helisys, and scanning laser sintering (SLS), which was invented at the University of Texas and commercialised by DTM. Within a few years ink-jet deposition methods were developed to build objects from powders (the 3D Printing method at M.I.T.) and waxy materials (Sanders Prototyping Inc., Wilton, NH). By 1994, sophisticated commercial systems were available to build quite accurate plastic parts from epoxy acrylates (SLA), engineering thermoplastics (FDM, SLS), casting wax (Sanders), or cellulose paper (LOM). The current capability for ceramic SEF came by adapting these methods for building polymers so they could instead build ceramic green bodies. A notable exception to this is 3D Printing, initially developed at MIT to produce ceramics. An excellent presentation of these methods, and many others, is in Beaman *et al.*[1]

What led to the boom in 1984–86? One decisive factor was developments in CAD systems with improved solid modelling technology. This is crucial for SFF, which must have an unambiguous mathematical description of the object's geometry. Indeed, the computer software is perhaps the most important component of an SFF system.

The starting point for SFF is the Solid Model or Surface Model. It defines what we want to build. A model for mechanical parts is created by a designer with a standard CAD package. Models for anatomical objects can be generated from medical CT files using appropriate software. The Solid Model is then put into a more easily-handled form— an 'interprocess model'. Most commonly, the surface of the CAD model is approximated by a tessellation of triangles, in an 'STL format' which is the de-facto standard for many SFF systems. Some information is lost in this approximation, but the relative simplicity of the STL triangle format makes is easier to do the subsequent calculations.

The SFF software uses the STL surface to create a set of 'slice files', whereby the solid object is converted into many slices, each representing a layer to be built. The slice files are then used to generate a set of 'build files' which have the instructions for building each layers. These build files are different for each SFF method. For methods such as SLA, FDM, SLS, and ink jet, the build files have detailed instructions to: a) define the borders of each slice by scanning vectors or contours; (b) filling the interior of the slice with patterns unique to each technique; and (c) creating the support structures to fix the object in space. These build files control the deposition device— they define how the laser will scan, how the ink-jet will move and fire, or the 'toolpath' of the extruder. The particulars of the materials engineering arise here: issues of heat transfer, shrinkage, curing, joining of adjacent material, etc. Adapting a commercial SFF method to build ceramic often requires careful attention to create appropriate build files. Another key issue are the support structures. These are separate fixtures which allow unsupported features such as overhangs, to be built. Support structures are critical with SLA, FDM, and some ink-jet methods, but not necessary if the object is supported by a powder bed (SLS and 3DP) or excess tape (LOM).

3. CERAMIC SFF

This is a rapidly moving field. By the time this review is published, it will be badly out of date. So consider this review to be only a limited survey as of early 1998. Table 1 presents an outline of six of the major methods, with reference to their use for SFF of ceramics. They will be discussed in turn.

Table 1 Survey of solid free-form fabrication methods for ceramics.

Commercial Method	Ceramic Method	Refs
SLA-stereolithography (3D Systems, Inc.) UV laser cures slices of epoxy resin	conventional SLA with ceramic-containing resin Novel photocuring of layers	21-24, 26-28 20
LOM-Laminated Object Manufacturing (Helysis, mc) builds solid from laser-cut sheets of paper	commercial LOM with ceramic green tape material modified green tape cut-and-place	3,4,5 6,7
FDM-Fused Deposition Modeling (Stratysis, Inc.) robotic extrusion of thermoplastic	commercial FDM with ceramic-thermoplastic material modified extrusion of ceramic slurries or wax mixtures	16.17 18-21
SLS -scanning laser sintering (DTM, Inc.) high power laser sinters layer of powder	sinter ceramic-polymer mixtures	30-31
3DP - three dimensional printing (MIT) build layers by ink-jet printing of binder on powder bed	3DP-printing binder on ceramic powder bed	8-9
SMM- Sanders Model Maker (Sanders Prototype,Inc.) solid hot melt ink-jet deposition of waxy build materials	SMM with wax-ceramic material Direct Ceramic Ink Jet Printing – print dilute ceramic slurry from printer	15 10-14

3.1 Methods Related to Laminated Object Manufacturing

Laminated Object Manufacturing (LOM) is the most literal version of 'layered manufacturing', as the object is actually built from individual sheets of stock material. The commercial version of the LOM process was introduced by the Helisys Corporation to produce models for metal casting patterns. The patterns are built from adhesive-bonded paper, with each layer of the model consisting of a single sheet of paper fed from a roll. A carbon dioxide laser quickly cuts the outline of each layer of the object. A heated roller, which activates the adhesive, laminates the sheet to the previous layer. The outside of the sheet is cut into small blocks by the laser to become the support structure. The process repeats, generating a solid object enclosed within and supported by blocks of the excess material. The object can then be removed from the support ('de-cubed'). Finished LOM objects often have a beautiful appearance of polished wood, with the 'grain' formed by the layers, almost as if the paper pulp had be assembled into wood again. Dimensional tolerances within 0.25 mm can be achieved.

Creating ceramic objects by LOM is easy in concept: replace the paper with ceramic green tape. Although it differs in detail, much of the prior art of tape cast and laminated ceramic technology is applicable. Ceramic LOM was first done with alumina by Griffin *et al.*[3,4] at Lone Peak Engineering (Draper, Utah, US), which now produces commercial ceramic prototypes made with LOM. The quality of LOMceramics is comparable with conventional ceramic materials. For example, high-toughness zirconia-alumina multilayer composites have been made with LOM.[5]

A recent variant of LOM has been developed by the CAM-LEM, Inc. and Case Western Reserve University[6,7] (Cleveland, Ohio, US). CAM-LEM also laminates green ceramic tape, but cuts each layer on a separate table, robotically picks up the cut layer, and places it onto the stack to be laminated. The separate cut-and-place method allows several advantages. The sheets can be cut at an angle by 5-axis laser cutting, allowing a much better approximation to vertical curved surfaces (terraces rather than stair-steps), via tangent cutting and trajectory smoothing. Internal voids within the layer can easily be produced, making hollow structures possible. Different materials can be used in separate regions of a single layer, creating composite structures. Alternately, fugitive materials can be introduced to produce well-defined internal passages. For example, a fluidic structure was built from silicon nitride tape and graphite fugitive tape. After lamination, binder removal, and burnout of the graphite fugitive, the silicon nitride was sintered to create a dense ceramic with a complex interior gas channels, for use as a fluidic circuit for an aerospace application.

3.2 Ink jet methods

There are two varieties of ink jet SFF techniques. The most mature (and commercial) variety is 3D Printing, which uses an ink jet to apply a binder to ceramic powder. Another variety directly deposits ceramic suspensions via an ink jet nozzle.

(a) Binder deposition

Three Dimensional Printing (3DP) also builds parts in layers. First, a thin layer of loose powder is spread over the surface of a powder bed. Next the 'green layer' is created by printing a binder from an ink-jet printer in the pattern defined by that particular slice file. This can be a refractory binder, such as colloidal silica, or a temporary polymer binder. A piston then lowers so that the next powder layer can be spread and the binder is printed again, in the pattern of the next layer. This layer-by-layer process repeats until the part is completed. The fabricated part is removed from the bed of unbound powder and sintered. 3DP is the one of the most flexible rapid prototyping technologies. The process can create parts of any geometry, including undercuts, overhangs, and internal volumes from any powdered material.

3D Printing has been used to prepare dense alumina components[8] from submicron alumina powders by printing a latex binder. The green density of the as-sintered part is too low

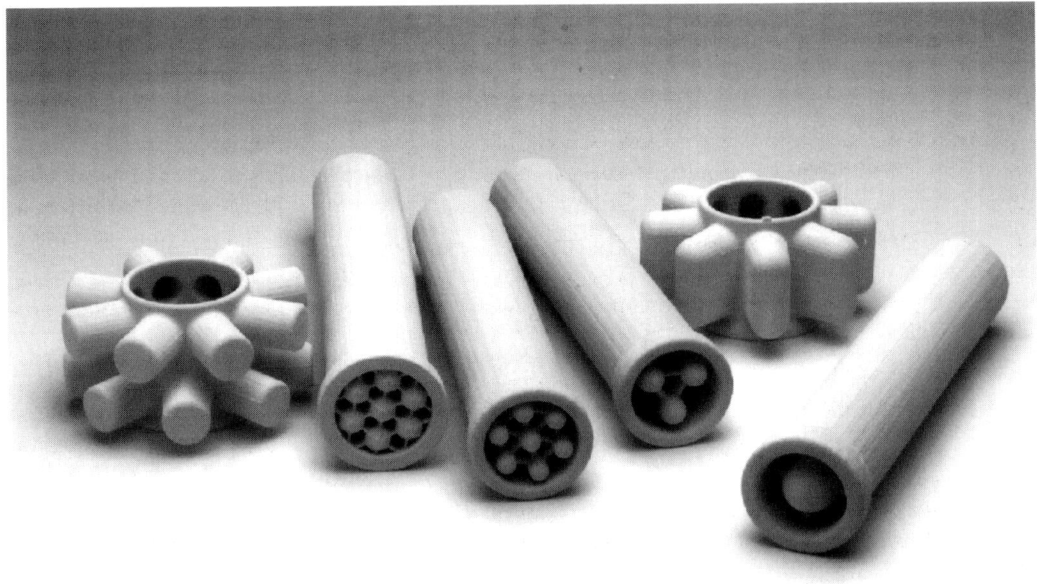

Figure 1 Ceramic candle filters produced by 3DPrinting. Courtesy M. Parish.

to directly sinter, but can be densified by isostatically pressing. After binder burnout and sintering, dense ceramics are achieved with properties which are comparable to the conventional processed material. Much of the 3DP activity involves refractory ceramics, printed with a colloidal silica binder. Ceramic moulds for metal casting are commercially by 3D printed by Soligen, Inc. (Northridge, CA), which is perhaps the most successful SFF commercial firm at present. Soligen has developed very rapid commercial 3DP machines with 100 ink jet nozzles. Specific Surface (Franklin, MA, US) uses a Soligen DSPC-300 to build complex ceramic filters for power plants.[9] Several examples are shown in Fig. 1.

(b) Ink jet deposition of ceramic suspensions

The 'direct ceramic jet printing' technique has been pioneered at Brunel University by Evans and Edirisinghe[10] for ceramic freeforming. A dilute suspension of ceramic powder in a volatile liquid is used as a ceramic ink.[11] The ceramic inks, containing 5 vol.% of 100 nm zirconia powder, have been formulated to be printed with a drop-on-demand ink jet printer[12] or with a continuous ink jet printer.[13]

Repeatedly printing with the zirconia ink on filter paper produced films of high green density zirconia, which could be sintered to full density. Kim and McKean[14] recently demonstrated ceramic patterning by ink jet deposition of a 10 vol.% titania suspension.

But conventional ink jet printers are designed to print small quantities of ink on paper, not to build up a significant deposit. Hot melt ink jet techniques, however, can create large

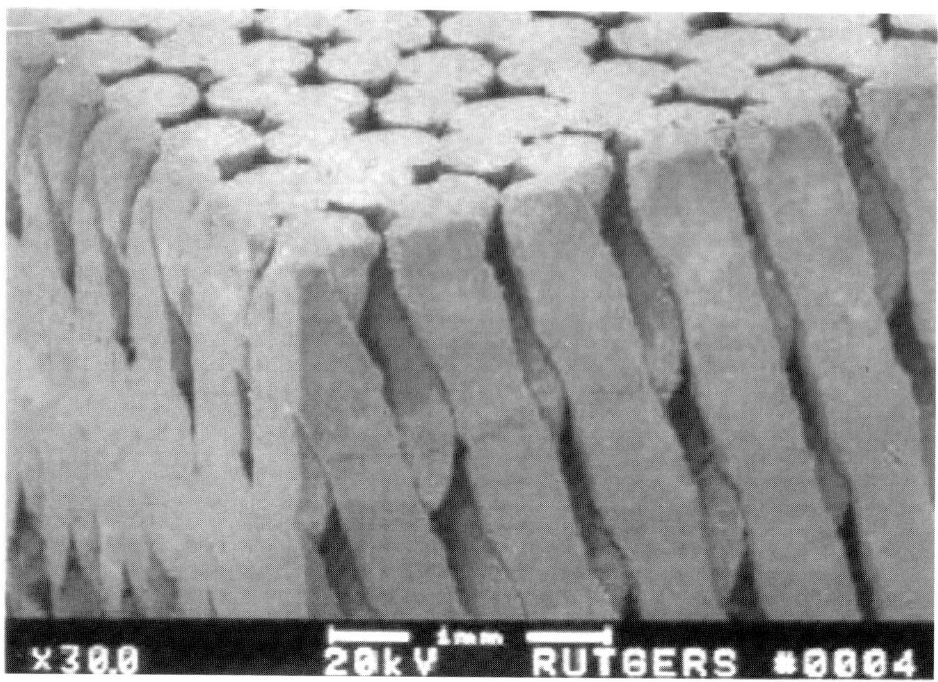

Figure 2 Piezoelectric Composite transducers from PZT produced by fused deposition modelling. Courtesy A. Safari.

deposits of waxy thermoplastic materials. The Sanders Model Maker (SMM) is a very accurate SFF device which uses has two hot melt materials deposited from drop-on-demand jets. The two materials are a thermoplastic build material and a waxy support material. Research is underway at Oxford and Brunel to do ceramic SFF with the Sanders Model Maker, using a suspension of ceramic powder in hot wax. Early results are encouraging using a wax suspension with 20 vol.% alumina powder.[15]

3.3 Extrusion methods

The principal commercial method is Fused Deposition Modelling, by Stratasys. The FDM system extrudes a thermoplastic build material from a small nozzle directed by an x-y positioning device. The build material is supplied on spools as a stiff filament about 2 mm thick. The filament is fed from the spool to the moving 'liquifier' head, where it is melted and laid down 'roads' of material with the desired width and thickness. The SFF object is built from these roads, following a prescribed tool path to create the each layer. Several groups have adapted the Stratasys FDM system by replacing the thermoplastic polymer filament with a ceramic powder-filled filament with the correct combination of properties.

The most notable effort has been by a team led by Rutgers University. They have produced accurate models from silicon nitride[16] using a filaments loaded with 55 vol.% powder in a thermoplastic resin blend. Specimens fabricated from Allied Signal's G54415R silicon nitride grade had microstructure and properties comparable to conventionally -processed material. The FDM silicon nitride had a room temperature strength of 900 MPa (4 point bend) with a fracture toughness of 8.5 MPa $m^{1/2}$ (chevron notch). The properties were similar when measured transverse and longitudinal to the direction of the extruded 'roads', demonstrating that it is possible to build structural ceramics from layers and distinct roads without leaving an undesirable texture. The FDM process can also be used with piezoelectric ceramics[17] to build advanced transducers and actuators. Figure 2 is an example of a complex structure made from PZT-5H. This oriented angular composite is designed to exploit the high piezoelectric shear coefficient, enabling a unique actuator design.

Other modifications of the Stratasys method have been done Lombardi *et al.*[18] at ACR, Inc. (Tucson, AZ, US), using small high pressure extruder in place of the Stratasys nozzle, which permits a wider variety of materials to be used. There have been several other extrusion methods. The 'Multiphase Jet Solidification' (MJS) process, developed Fraunhofer-IFAM[19] uses a thermoplastic-powder mix which is similar to conventional injection moulding material. The process can be used with any injection mouldable ceramic or metal system. The MJS consists of a robotically controlled nozzle which builds parts from small beads of extrudate. Subsequent binder removal, sintering, and properties are similar to the injection moulded version. The 'Robocast' process,[20] being developed at Sandia Laboratory extrudes a highly concentrated aqueous suspension of ceramic powders, which solidify rapidly by drying. Similar techniques have been developed by Calvert[21] based on reactive systems which solidify by gelation.

3.4 Photocuring

Photocuring is the basis of stereolithography[2] (SLA), one of the most popular and most accurate SFF techniques. Commercial SLA machines (3D Systems, CA) produce plastic prototypes from epoxy resins by photopolymerisation of a liquid monomer with a UV laser. The first step is curing a thin (250 microns) layer by drawing the cross section on the surface of the resin with the laser. The part is attached by supports to an elevator platform beneath the surface of the resin. After curing the layer, the elevator platform dips into the liquid resin allowing the liquid resin to flow over the cured portion of the part. A doctor blade sweeps over the part, leaving a layer of fresh liquid resin which becomes the next cured layer as the laser draws the next cross section. Repeating these steps many times builds up the three dimensional plastic part.

It is possible to use the photo-cured polymer as a binder to build ceramics by stereolithography with a 'ceramic resin' – a very concentrated suspension of ceramic powder in a photocurable monomer liquid. The green body is build layer-by-layer on the SLA machine, and is later fired to obtain a dense ceramic object. The original work by Griffith[22] demonstrated SLA using several powders in acrylate monomers and in aqueous acrylamide monomer. Brady[23] developed this into a practical ceramic SFF process, and outlined the physical parameters of curing and part building. Hydroxyapatite bioceramics have been built as with SLA from files generated from medical CT data.[24] Figure 3 is an example of a ceramic SLA object built by Brady from a suspension of 55 vol.% alumina powder in hexane-

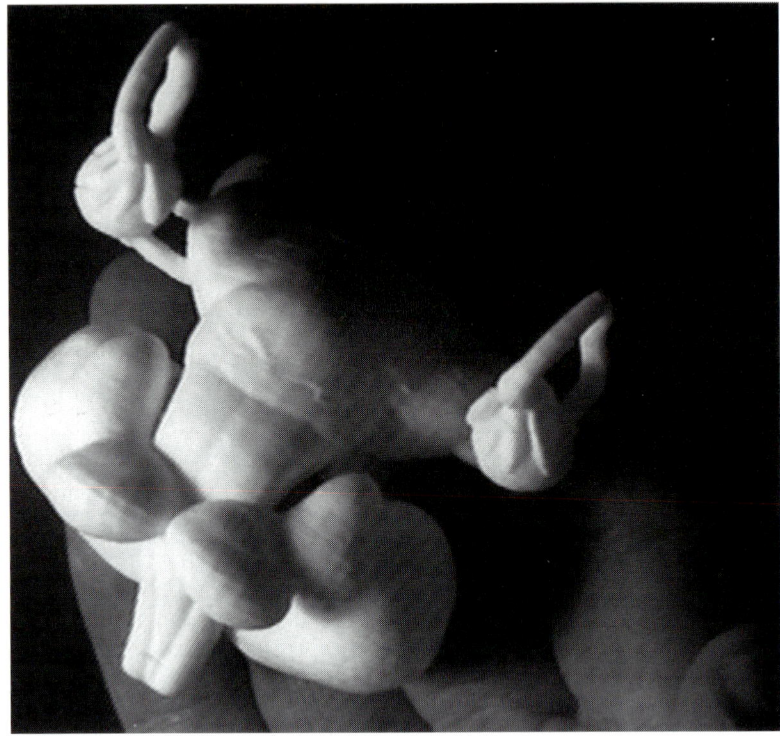

Figure 3 'Anja Test', a digital sculpture by Michael Rees, built in alumina by stereolithography. Photo by permission of the artist.

diol-diacrylate. This is the sculpture 'Anja Test', by Michael Rees, a pioneer of 'digital sculpture' realised by SFF.[25] The CAD file for Anja Test was sent to Brady by FTP, built on an SLA machine, and sintered. It is a good example of the fine detail and freedom of shape provided by SLA.

Lio and Coyle[26] and Chartier and Chaput[27] have explored photopolymerisable suspensions for ceramic SLA, and Zimbeck[28] have reported on some properties of photocured ceramics. Ventura and her colleagues at SRI has developed a novel photoforming process[29] that does not involve UV laser scanning. Rather, layers of a photosensitive ceramic suspension, with photosensitivity in the visible range, is cured through a mask. The mask is created by a computer display screen.

3.5 Scanning Laser Sintering

Scanning Laser Sintering (SLS) is one of the pioneering SFF methods. The process is described in detail by Bourell and his colleagues.[1] The feedstock for SFF is loose powder, which is laid down in thin layers. A high power CO_2 laser is scanned over the powder bed to define each layer by sintering the powder. For plastic parts, the laser fully sinters the polymer powder. The temperatures required to sinter ceramics are inconveniently high to directly sinter. Instead, refractory powders (ceramic or metal) are coated with a polymer binder. The laser melts the binder to 'sinter' the layers and build a green object. The green object is simply removed from the loose powder bed, and then undergoes a conventional firing operation to sinter the part. Perhaps the first SFF ceramics were made by SLS, from SiC[30] and alumina.[31]

4. APPLICATIONS OF CERAMIC FREE FORM FABRICATION

The current application is to create ceramic prototypes by allowing RP methods to also make ceramics. This was one of the motivations of the US DARPA program, aimed at reducing the barrier to the use of ceramics by shortening the long lead time and high cost of ceramic prototypes. Producing one-of-a-kind ceramic products, or short-run ceramic products is an obvious commercial outlet. This is anticipated to be a viable market, and at least one firm (Lone Peak) is committed to exploit it.

Ceramic SFF methods can also be used to fabricate otherwise unobtainable objects. For example, Specific Surface Inc. sells special ceramic filters, whose function derives from a configuration that can only be grown by an SFF technique. Other examples could be actuators or sensors with embedded electrodes or other functionality. Many of these are made today by multilayer (ML) technology— but could be extended to the third dimension with SFF. Biomedical objects, built to fit a particular anatomical defect, could be grown to shape defined by a medical image, perhaps with a sophisticated interior configuration (as with bone tissue engineering scaffolds).[32]

Could the techniques of SFF ultimately be used for high-volume manufacturing of ceramics? I think so. Building ceramic products from many thin layers is already an important segment of industrial ceramics. Multilayer electronic packages, capacitors, varistors, etc. are routinely manufactured form many layers of ceramic tape, sometimes with complex patterns of metal traces and vias. The raw materials and physical processes are much the same, but traditional ML processing employs 'hard' tooling and 'hard' automation. SFF is the ultimate in 'soft' automation: one device makes any shape with no special tooling. Several entirely different objects could be made simultaneously on the same build platform.

Now the SFF process is very slow— typically it takes a day to fabricate a ceramic object (it is 'rapid prototyping' in name only). But increasing the speed by only a hundred fold brings the cycle time down to a fraction of an hour, which may be acceptable for complex shape if no subsequent machining is required. Increasing the speed by a factor of a thousand brings the cycle time down to about a minute. This may well be fast enough for complex objects.

Is it reasonable to expect the SFF methods to be speeded up by two or three orders of magnitude? Why not? SFF is heavily dependent on information processing, which will certainly increase a thousand fold in a few years. The physical part of current SFF machines has one active device (one laser, one ink jet, one extruder), so are quite slow. Clearly this will be done much faster with multiple devices. Green tape as a common feedstock for conventional multilayer technology, where it is used for products with very low unit cost (e.g. capacitors). This suggests the LOM methods could readily be developed for practical production. Tape handling techniques already exist at commercial quantities. Tape cutting for LOM could be done much faster with more powerful lasers, faster scanners, and the use of multiple beams.

Photochemistry is inherently fast, since the curing reaction occurs in a fraction of a second. Stereolithography-based methods could be accelerated with faster scanning or multiple beam laser systems, and the current recoat method could be replaced with faster curtain coating or spraying techniques. Extrusion systems could employ multiple nozzles for faster FDM techniques. Ink Jet methods have already been accelerated for lower-accuracy 'concept modeller' SFF machines that build from polymers with multiple jets. For example, the 3D Systems 'Actua-Multi Jet Modeler' has a 96-jet printhead, and recently Z-Corporation (Somerville, MA) released a machine based on the MIT 3D Printing method with 128 jets. These are relatively few jets, compared commercial graphics printers, which have ink jet printheads with 1000 nozzles (Xaar, Ltd., Cambridge). A thousand-fold improvement in ink-jet SFF machines is quite reasonable.

5. CONCLUSIONS

High quality, sophisticated ceramic objects have been produced by a variety of solid freeform fabrication (SFF) techniques. The full range of powder-processible ceramics have been demonstrated, including aluminum oxide, silica, hydroxyapatite, silicon nitride, and lead zirconate titanate. As of early 1998, each of the prominent commercial SFF methods have been applied to ceramics, involving processes such as extrusion, ink-jet deposition, green tape lamination, and photopolymerization. The quality of SFF ceramics compares well with conventionally processed ceramics.

6. REFERENCES

1. J.J. Beaman, J.W. Barlow, D.L. Bourell, R. H. Crawford, H.L. Marcus, and D.P. McAlea, *Solid Freeform Fabrication: A New Direction in Manufacturing,* Kluwer Academic Publishers, 1997.
2. Paul F. Jacobs, *Rapid Prototyping & Manufacturing – Fundamentals of Stereolithography,* 1st edn, Society of Manufacturing Engineers, Dearborn, MI 1992.
3. C. Griffin, J. Dautenbach, and S, McMillin, 'Desktop Manufacturing: LOM vs. Pressing' Ceramic Bulletin, 1994, **73** (8), 109–114 (1994).
4. C. Griffin, J. Dautenbach, and S. McMillin, 'Solid Freeform Fabrication of Functional Ceramic Components using a Laminated Object Manufacturing Technique', *Solid Freeform Fabrication Symposium Proceedings,* H.L. Marcus, J.J. Beaman, J.W. Barlow, D.L. Bourell, and R.H. Crawford, eds, University of Texas at Austin, 1995, 25–30.
5. E.A. Griffin, D.R. Mumm,and D.B. Marhall, 'Rapid Prototyping of Functional Ceramic Composites' *Ceramic Bulletin,* 1996 **75** (7), 65–68.
6. J. D. Cawley, A. H. Heuer, W. S. Newman, B. B. Mathewson, 'Computer-Aided Manufacturing of Laminated Engineering Materials' *Ceramic Bulletin,* May 1996, **75** (5), 75–79.
7. ZE. Liu, P. Wei, B. Kernan, A. H. Heuer, J. D. Cawley, 'Metal and Ceramic Components made via CAM–LEM Technology', *Solid Freeform Fabrication Symposium Proceedings,* H.L. Marcus, J.J. Beaman, J.W. Barlow, D.L. Bourell, and R.H. Crawford, eds, University of Texas at Austin, 1996, 377–384.

8. J. Yoo, K. Cho, W.S. Bae, M. Cima, and S. Suresh, 'Transformation-toughened Ceramic Multilayers with Compostional Gradients', *J. American Ceramic Soc.,* 1998, **81** (1), 121–32.
9. Specific Surface, Inc., Franklin, MA website: www.specsurf.com
10. M. Edirisinghe, 'Freeforming Ceramics', *Materials World,* 1997, **5** (3), 138–140.
11. P.F. Blazdell, J.R.H. Evans, M.J. Edirisinghe, P. Shaw, and M. J. Banstead, 'The Computer Aided Manufacturing of Ceramics using Multilayer Jet Printing', *J. Materials Sci. Letters,* 1995 **14**, 1562–565.
12. Q.F. Xiang, J.G. Evans, M.J. Edirisinghe, P. F. Blazdell, 'Solid Freeforming of Ceramics using Drop on Demand Jet Printer' *J. Eng. Manuf.* (in press)
13. W.D. Teng, M.J. Edrisinghe, and J.R.H. Evans, 'Optimization of Dispersion and Viscosity of a Ceramic Jet Printing Ink', *J. American Ceramic Soc.* 1997, **80** (2), 486–94.
14. S.J. Kim and D.E. McKean, 'Aqueous titania suspension for novel applicaton of ink ject printing for ceramic patterning', *J. Mat. Sci. Letters,* 1998, **11**, 141–144.
15. K.Seerden, N. Reis, J.R.H. Evans, J.W. Halloran, and B.Derby, unpublished research 1998
16. M. K. Agarwala, A. Bandyopadhyay, R. vanWeeren, A. Saran, S.C. Danforth, N. Langaran, V.R,. Jamalabad, P.J. Whalen, 'FDC, Rapid Fabrication of Structural Components', *Ceramic Bulletin,* 1996, **75** (11), 60–65.
17. A. Bandyopadhyay, R. K. Panda, V. F. Janas, M. K. Agarwala, S.C. Danforth, and A. Safari, 'Processing of Piezocomposites by Fused Deposition Technology', *J. American Ceramic Soc.,* 1997, **80** (6), 1366–72.
18. J.L. Lombardi, R.A. Hoffman, J.A. Waters, D. Popovich, C. Souvignier, and S. Boggavarpu, 'Issues Associated with EFF & FDM Ceramic Filled Feedstock Formulation', *Solid Freeform Fabrication Symposium Proceedings,* D.L. Bourell J.J. Beaman, R.H. Crawford, H.L. Marcus and J.W. Barlow, eds, University ofTexas at Austin, 1997, 457–64.
19. M. Greulich, M. Greul, and T. Pintat, 'Fast Functional Prototypes via Multiphase Jet Solidification', *Rapid Prototyping Journal,* 1995, **1** (1), 20–25.
20. J. Cesarano III, T.A. Baer, and P. Calvert, 'Recent Development in Freeform Fabrication of Dense Ceramics from Slurry Deposition', to appear in *Proc. Solid Freeform Fabrication Symposium,* D. Bourell *et al.,* eds, University of Texas at Austin, Austin, Texas, 1997.
21. P. Calvert, R. Crockett, J. Lombardi, J. O'Kelly, K. Stuffle, 'Extrusion methods for solid freeform fabrication', *Solid Freeform Fabrication Symposium Proceedings,* Univ. of Texas, Austin, 1994, 50–55
22. M.L. Griffith and J. W. Halloran, 'Free Form Fabrication of Ceramics via Stereolithography', *J. American Ceramic Soc.,* 1996, **79** (10), 2601–2608.
23. G. Allen Brady and John W. Halloran, 'Stereolithography of Ceramic Suspensions', *Rapid Prototyping Journal,* 1997, **3** (2), 61–65.
24. R.A. Levy, T.M. Chu, J.W. Halloran, SE. Feinberg, S.J. Hollister, 'CT-Generated Porous Hydroxyapatite Orbital Floor Prosthesis as a Prototype Bioimplant', *American J. Neuroradiology,* 1997, **18**, 1522–1525.
25. Michael Rees, 'Rapid Prototyping: Realizing Convoluted Form and Nesting in Sculpture', *Prototipazione e Produzione Rap ida,* May 1997, (www.sound.net/~zedand00/).
26. Hongmei Lio and T.W. Coyle, 'Photoactive suspensions for stereolithography of ceramics', *J. Canadian Ceramic Soc.,* 1996, **65** (4), 254–262.
27. C. Hinczewski, S. Corbel, T. Chartier, 'Ceramic suspensions suitable for stereolithography', *J. European Ceramic Society,* **18** (6), 583–590.
28. W. Zimbeck, M.Pope, and R.W. Rice, 'Microstructures and strengths of metals and ceramics made by photopolymer-based rapid prototyping', *Proc. Solid Freeform Fabrication Symposium,* D. Bourell et al., eds, University of Texas at Austin, Austin, Texas 1996, 411.
29. S. C. Ventura et al., 'A New SFF Process for Functional Ceramic Components', *Proc. Solid Freeform Fabrication Symposium,* D. Bourell et al., eds, University of Texas at Austin, Austin, Texas, 1996, 327.
30. C.J. Nelson, N.K. Vail, J.W. Barlow, J.J. Beaman, D. L. Bourell, and H.L. Marcus, 'Selective Laser Sintering of Polymer-coated Silicon Carbide Powders', *Ind. Eng. Chem. Res.,* 1995, **34** (5) 1641–1645.

31. K. Subramanian, N.K. Vail, J.W. Barlow, and H.L. Marcus, 'Selective Laser Sintering of Alumina with Polymer Binders', *Rapid Prototyping Journal,* 1995, **1** (2), 24-35.
32. A.I. Caplan and S.P. Bruder, 'Cell and Molecular Engineering for Bone Regeneration', *Principles of Tissue Engineering,* R.P. Lanza, R. Langer and W.L Chick eds, Harcourt Brace & Co., 1997, 603–618.

MECHANICAL PROPERTIES

Biaxial Disc Flexure – Modulus and Strength Testing

R. MORRELL, N. J. McCORMICK, J. BEVAN and M. LODEIRO

National Physical Laboratory, Teddington, Middlesex, TW11 OLW, UK

J. MARGETSON

Defence Research Consultancy, Westcott, Aylesbury, Bucks, HP18 ONZ, UK

ABSTRACT

There is widespread use of biaxial testing of ceramics, but relatively little information concerning the accuracy of testing, and the pitfalls that can arise unless the test is carefully prescribed. This paper reviews recent work in measuring elastic moduli using a biaxial natural frequency method, and in investigating some the potential sources of error in strength testing.

1. INTRODUCTION

Biaxial strength testing has been used for many years, and there is a wide variety of test geometries described in the technical literature and other possible derivatives (Table 1). Typically, there are several advantages claimed for biaxial flexure testing of discs compared with uniaxial testing of bars, including ease of test-piece preparation, use for thin sheet materials, ability to test a large effective surface area free from edge finishing defects, and an equibiaxial stress distribution which is more searching for defects than a uniaxial distribution as found in a bar test. One test geometry, three-ball support/guided punch loading, has been standardised for some time as ASTM F394, aimed at electronic substrate materials. The concept is that out-of-flatness can be tolerated by supporting only at three points. Another, the ring supported/ring loaded geometry, has recently been agreed as a revision in ISO 6474 for orthopaedic alumina ceramics, and the same test is expected to be used for orthopaedic zirconia materials. There is also a literature on testing glass sheet, again the main argument being the avoidance of stressing cut edges.

Biaxial strength testing also has some disadvantages compared with uniaxial testing. Firstly, friction between the test-piece and the loading and support parts of the test jig has been widely recognised in uniaxial flexural testing, and has been eliminated in recently standardised tests (CEN EN 843-1, ASTM C1161) by requiring the use of freely rolling contact rollers. In biaxial testing, friction is usually ignored in the analysis, but will clearly play a similar if not larger role than in uniaxial testing. There is no simple equivalent to rolling rollers, and the nearest equivalent is to use freely rolling balls captive in a ring. Secondly, it is unclear what the optimum test geometry in terms of test-piece aspect ratio and edge overhang should be. The use of thin plate analytical solutions for the stress field may have limitations, especially regarding thin materials, such as electrolyte membranes, which may flex excessively, and thick materials, which may require correction for shear effects. Thirdly,

Table 1 Example biaxial strength testing geometries.

Circularly uniform	Reference	Circularly non-uniform	Reference
Ring on ring	ISO6474, Soltesz[1] Fessler[2]	Punch on three balls	ASTM F394, Kirstein[4]
		Ball on three balls	Godfrey[5]
Ball on ring		Ball on ring of balls	
Punch on ring		Ring of balls on ring of balls	Godfrey[6]
Ball with flat on ring		Ball with flat on three balls	Morrell[7]
Fully or partly pressurisation of ring supported disc	Field[3]	Ring on ring, square test-piece	Entwhistle[8]

all the analytical equations for fracture stress require the insertion of a value for Poisson's ratio, which may not be independently available. Finally, there is frequently a need to relate the strengths of materials in uniaxial and biaxial conditions, and the relationship is unclear and may depend on the failure criterion employed and the surface finishing to which the respective test-pieces have been subjected.

In contrast to testing rectangular or round section test bars in flexure, there has been very little critical review of testing methodology. This paper commences this evaluation by dealing with biaxial elastic moduli measurement, and finite element stress analysis of the ring supported/ring loaded disc test geometry.

2 BIAXIAL ELASTIC MODULUS MEASUREMENT BY IMPACT EXCITATION

2.1 Background

The value of Poisson's ratio required for biaxial strength testing is seldom known accurately, and is not a readily measurable parameter. Normally, a bar test-piece assumed to be isotropic in properties is induced to vibrate flexurally and in torsion, from which separate values for Young's modulus, E, and shear modulus, G, are obtained. Poisson's ratio, ν, is deduced from the relationship $\nu = (E/2G) - 1$, assuming elastic isotropy. Small errors in E and G lead to large proportional errors in ν. Such facilities are not generally available, and a skilled operator is required. For these reasons, the ability to measure Poisson's ratio on the same test-pieces as being used for strength testing would be a considerable advantage. One method, which has been known for some time (Martincek[9], Glandus[10]) but is not in widespread use, is to employ disc resonance or natural impact frequency determination. Glandus applied the natural frequency analysis to thermal shock cracking detection in ceramics. The

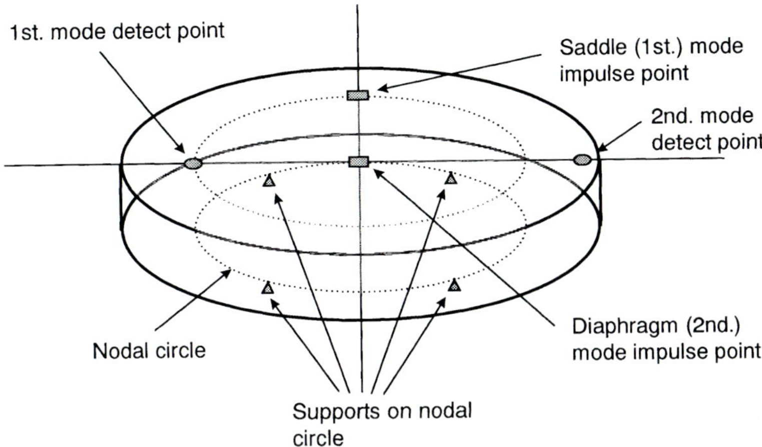

Figure 1 Positions of impact for exciting the first two flexural modes in a disc supported on the nodal circle. The detection points give maximum signals for the respective frequencies.

method has been suggested in an equipment manufacturer's handbook, but only recently has there been a proposal to standardise it as an annex to ASTM C1259.

2.2 Principle

If a disc is supported on the nodal circle (of radius 0.681 of the disc radius) and is struck centrally or resonated by central driving, a symmetrical "diaphragm' mode of vibration is induced, with the disc edge being an antinode (Figure 1). This is the fundamental frequency. The second frequency is obtained if the disc is struck or driven at a point on the nodal circle midway between the support points. The disc then vibrates in a 'saddle' mode at a higher frequency than the fundamental. By taking the ratio of these two frequencies, together with disc thickness to radius ratio, Glandus[10] provided a look-up table for estimating Poisson's ratio to an accuracy of about ± 0.002 assuming perfect isotropy in the test material. Each frequency could then be used independently to calculate *E,* the average of which with ν could be used to calculate *G*.

2.3 Experimental issues

In order to ensure repeatability of measurement, limits have to be laid down for the disc geometry. These are recommended as:

1. Circularity of the disc: better than *d*/500, where *d* is the diameter, is required to avoid obtaining two closely separated natural frequencies due to ellipticity.
2. Flatness of the disc faces: experiments with progressively curving a metal disc using a ring on ring test jig showed that an out-of-flatness of about d/100 produced a measurable change in frequencies, implying that discs should be flatter than this.

3. Parallelism of disc faces: disc thickness enters the calculation equation for E as the third power. Non-parallel faces lead to two closely spaced natural frequencies, similar to the ellipticity effect. The practical limitation in out-of-parallelism is likely to be typically $t/100$, where t is the thickness.

In addition, inspection of the look-up tables shows that results for ν become much more sensitive to disc geometry when the thickness to diameter ratio, t/d, is greater than 0.08. For testing, large discs can be placed concentrically on four equally spaced points lying on the nodal circle. This is most conveniently achieved by defining the nodal circle on a piece of paper, and positioning four equal sized supports, preferably hard rubber or polymer, on this circle. Positioning does not need to be particularly precise. Small discs are most easily supported by ceramic or glass fibre blanket. There is a lower limit on practical disc size controlled by the natural frequencies produced. These need to be less than 50 kHz in order to be readily detected. Low modulus materials can be tested in smaller sizes than higher modulus materials.

In the impact excitation method, the disc is struck at the centre for the first mode, and on the nodal centre between supports for the second mode, typically 1.5 times the first mode. A ceramic ball glued to the end of a flat flexible plastic strip makes a convenient striker which can give single impacts normal to the test surface. The frequencies can be detected by microphone or piezo-transducer contacting the disc at the appropriate positions (see Figure 1). Measurement of the natural frequencies, which are typically in the range of 5 to 50 kHz for practical geometries, can be achieved by using a proprietary instrument, such as the 'Grindosonic' (Lemmens Elektronika BV, Leuven, Belgium), or by using a Fourier transform frequency analyser. The latter has some advantages in that the consistency of the frequency spectrum can be seen, and a better assessment of the repeatability can be made than with the former instrument. In either method, the accuracy of determining the natural frequencies is typically ± 2 Hz, with a similar level of repeatability.

2.4 *Calculations*

Having obtained ν from the ratio of frequencies and the t/R (where R is the disc radius) ratio in the look-up table, the Young's modulus E can be calculated separately from both modes of frequency f_1, f_2 as follows:

$$E_{1,2} = \frac{37.6991\, f_{1,2}^{2}\; d^{2}\; m\,(1-\nu^{2})}{K_{1,2}^{2}\; t^{3}} \tag{1}$$

where d is the disc diameter, m is the disc mass, and K_1, K_2 are dimensionless constants which are established from further look-up tables as functions of Poisson's ratio and t/r. The mean of the two results can be taken as the best value.

2.5 *Error analysis*

An analysis of the propagation of errors is complex because of the use of look-up tables, but it can be shown that the precision of the test disc is the most critical aspect. The tolerances on the test-piece described above should permit Poisson's ratio to be measured to ± 0.01 or

Table 2 Disc moduli test results.

Material	Diam, *d*, mm	Thickness, *t*, mm	Mass, g	Density, kg/mm^3	f_1,Hz	f_2, Hz	*E*, GPa	G, GPa	ν
99.9 Al_2O_3	73.85	2.79	47.92	3957	8533	5320	396.6	158.7	0.25
95 Al_2O_3	98.86	4.16	119.87	3754	6353	3969	302.2	121.1	0.25
95 Al_2O_3	99.80	4.17	120.59	3697	6379	3984	310.0	124.2	0.25
Alumina	77.24	8.71	150.61	3690	21045	13384	312.2	125.8	0.24
Y-TZP1	79.29	6.69	197.97	5993	10500	6220	207.3	78.7	032
Y-TZP2	79.62	6.73	201.35	6012	10508	6220	209.2	79.4	0.32
RBSN1	100.16	4.77	92.79	2469	6321	3989	157.5	63.8	0.23
RBSN2	100.13	4.88	93.63	2437	6405	4047	156.1	63.3	0.23
SSN	88.42	7.76	150.29	3154	15375	9433	282.8	110.7	0.28
RBSC	100.04	5.00	125.89	3203	7385	4524	248.4	97.8	0.27
97 Al_2O_3	29.68	2.25	5.37	3454	36704	24072	285.2	119.7	0.19

better, which is likely to give more consistent results than attempting separate measurements of *E* and *G*. The error in *E* is less than 2%. These estimates also assume that the material is isotropic and has homogeneous density. Any tendency for inhomogeneity will lead to doubling of frequencies.

2.6 Example tests

A series of large, accurately machined ceramic discs (Table 2) has been evaluated, and the results calculated according to the look-up tables. The values for Poisson's ratio are consistent with those in the literature for similar materials. In each case, the impact frequency determinations were repeated at least five times and until consistent results were obtained. Also in Table 2 are the results for a smaller disc 30 mm diameter and 2.25 mm thick, where it can be seen that the frequencies are 24 and 37 kHz for the two modes. These results would suggest that for a high-E material, these dimensions are close to the lower limits for the test to be within convenient audio detection range.

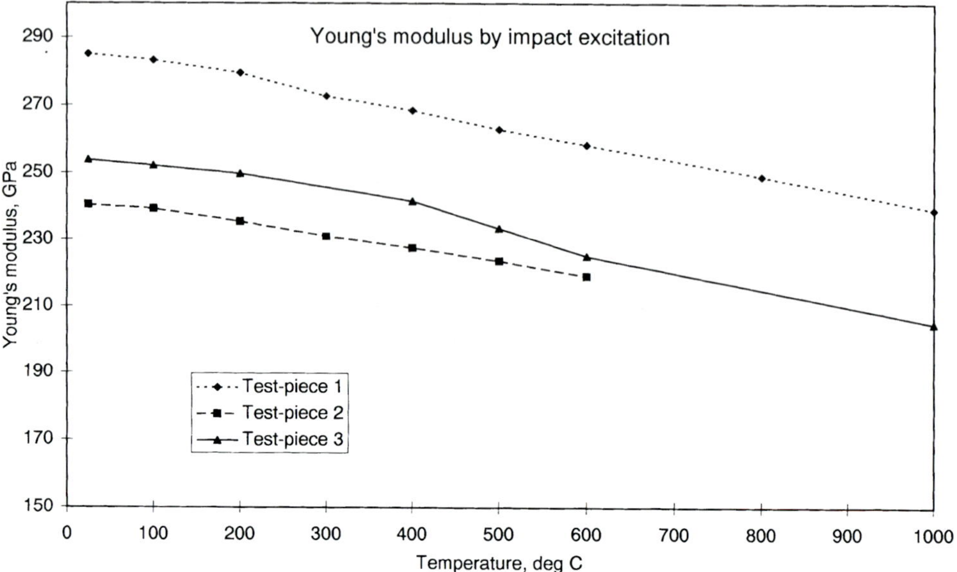

Figure 2 Young's modulus as a function of temperature determined using impact excitation of disc test-pieces.

2.7 Extending the test to elevated temperatures

In common with bar flexure testing, the method can be extended to elevated temperatures, for example, by supporting the test-piece on a resilient fibre blanket material inside a heated enclosure. Impacts can be made by dropping small ceramic balls (spherical bearings or grinding media) down a guide tube onto the surface of the test-piece. The guide tube needs to be angled to the vertical so that the balls bounce off the test-piece, and do not reside on its surface, which could affect the natural frequencies produced. An alternative is to fire the ball upwards at the test-piece supported over a tube. The use of a frequency analyser is invaluable because precision of striking is reduced and both natural frequencies can be excited and recorded simultaneously. A piezo-transducer can have a refractory extension piece added, which can be inserted through a side port in the furnace to rest on the disc surface. A microphone can have a ceramic wave-guide also passing through a side port. In contrast, the 'Grindosonic' may switch between the required vibration modes, or detect other modes, in an inconsistent fashion if the strike and detect positions are not appropriately located. This can occur due to test-piece movement within the furnace with repeat striking. Figure 2 shows results for E on three alumina discs, indicating similar trends with increasing temperature.

2.8 Conclusions

The method of impact excitation of discs for determining elastic properties appears very straightforward, but requires attention to the dimensions of the test-piece and the precision of its geometry. Experimental trials have shown that a Fourier transform frequency analyser is the most convenient method of identifying the natural frequencies of the disc, especially for work at elevated temperatures.

3. BIAXIAL STRENGTH TESTING

3.1 Background

As identified in the Introduction, all the test geometries listed in Table 1 have assumptions in their mathematical analyses and associated experimental unknowns which limit understanding of testing requirements for reliable data. Two principal ones are friction at the loading and support positions, and thickness effects. In addition, the accuracy of positioning of the loading and support positions relative to the disc diameter will inevitably play a role in the achievement of cylindrical symmetry and homogeneity of the stress field.

The geometry that is perhaps most widely reported in the literature, and the most widely used in laboratories, is the ring supported/ring loaded disc. The analytical equations for the stress distribution in an isotropic body are well known, and are derived from Timoshenko's thin plate analysis for small deflections. Inside the inner loading ring:

$$\sigma_r = \sigma_\theta = \frac{3P}{4\pi t^2}\left[2(1+\nu)\ln\frac{a}{b} + \frac{(1-\nu)(a^2-b^2)}{R^2}\right] \tag{2}$$

Between the load rings:

$$\sigma_r = \frac{3P}{4\pi t^2}\left[2(1+\nu)\ln\frac{a}{r} + \frac{(1-\nu)(a^2-b^2)}{R^2} - \frac{(1-\nu)(r^2-b^2)}{r^2}\right] \tag{3A}$$

$$\sigma_\theta = \frac{3P}{4\pi t^2}\left[2(1+\nu)\ln\frac{a}{r} + \frac{(1-\nu)(a^2-b^2)}{R^2} + \frac{(1-\nu)(r^2-b^2)}{R^2}\right] \tag{3B}$$

where R disc radius, a = support ring radius, b = loading ring radius, t = disc thickness, P = applied force, and v is Poisson's ratio. These equations show that there should be a uniform biaxial stress within the inner loading ring which as in beam flexure is a maximum at the surface. The radial and tangential components are different between the loading and support rings, but both drop to zero at the support ring position. Outside the support ring, the stresses increase again. The parts of the test-piece outside the support ring have a stiffening effect on the test-piece, and to avoid the stresses in this region becoming of significant magnitude, the overhang of the support ring should be neither t0o large nor too small such that there is a risk of edge damage affecting the results.

3.2 2D finite element analysis (FEA) of the ring supported/ring loaded test

A systematic study was made using Rockware Microfield software used in the axisymmetric mode. Precautions were taken concerning the solutions for the stress fields at critical points, and higher densities of elements were inserted in regions close to the loading and support rings.

There have been a number of FEA studies reported in the literature, and a common conclusion is that there is a small stress concentration on the tensile surface opposite and just

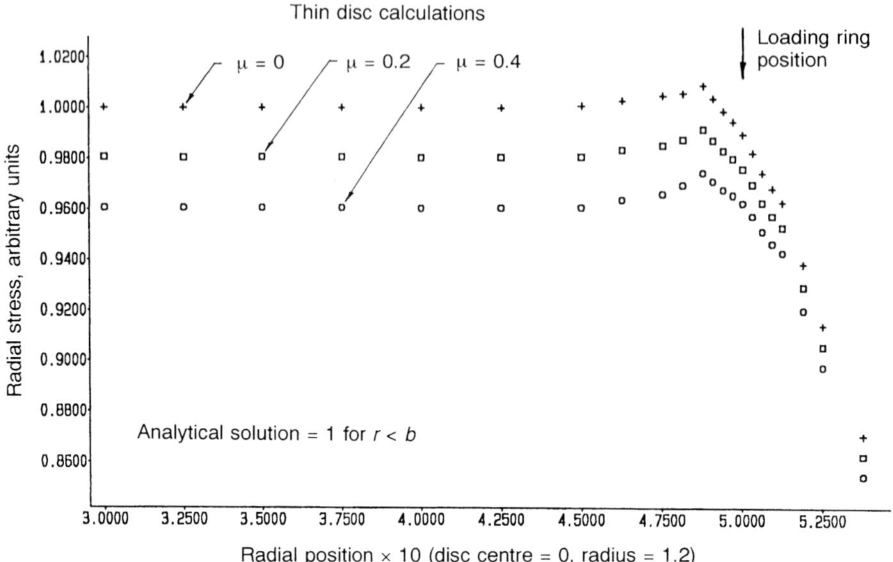

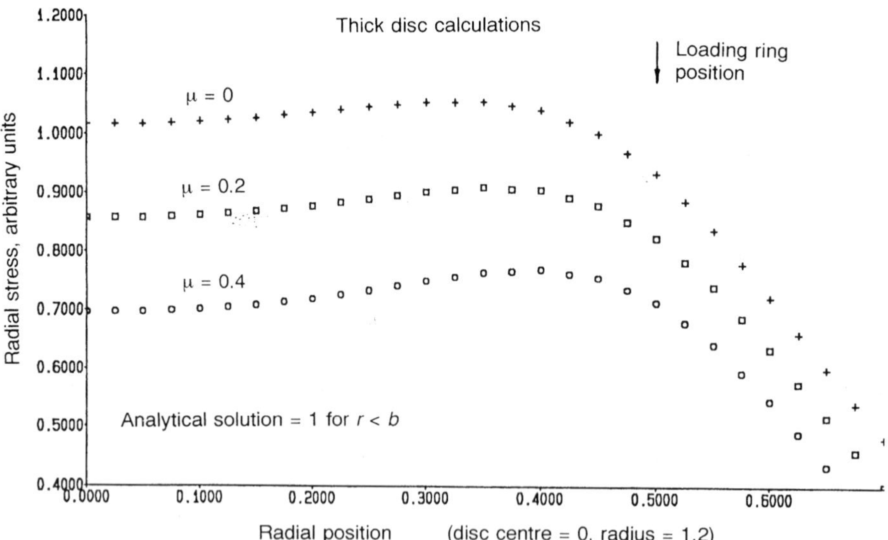

Figure 3 2D axisymmetric solutions for the stress distribution in discs with frictional loading contacts for (a) thin disc solution, $t/R = 24$, (b) thick disc solution, $t/R = 0.33$.

inside the loading ring position. The results of this study are shown in Table 3. For a disc of diameter to thickness ratio of about 40, the error in ignoring this effect is small, about 0.2%, but as the disc thickness or the overhang outside the support ring is increased, the error increases, becoming 7% at a diameter to thickness ratio of 6. The stress at the disc centre also departs from the analytical value for the thicker geometry, being enhanced by 1.6%.

Table 3 - Biaxial flexural strength testing - summary of 2D FEA results

Support ring diam., units[a]	Loading ring diam., units[a]	Disc diam., units[a]	Disc thickness, units[a]	Coeff. of friction	Radial stress error[b] relative to analytical solution, %	
					Centre	Near loading ring
2	1	2.02	0.05	0	< –0.1	+0.4
					< –0.1[c]	+0.2[c]
2	1	2.1	0.05	0	~0	+0.5
2	1	2.4	0.05	0	~0	+1.1
				0.2	–1.9	–0.8
				0.4	-4.0	–2.5
2	1	2.4	0.4	0	+1.6	+7
				0.2	–14	–9
				0.4	–30	–27

a Arbitrary units to provide appropriate aspect ratios
b Hoop stress values are similar to radial stress values at this point
c Solution with refined mesh in region of support and loading points.

The added effect of friction (equal at support and loading rings) was also estimated for discs with an overhang of 10% of the support diameter, a typical experimental geometry. The effects of friction are more marked than changes in geometry (Table 3), and cause a reduction in the actual stress over the disc central region compared with the frictionless cases (Figure 3a). The effect is very marked with thicker discs of aspect ratio 6 (Figure 3b). Small stress concentrations still exist near the loading ring which are of similar magnitude to the frictionless case.

3.3 3D FEA of ring supported/ring loaded disc

In order to study the effects of misalignment of the loading rings, it was necessary to use 3D rather than 2D FEA, because the latter could cope only with the axisymmetric case. An ANSYS package was used on a Silicon Graphics workstation. The mesh was established basically by devising an in-plane array, and then extruding this in the z-direction. The first problem was to set up a satisfactory element mesh which did not give excessive risks of errors. The initial use of sector elements at the disc centre was found to be unsatisfactory. Other designs for the central region were examined and evaluated by employing the quadrature and skew procedures in the software to test element acceptability. The axisymmetric condition was first examined for disc aspect ratios of 40 and 20 with a 5.6% of disc diameter overhang of the support ring. Uniform normal line forces were applied around the loading and support rings. The calculations proved to be very computer intensive because of the number of elements involved and the need to satisfy the mesh verification tests. No mesh refining techniques were employed because of the complexity of the problem and the need to achieve high precision in the calculations.

Table 4 - Biaxial flexural strength testing - summary of 3D FEA results

Support ring diam., units	Load-ing ring diam., units	Disc diam., units	Disc thickness, units, friction coeff.	Eccentricity, units	Stress errorrelative to analyticalsolution, %	
					Centre	Near loading ring
1.8	0.8	2.0	0.05, μ=0	0 0.05	~0 ~0	+ 1.1 ±12
			0.05, μ>>1*	0 0.05	-14 -14	−12 −12±7
			0.05, μ=0	0 0.05	+0.5 + 0.5	+1.3 ± 2.6
			0.05, μ>>1*	0 0.05	−8 −8	–4.5 –4.2 ± 5.0

*Equivalent to fixed boundary conditions rather than sliding with a given friction coefficient

For the frictionless condition, the central region radial stress was found to be raised by only 0.5% compared with the analytical solution for the aspect ratio (*t/d*) of 0.05 and the stress concentration near the loading ring was about 1% (Table 4), figures broadly in line with the 2D axisymmetric results. For boundary conditions in which no sliding across either the support ring or the loading ring is permitted, which is equivalent to a very high friction coefficient, certainly greater than 1, a significant stress reduction in the central region is produced, in line with the 2D solution, although the reduction is lower for the thicker disc.

In order to examine the effect of eccentricity of the loading ring relative to the support ring, the mesh was distorted by dragging the nodes under the loading ring by 5% of the disc radius (Figure 4). The principal effect of doing this appeared to be the tilting of the stress distribution in the direction of loading ring displacement (Fig. 5), with the maximum stress being enhanced opposite the part of the loading ring which is nearer the disc centre. This situation accords with the situation in four-point beam flexure (Baratta *et al.,*[11] Morrell[12]) in that the bending moment resulting from uniform load applied to the loading ring must have increased since the distance from the support ring has increased. In Table 4, the stresses at the centre appeared to be virtually unaltered by the displacement, but near the loading ring the stress increase could be large. The results shown in the last column of Table 4 indicate that the magnitude of the stress enhancement declined with increasing disc thickness, and varied with increasing friction. The continuity of the solution on the tensile face was found to be good except in the region of the support ring because point forces had been assumed, which causes a stress singularity. In practice such stresses can be ignored since contacts are spread. Figure 6 shows a plot of the stress error defined as the difference in maximum principal stress between adjacent elements divided by the peak stress at that location. The possible error levels are generally less than 2% within the central region

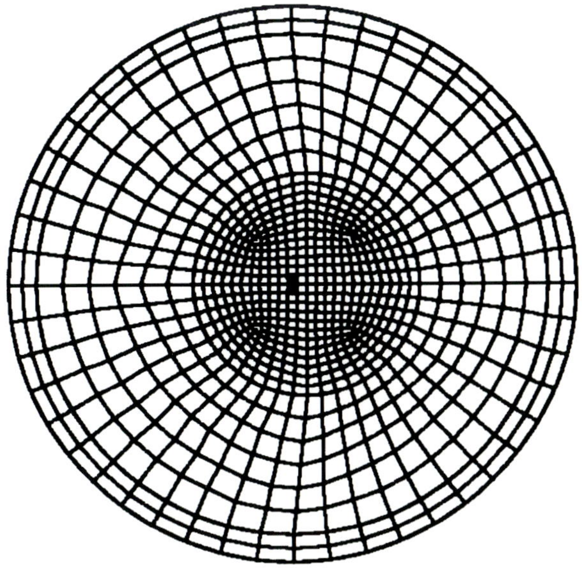

Figure 4 Finite element mesh used for 3D analysis of eccentric disc.

except at the four areas where wedge elements were used, but these are remote from the diameter along which the loading ring displacement was made and the maximum stress changes were calculated.

As the complexity of the precautions taken has increased, so has the confidence in the accuracy of the solutions. The key points in making this successful analysis are in the checks made in optimising the element meshing and in ensuring that the necessary inclusion of wedge elements is made remote from the areas of most importance for the stress determinations.

3.4 Discussion

Both 2D and 3D solutions show that there is a stress concentration on the tensile face opposite the loading ring. In the axisymmetric frictionless case, the stress is enhanced by between 1 and 2% for a thin disc of diameter to thickness ratio typically 40:1, the uncertainty being typical of the level achievable by the FE technique employed, notably the density of meshing close to this point, and the assumptions made concerning the application of point forces. For thicker discs the stresses in the central region increase a little, and the stress concentration also increases compared with thin discs. From this it can be deduced that the diameter to thickness ratio should be greater than about 20 : 1 to ensure that stress errors are kept below about 2%. With thicker discs, the stress is underestimated to a level which may be significant.

The 3D FEA shows that co-axiality of the loading and support rings is important. For the thin frictionless disc, the displacement between the ring axes should be less than 0.01 radius units to keep the stress errors below about 2%, or for a 30 mm diameter disc, to better than 0.15 mm. Time did not permit 3D solutions for making the disc eccentric to the loading ring axes, but it is postulated that similar concentricity requirements would be found.

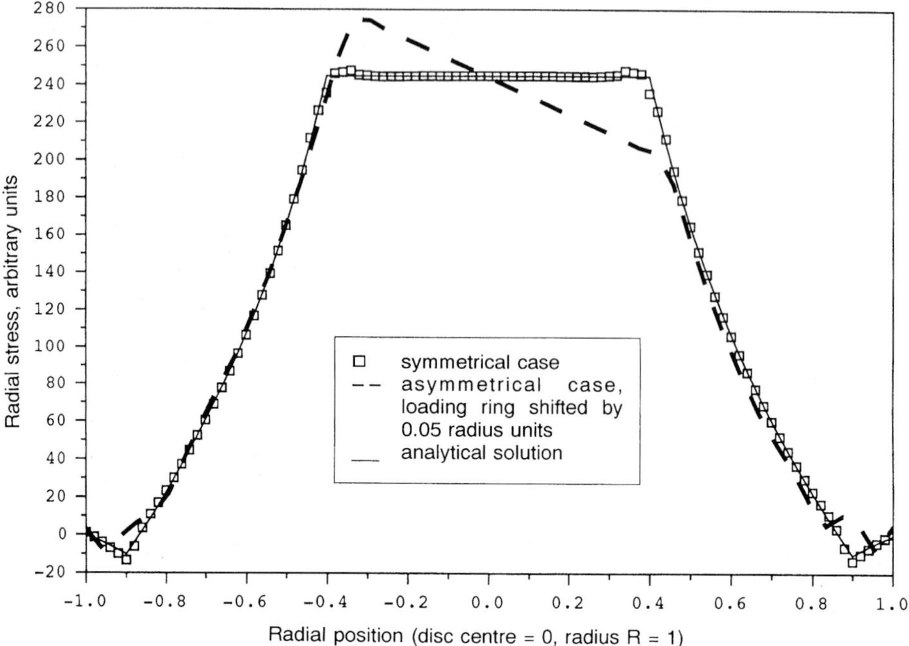

Figure 5 Radial stress distribution in eccentric ring loading of a symmetrically supported disc.

When frictional effects are added, the stress in the central region of the disc becomes significantly less than the analytical solution. For a typical coefficient of friction of 0.4 between a ceramic test-piece and a metal loading ring, the stress reduction due to friction is much greater than the stress enhancement due to stress concentration or thickness effects. By ignoring friction, the stress field is seriously overestimated, leading to higher apparent strength than the true values. This points to a need to minimise friction effects as far as possible, especially for thicker discs. For the ring-on-ring geometry this is not easy to solve. The use of rubber and paper as interface materials is claimed not only permit slight out-of-flatness but also possibly permit shearing at the loading ring to reduce friction effects. However, it can readily be surmised that this is not a perfect solution. Godfrey's[6] ring-of-balls/ring-of-balls geometry, which is close approximation to the ring-on-ring geometry, might eliminate friction completely provided that the balls in the support ring are allowed to roll freely outwards and the balls in the loading ring to roll inwards (the parallel with rolling rollers in the beam test). However, the means of retaining the balls in a ring might constrain them to rotate without rolling, and thus still result in a frictional effect.

3.5 Biaxial testing experiments

There are two possible approaches to establishing experimentally the magnitude of these FE predictions of errors: testing large numbers of nominally identical, conventionally sized test-pieces with systematic deviations of practice; or applying strain gauges to large test-pieces and evaluating stress/strain behaviour without fracturing them. Because of perceived

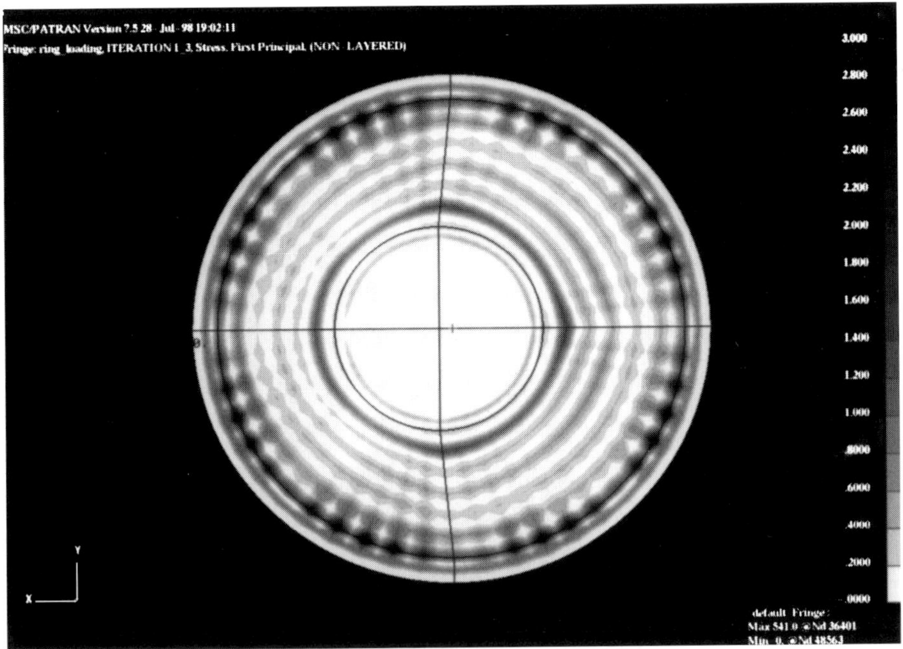

Figure 6 Stress error plot for eccentrically loaded disc.

difficulties in eliminating experimental variables such as geometrical precision and the need to test large numbers to obtain statistical confidence in comparing results, the second approach was chosen. Biaxial strain gauges were applied to large disc test-pieces along a diameter corresponding to a loading ring. Fixed ring jigs were made from brass and designed to allow adjustment of the lateral position of the test-piece relative to the rings, and to permit the strain gauge leads to exit. The support ring plate was placed directly onto the platen of a testing machine, and the upper loading plate was loaded through a large ball bearing located at the centre of the loading ring.

Cyclic loading and unloading tests were performed using either direct disc/ring contact, or with combinations of paper and rubber interlayers, or with the disc surface greased. Initial results have not been particularly conclusive possibly because of the sensitivity of results to precise position of the strain gauge relative to the loading rings. However, it was clear that at low applied forces, the load/strain behaviour was non-linear due geometrical imperfections. Small out-of-flatnesses in any of the components in the system could cause this. At higher forces when all surfaces are in good contact, all strain gauge outputs became linear and parallel within the accuracy of matching of the gauges. For the concentric case, the slope of the line was computed to correspond to the flexural modulus of the disc. For the eccentric case the strain outputs of the gauges deviated broadly in accordance with the FEA expectations, with a tilting of the stress field along the direction of eccentricity. However, it was not at this stage possible to obtain accurate numerical measurements because of experimental difficulties, and further work is needed to improve the test jig design.

4. SUMMARY

Some of the issues of concern in biaxial flexural testing have been reviewed and examined in more detail. Some sources of potential error have been identified. Poisson's ratio, which is required for evaluating disc flexure tests, can be measured directly and reliably on larger discs using the impact excitation method. The use of thin plate theory to calculate flexural strength incorporates assumptions which are not always borne out experimentally, notably friction effects and geometrical precision, both of test-piece and the loading arrangements. The magnitudes of some of the errors have been identified by finite element analysis, but confirming these factors experimentally has proved to be difficult.

REFERENCES

1. U. Soltesz, H. Richter and R. Kiezler, *Proc. Conf. Ceramics in Clinical Applications,* Milan, June 1986, *High Tech Ceramics,* P. Vincenzini, ed., Elsevier, Amsterdam, 1987, 149–58.
2. H. Fessler, and D. C. Fricker, *J. Amer. Ceram. Soc.,* 1984, **67** (9), 582-8, and correction: 1988, *ibid* **71**(10) 904.
3. M. J. Matthewson and J. E. Field, *J. Phys. E (Sci. Instr.),* 1980, **13** (3), 355-9.
4. A. F. Kirstein and R. M. Woolley, *J. Res. NBS,* 1967, **71C** (1), 1–10.
5. D. J. Godfrey, *Mater. Sci. Technol.,* 1985, **1** (7), 510-5.
6. D. J. Godfrey, and S. John, *Proc. Second Int. Conf. Ceramic materials and Components for Engines,* Lübeck-Travemünde, 14–17 April 1986, W. Bunk and H. Hausner, eds, Verlag Deutsche Keramische Ges., 1986, 657-65.
7. R. Morrell, unpublished results from a UK round robin.
8. K. M. Entwistle, *J. Mater. Sci.,* 1995, **30**, 3773–81.
10. J. C. Glandus, 'Rupture fragile et résistance aux chocs thermiques de céramiques à usage méchaniques', Thesis, University of Limoges, France, 1981.
11. F. Baratta, G. D. Quinn, and W. T. Matthews, *Errors associated with flexure testing of brittle materials,* Technical Report TR 87-35, July 1987, US Army Materials Technology Laboratory, Watertown, Mass., USA (reprinted March 1997).
12. R. Morrell, *Flexural strength testing of ceramics and hardmetals,* NPL Measurement Good Practice Guide No. 7, 1997, National Physical Laboratory, UK.

Hertzian Testing of Ceramics

S. G. ROBERTS

Department of Materials, University of Oxford, Parks Road, Oxford OX1 3PH, UK

ABSTRACT

The paper reviews recent work on the fracture mechanics of ring crack formation, from pre-existing small surface cracks, when a hard sphere is pressed elastically against a hard surface. It summarises the ways in which data from these 'Hertzian' tests may be used to determine the extent of surface cracking damage, the materials fracture toughness, and the strength of any residual stress in the surface. Examples are given of experiments applying these test methods.

INTRODUCTION

If a hard sphere is pressed against a hard surface, the contact is purely elastic at low and moderate loads. If the load on the indenter is P and the indenter radius is R then the radius of the contact area, a, is given by:[1,2]

$$a = \left(\frac{3RP}{4E^*}\right)^{1/3} \tag{1}$$

where

$$\frac{1}{E^*} = \frac{1-\nu_1^2}{E_1} + \frac{1-\nu_2^2}{E_2} \tag{2}$$

ν_1, ν_2 and E_1 E_2 are Poisson's ratio and Young's modulus for the sphere and substrate respectively. The peak pressure under the contact is p_0:

$$p_0 = \frac{3P}{2\pi a^2} = \frac{3}{2\pi} P^{1/3}\left(\frac{4E^*}{3R}\right)^{2/3} \tag{3}$$

The resulting stress field in the test specimen, though complex, has been completely characterised.[1,2] The principal feature, from the point of view of fracture mechanics, is a radial stress that is tensile close to the surface. This tensile stress decays with distance, r, from the contact centre:

$$\sigma_{rr} = \frac{1-2\nu}{2}\frac{P}{\pi r^2} \tag{4}$$

The radial stress decreases rapidly with depth and eventually becomes compressive (see Fig. 1). At sufficiently high indenting loads, the near-surface radial tensile stress can propagate a pre-existing crack in the surface. The initial stage of propagation is to form a ring around the contact centre. Subsequent increases in loading may then stably drive the ring deeper; the crack follows the principal tensile stress trajectory, to form a characteristic 'cone crack'.

This paper reviews recent work on the fracture mechanics of ring crack formation in brittle materials, and summarises the ways in which data from Hertzian tests may be used to determine the extent of surface cracking damage, the material's fracture toughness, and the magnitude of any residual stress in the surface. Examples are given of experiments applying these test methods.

SURFACE FLAW TESTING

The test method for the determination of the populations of surface-breaking cracks in brittle materials using Hertzian indentation is to perform a number of tests on the surface, to the point at which a ring crack is formed in each test. (In our test instrument, the formation of ring cracks is detected by the associated acoustic emission.) For each test, one then measures the diameter of the ring crack. It is then possible, from the ring crack diameter and the ring crack formation load, to determine the size of the surface flaw that was the origin of the ring crack. To do this, it is necessary to find the depth of the crack at the radial position of the ring that at the fracture load experiences a stress intensity equal to the critical stress intensity for fracture, K_{Ic}.

There have been a number of papers calculating the stress intensity factors associated with cracks driven by Hertzian indentation.[2–7] Early calculations of this type used only the near-surface radial tensile stress component of the stress field, effectively assuming that this stress was uniformly applied to the crack. However, the stress field actually decays rapidly with increasing depth, and most of the crack surfaces are subjected to a lower tensile traction than that resulting from the near-surface stress (see Fig. 1). If the crack is deep enough, the stress on some parts of the crack faces may even be compressive.

Cracks will thus experience a stress intensity lower than that calculated assuming that they are in a uniform stress equal to the surface radial stress. Recently, Warren *et al.*[8] applied the distributed dislocation technique[9] to Hertzian fracture. This method allows efficient evaluation of the stress intensity factors in strongly varying stress fields. We have used these stress intensity calculations in the interpretation of the experimental data and in the simulation of Hertzian testing described in this paper.

The stress intensity factors used in this paper are calculated for the relatively simple case of a planar crack of depth c, normal to and intersecting the free surface. This is an approximation to the real situation. The cracks are likely to be 'thumbnail' cracks of a length comparable to their depth, though on a scratched or abraded surface such cracks may lie in closely connected or overlapping 'chains', thus resembling planar cracks. Two questions then arise: (1) what type of crack, linear or thumbnail, should one use to calculate stress intensity factors)? (2) if a thumbnail crack, what aspect ratio should be used, and where on the crack front should K be calculated? A detailed analysis by Lin *et al.*[10] has compared the stress intensity factors associated with elliptical thumbnail cracks, of various aspect ratios, in the Hertzian stress field with those associated with planar cracks. They found that for all thumbnail cracks, while the stress intensity at the deepest point of the cracks is rather less

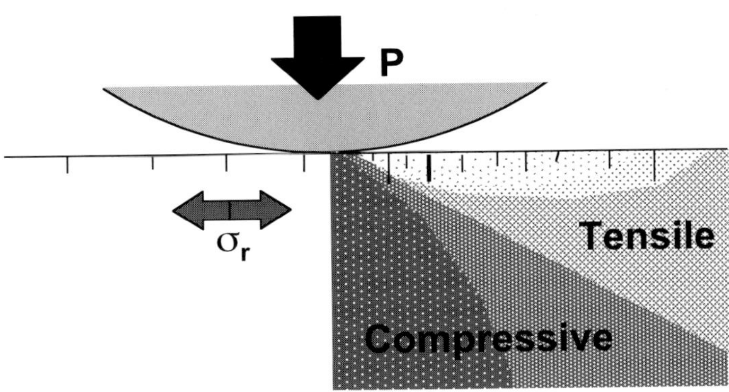

Figure 1 Radial tensile/compressive stresses around a Hertzian contact. There is a shallow region of strong radial stress near the contact edge, which decreases rapidly with depth and eventually becomes compressive. Surface breaking cracks are generally within a strongly varying stress field.

than for a planar crack of the same depth, the stress intensity factor near the free surface was always very close to that of a planar crack of the same depth, and within 2% of the planar crack value for all elliptical cracks wider than their depth. Thus the assumed stages of crack propagation for a thumbnail crack are (1) K exceeds K_{Ic} only near the surface of the crack – it extends rapidly 'sideways' to form a ring crack (2) K now exceeds K_{Ic} at the base of the ring crack, and this may extend rapidly in depth until K at the base of the crack drops below K_{Ic}. In practice these stages may be near-simultaneous. Swindlehurst and Lawn[11] could not distinguish these two stages in their acoustic emission studies of ring crack formation. Whatever the initial crack shape, however, the study of Lin *et al.* indicates that the stress intensity calculations outlined below do give the conditions for unstable crack extension to occur, eventually forming the ring system.

FLAW SIZES AND FLAW DENSITIES

In a given test, a ring crack forms at indentation load (P); the radius of the ring crack generated (r) gives the distance of the originating flaw from the contact centre. The depth of the original (surface-breaking) flaw (c) is calculated by finding the depth of the crack at position r which has $K = K_{Ic}$ at load P. For surface with other than very sparse flaw distributions, use of the following approximation expression[8] for K as a function of P, c and r gives results very close to those using the full calculation:

$$K_I = 1.12\, p_0 \sqrt{\pi a}\, \sqrt{c/a} \left(\frac{(1-2\nu)}{3(r/a)^2} - \frac{2\alpha(c/a)}{\pi} \right), \qquad (5)$$

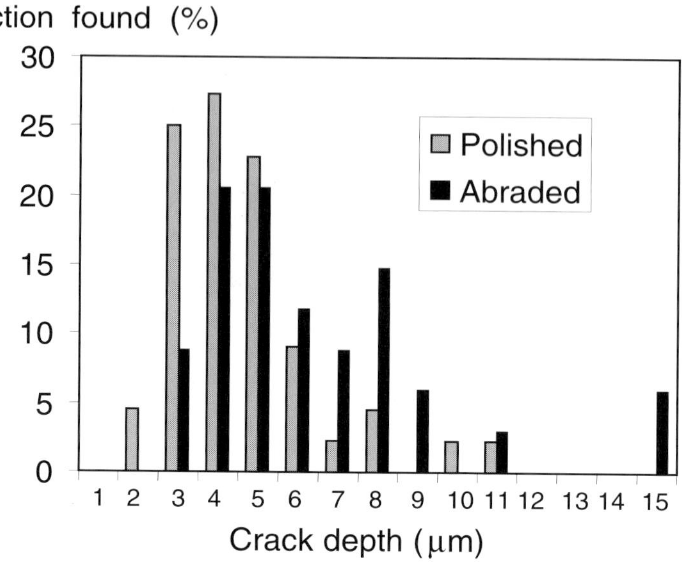

Figure 2 Surface quality in differently polished alumina specimens, as percentage of cracks found with 1 μm size bands (i.e. '2 μm' implies crack depth between 2 and 3 μm) 'Abraded' surface was ground with 600 grit SiC, 'Polished' surface was ground with 600 grit SiC and then polished with 6 μm diamond on a soft cloth until grinding marks were removed. Results of Hertzian tests with alumina ball.

where a and p_0 are calculated as in equations 1 and 3, and a is given by:

$$\alpha = \frac{-1}{\sqrt{u}}\left[u\left(\frac{1-\nu}{1+u}\right) - 2(1+\nu)\sqrt{u}\tan^{-1}\left(\frac{1}{\sqrt{u}}\right)\right] \tag{6}$$

with:

$$u = \left(\frac{r}{a}\right)^2 - 1 \tag{7}$$

Thus, each test gives a value of r (the ring crack radius) and P (the indenting load to produce the ring crack). Equations 1–3 are then used to calculate a and p_0 for each test, then either the full numerical solution for K as a function of crack size and position, or the approximate solution above, is used iteratively to find the value of crack depth for which $K = K_{Ic}$. This is taken as the size of the original flaw. The simplest way to present data derived from a series of tests is to examine the distribution of flaws found within a given size range. Figure 2 shows the results of such tests on two alumina samples, identical apart from their surface finish. The mean flaw depth for the 'abraded' surface (ground with 600 grit SiC) was 6.4 μm, and that for the 'polished' surface (ground and then finished with 6 μm diamond paste) was 4.6 μm; the figure shows the increased spread in flaw size produced by the grinding treatment.

A more sophisticated treatment of the data is to calculate the density of flaws found in the surfaces; i.e. the number per unit area of flaws in a given size range. This is done by invoking the concept of 'searched area', originally proposed and used by Wilshaw.[2] The searched area for cracks within a given size range is evaluated as follows: for each test, calculate the area of the test surface around the indentation within which, had there been a crack of a size within the range of interest, it would have formed a ring crack at a load below that actually reached during the test (usually the calculation is done for a crack of a size midway in the range). These searched areas are annuli surrounding each test site, the inner and outer radii of which vary with the depth of the crack being 'searched' for. Then, in a series of tests, the total searched area for cracks within a given size range is simply the sum of all the searched areas for individual tests. The crack density, $\rho(c)$ (cracks per unit area of test surface) is then calculated as:

$$\rho(c_1 < c < c_2) = \frac{\text{Number of cracks found } (c_1 < c < c_2)}{\text{Total searched area } (c = (c_1 + c_2)/2)} \quad (8)$$

Some complications in the calculation of searched areas were not taken into account in Wilshaw's use of the method.[2] The equations used in his method effectively assume that (a) all cracks are oriented circumferentially with respect to the indentation centre; (b) the stress acting on each crack is the surface skin stress. Both of these factors overestimate the stress intensity acting on an 'average' flaw, and thus overestimate the searched areas.

The first effect leads to a very large error in the searched area and calculated crack densities, as only a very small fraction of the cracks present in a randomly damaged surface will have orientations close to the circumferential direction; cracks even a few degrees from this orientation will have very substantially lower stress intensities acting on them, and will be effectively 'invisible' to the Hertzian test. The second effect becomes more important as deeper cracks are considered, and can lead to the inner radius of the searched area varying its position in a complicated way with increasing load, making the calculations of searched area more complex. However, all these effects can be taken into account[12] and we have developed computer codes to calculate real searched areas and thus crack densities (these are two or three orders of magnitude higher than those calculated using Wilshaw's original searched area equations).[2]

Figure 3b shows the crack densities calculated using this extension of Wilshaw's method for the tests shown as 'fraction of cracks found' in Figure 3a. The data are from tests done on alumina/SiC 'nanocomposite' ceramics[13] and show the improved surface finish produced in the nanocomposite for the same polishing treatment.

Clearly, a certain minimum number of tests need to be done to find a representative number of the cracks actually present in a given surface; to get any data on the rarer cracks, enough tests need to be done to find at least one, preferably more. To see how many tests need to be done to gain a reasonably accurate picture of the true flaw densities, we performed computer simulations of Hertzian testing.

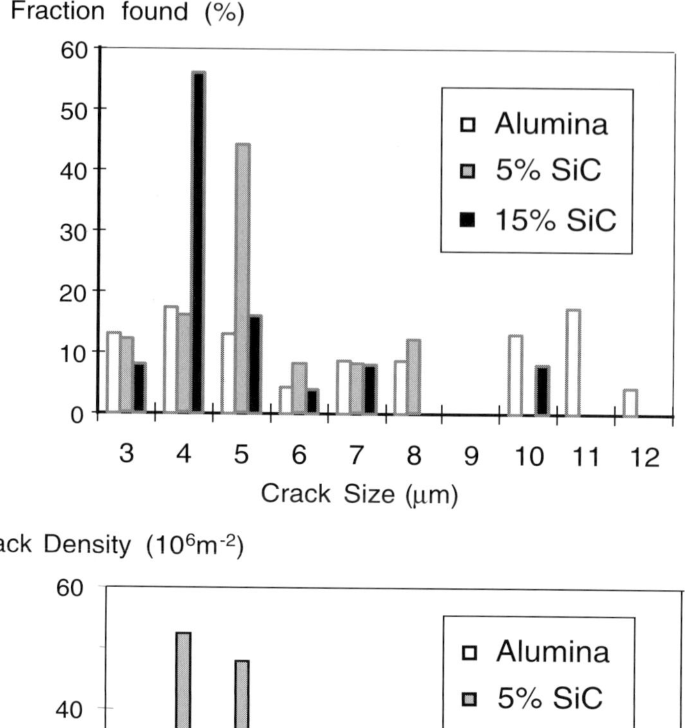

Figure 3 Hertzian tests for surface condition on alumina/SiC 'nanocomposites', all polished using the same treatment (6 μm diamond powder finish). (a) number of cracks found in each 1 μm size band; (b) the same data converted to area densities of cracks in each size banand, using the 'searched area' method.

SIMULATED TESTS

The method used was to generate a distribution of flaws random in position and orientation, corresponding to operator-defined flaw densities for each crack size, around a point at which a simulated test was performed. The test steadily increases the simulated load until the first

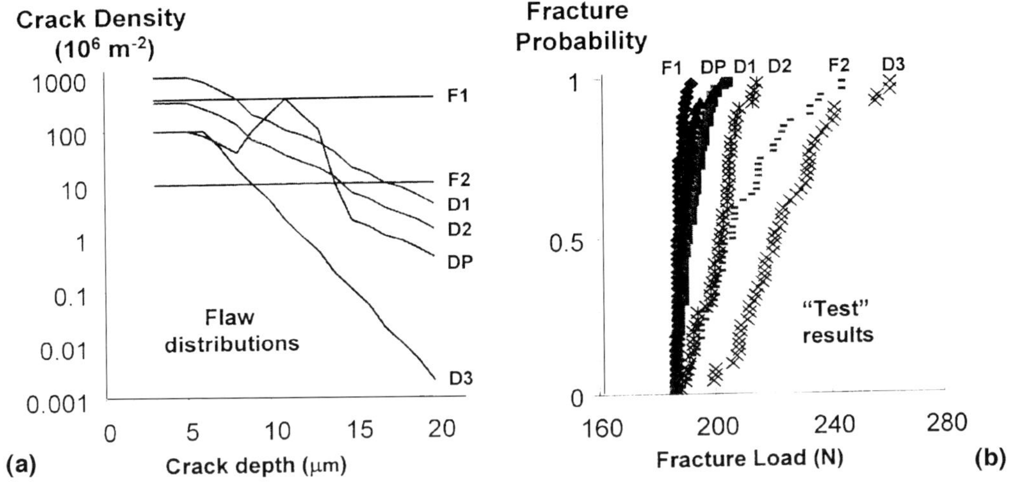

Figure 4 Simulations of Hertzian testing on alumina. The flaw distributions in (a) were used in simulations of a series of 50 tests with a 10 mm diameter ball. (b) shows the results, as probability of fracture as a function of indentation load.

flaw with $K = K_{Ic}$ is found, and the size, position and orientation of the critical flaw noted. Each 'test' in a series is performed in a newly-randomised flaw distribution.

Data from simulated tests on different types of flaw distribution are shown in Fig. 4 (test surface and test ball of alumina, 10mm diameter ball). The flaw distributions used as input data are shown in Fig. 4a; we used:

1. Two 'flat' flaw distributions, Fl and F2, with an equal density for all flaw sizes — while probably rather unrealistic, this should provide a simple base for comparison with more realistic distributions;
2. Three 'decreasing' flaw distributions, Dl, D2 and D3, where the flaw density is lower for longer cracks; this probably corresponds to the state of a well-polished surface;
3 One 'peaked' flaw distribution, DP, where the flaw density is a maximum at a given flaw size; this type of distribution might correspond to that in a surface abraded with a particular grit size.

Results are shown in Fig. 4b, as cumulative probability of (ring) fracture as a function of test load. This type of plot is used in experiments as a simple 'first view' of the data, before the rather lengthy ring crack size measurements and subsequent calculations to give flaw densities are done, and is was of interest to see what type of information could be extracted from such plots alone.

The plots have two main features of interest:

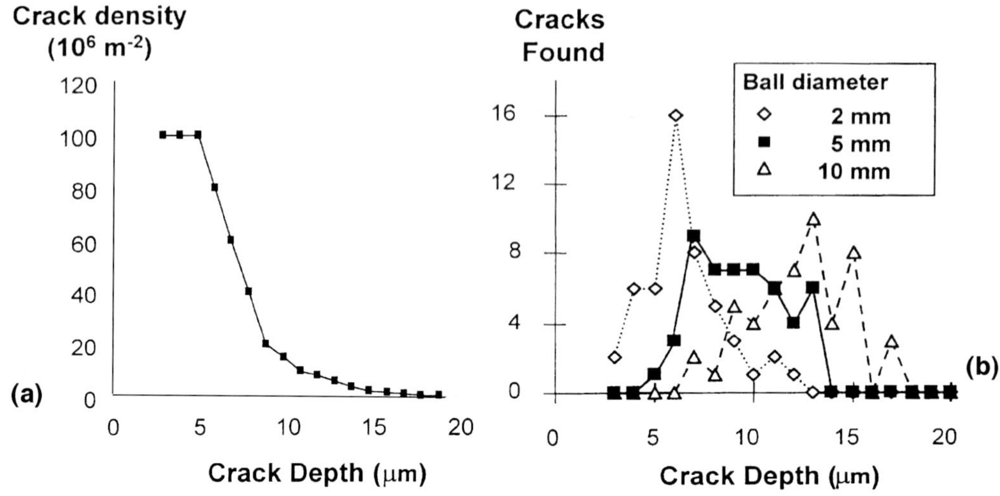

Figure 5 Simulations of Hertzian testing on alumina; effect of test ball size. The flaw distribution in (a) was used in simulations of a series of 50 tests with 2, 5 and 10 mm diameter balls. (b) shows the results, as number of cracks found of each size.

1. There is a definite minimum load for fracture, which does not vary significantly with the details of the flaw distribution. The reasons for this, and its uses to calculate K_{Ic} and residual stresses, are outlined later in this paper.

2. The slope of the plots is closely related to the flaw density around ~10 μm flaw size; 'flat', 'decreasing' and 'peaked' flaw distributions with high flaw density at this flaw size (Fl, Dl, DP) all have steeply rising fracture probability/load graphs. All flaw distributions with low flaw densities at this flaw size have more slowly rising, and scattered, fracture probability /load graphs.

This tendency for tests to 'focus' on flaws in a narrow size range is shown in Fig. 5. The flaw distribution deduced from the test results is peaked around a favoured flaw size, which increases with increasing test ball size. It is possible for cracks, even if present in large numbers, not to be detected at all if they are of a size very different from the favoured one for the size of ball used. The reasons for this are explored below.

MINIMUM FRACTURE LOAD

The reasons for the existence of the minimum fracture load shown in Figure 4 and in all experimental results from Hertzian tests are connected with the decrease in radial stress with depth and radial position illustrated in Fig. 1; this was first noted by Warren.[14] At any given load, the complexity of the stress field around the contact means that there is one position and depth of surface-breaking crack for which the crack has the maximum possible stress intensity. This crack is indicated by a heavy line in Figure 1. Cracks deeper than this have tips that find themselves in a decreasing or even compressive parts of the stress field, cracks closer to the contact centre again have crack tips in weaker or compressive parts of the stress fields, shorter cracks have lower stress intensity because of the shorter crack surface over which tensile tractions can act, and cracks further away from the contact centre are in a weaker tensile stress overall.

For a given test material/ball combination, the crack with the maximum stress intensity factor is found at constant *c/a* and *r/a* values (that is, the absolute depth and position of this flaw increase with increasing load); the stress intensity factor on this crack increases monotonically with increasing applied load. The minimum fracture load corresponds to the load at which this 'most favoured' flaw first has $K = K_{Ic}$. Below this load, there are no surface-breaking cracks, of any depth or position, that have $K \geq K_{Ic}$; ring crack formation is not possible at such loads.

The *r/a* value at which the 'most favoured' flaw is found is typically between 1.05 and 1.2. This is as observed experimentally.[4] Previous explanations have been in terms of the statistical distribution of flaws on the test surface[4] or the effects of interfacial friction;[13] such explanations are unnecessary, as the fracture mechanics of surface breaking cracks in the full Hertzian stress field, as outlined here, give this result.

FRACTURE TOUGHNESS FROM HERTZIAN TESTING

The minimum fracture load in Hertzian testing depends only on the elastic properties of the test ball and substrate and the fracture toughness of the test materials. If the ball and test material have different elastic constants, then interfacial friction between the two will also influence the minimum fracture load.[4] However, we will consider here only the case where the ball and test material are identical, when frictional effects play no part. In this case, relation between K_{Ic} and the minimum load for fracture, P*, is given by Ref. 14:

$$K_{Ic} = \left(\frac{E^* P^*}{CR}\right)^{1/2} \quad (9)$$

where the constant *C* depends only on the Poisson's ratio of the test material and test ball (as listed in Table 1), and *R* is the radius of the test ball. Thus, to determine fracture toughness, all that is required is to establish a well-defined value for the minimum load to create a Hertzian ring crack. In practice, we find that between 20 and 40 tests are usually sufficient. Plots of fracture probability against test load help to indicate whether the minimum load is well defined; a steeply rising curve above the minimum fracture load gives one confidence that a reliable value has been measured. Clearly, the method depends on there being a high probability in a series of tests that a flaw of the right depth and position will be found, in at least one test, so as to give fracture at or very close to the minimum fracture load. To this

Table 1 C **(eqn 9) as a function of** ν **(data calculated using a full numerical solution[14]).**

ν	C	ν	C
0.10	789	0.23	2490
0.11	850	0.24	2790
0.12	917	0.25	3131
0.13	991	0.26	3530
0.14	1074	0.27	4001
0.15	1167	0.28	4560
0.16	1270	0.29	5229
0.17	1386	0.30	6037
0.18	1517	0.31	7022
0.19	1665	0.32	8235
0.20	1883	0.33	9748
0.21	2025	0.34	11658
0.22	2247	0.35	14106

end, it is found experimentally to be useful to abrade the test surface lightly so as to introduce a moderately high flaw density. Since the constant C is very sensitive to ν, accurate determination of ν is necessary for best interpretation of data; use of a natural vibration frequency measuring instrument such as 'Grindosonic' is recommended.

The method has an advantage over techniques based on sharp indentation, (e.g. that of Anstis *et al.* [14]) as the contact is purely elastic at all times, giving rise to a fully characterised stress field in which the small flaws in the surface can propagate. For techniques based on sharp indentation, in contrast, the final length of the median/radial cracks used to determine fracture toughness is determined by the complex elastic/plastic stress field around these indentations, which is not very well characterised; the review by Ponton and Rawlings[17] gives 19 different formulae for determining factor toughness from such experiments, which give a wide variation in the calculated results from a given set of data.[18]

The detailed analysis[14] leading to eqn (9) also provides the basis for 'Auerbach's law'[17] that, in Hertzian fracture, the fracture load is proportional to the indenting ball radius. For tests on surfaces with a moderate or high flaw density, ring crack formation is likely to occur near to the minimum load, P^*, which is proportional to the ball radius, R. Further, since in ideally brittle materials:

$$K_{Ic} = (2\gamma E)^{1/2}, \tag{10}$$

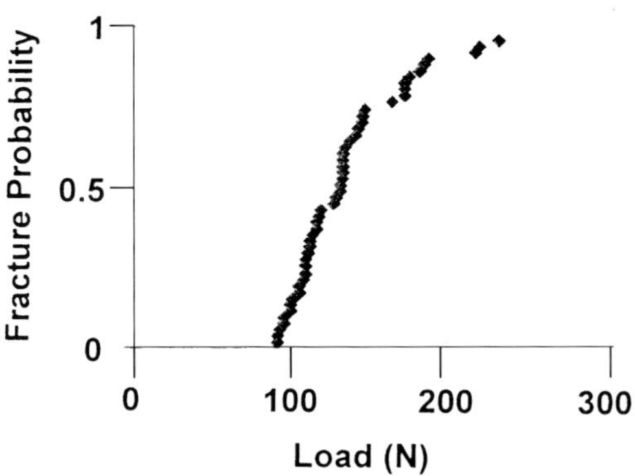

Figure 6 Typical experimental data for probability of fracture as a function of indentation load in glass. Note the well-defined minimum fracture load.

the constant of proportionality between P^* and R is linearly dependent on γ, the surface energy of the specimen material, as noted by Lawn.[20]

Fracture Toughness Results

Figure 6 shows a typical experimental curve of fracture probability as a function of load for tests on glass. Note that there is a well defined minimum load for fracture, from which one can determine a value for fracture toughness. Table 2 gives values for fracture toughness determined in this way[14] for float glass and polycrystalline alumina. The mean values and standard deviations are: glass, 0.78 ± 0.08 MPa m$^{1/2}$; alumina, 3.96 ± 0.27 MPa m$^{1/2}$. These values are in good agreement with those in the literature – for example Zeng *et al.*[7] found values of 0.8 MPa m$^{1/2}$ for glass and 3.77 MPa m$^{1/2}$ for alumina.

Table 2 Fracture toughness of glass and alumina determined by Hertzian testing.

Material	Surface treatment	Sphere radius (mm)	Minimum fracture load (N)	K_{Ic} (MPa m$^{1/2}$)
Glass	As-received	2.5	127	0.78
		5.0	340	0.90
	SiC grit	2.5	105	0.71
		5.0	231	0.74
Alumina	SiC grit	2.5	470	3.72
		5.0	982	3.80
	Diamond	2.5	540	3.99
		5.0	1273	4.33

EFFECT OF SURFACE STRESSES

The above treatment for the determination of fracture toughness assumes that the surface is initially stress-free. A residual stress in the surface will change the stress intensity experienced by surface flaws and thus change the minimum load for Hertzian fracture.[21] Stress intensities on cracks normal to the surface were calculated as:

$$K = K_{Hertz} + K_{\sigma} , \tag{11}$$

where K_{Hertz} is the stress intensity induced on the crack by the load applied to the indenting ball and K_{σ} is the stress intensity induced on the crack by the residual stress in the surface. K_{Hertz} was computed numerically using the full Hertzian stress field,[8] and K_{σ} was taken as:

$$K_{\sigma} = 1.12\,\sigma\left(\pi c\right)^{1/2} , \tag{12}$$

where c is the depth of the crack and σ is the surface stress, assumed to be biaxial and uniform with depth over a depth somewhat greater than that of the crack. A computer program was used to search for the crack with maximum K (K_{max}) for increasing values of indenting load P. The minimum fracture loads, P_{σ}*, which gave $K_{max} = K_{Ic}$ at were noted for a range of values of σ.

Figure 7 shows the predicted effects of compressive residual stresses on the Hertzian indentation of float glass (E = 70 GPa, ν = 0.25) with a 10mm diameter ball of the same material. Figure 7a shows the maximum K on cracks normal to the surface for a given indenting load, P; Figure 7b shows the minimum fracture load (P_{σ}*) as a function of residual stress (σ).

As the compressive surface stress increases in magnitude, the maximum K possible on any surface crack reaches K_{Ic} (taken as 0.80 MPa m$^{1/2}$) at loads increasingly greater than for the stress-free case. The shift in minimum fracture load in Hertzian testing can be therefore be used to determine residual stresses, provided that a sample known to be stress-free is available as a reference, or if the test material's K_{Ic} is known. (For other materials than glass, similar curves to those in Fig. 7 can easily be generated.)

Approximate Determination of Surface Residual Stress

A rough estimate of the residual stress can be obtained by a knowledge of the flaw size giving fracture at the minimum load, c*. We can calculate the apparent fracture toughness (${}^{\sigma}K_{Ic}$) from P_{σ}* for the stressed surface according to equation (9) above; the difference from the true K_{Ic} (${}^{0}K_{Ic}$) is then approximately:

$$ {}^{0}K_{Ic} - {}^{\sigma}K_{Ic} = 1.12\sigma\sqrt{\pi c_0^*} \tag{13}$$

where c_0* is the crack depth for fracture at the minimum Hertzian indentation load in a stress-free surface; (for a sphere of diameter 10mm, c_0* = 13.2 µm for glass, and 11.2 µm

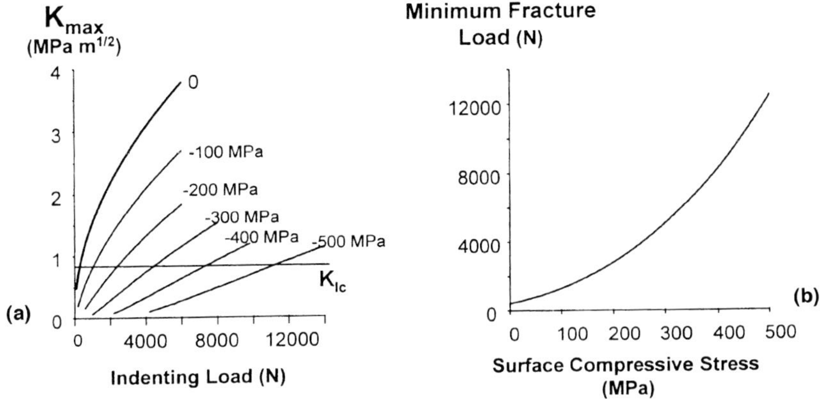

Figure 7 (a) Maximum stress intensity factor for any crack in the indented surface as a function of indentation load and surface stress; (b) minimum fracture load as a function of surface stress. Calculations are for glass, indented with a 10mm diameter glass ball.

for alumina). Calculations of the residual stress using c_0* give better estimates of the residual stress than those using c_σ*, the actual critical crack size and position in the stressed surface.[21] In simulated tests on stressed surfaces, results using this approximate method are within 20% of the input stress value, and often within a few %.

RESIDUAL STRESS – RESULTS

Strengthened Glass

We used the Hertzian method to obtain values for the residual stresses in the surface of chemically strengthened glass specimens. The strengthening process consists of immersing the glass specimens in a potassium nitrate bath heated to 460 °C for 42 hours.[22] Potassium ions are exchanged on a one-for-one basis with sodium ions near the surface, with the concentration of potassium ions falling off with depth. The potassium ions are larger than the sodium ions they replace, and so a compressive stress is generated in the surface layer of the glass. For the specimens tested here, the compressive stress profile has been determined by differential surface refractometry;[23] the surface compressive stress is approximately 450 MPa, and falls off steadily with depth; at 10 μm below the surface the stress is ~380 MPa. Below a depth of approximately 40 μm virtually no ion exchange has occurred.

Figure 8 shows typical cumulative fracture probability plots versus fracture load for annealed glass (abraded with 1000 grit SiC) and chemically strengthened glass, indented with a stainless steel sphere of diameter 5mm. Because of the compressive residual stress, the minimum load for fracture for the strengthened glass, 800N, is considerably larger than that for the annealed glass, 89N.

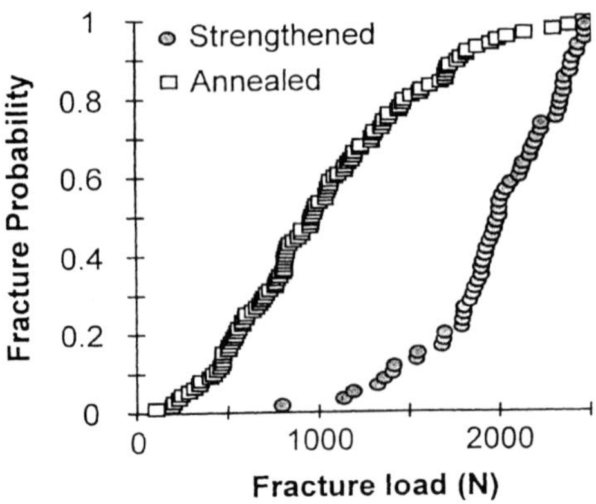

Figure 8 Cumulative probability of Hertzian fracture for annealed glass and for ion-strengthened glass. The threshold fracture load is much higher (800 N) for the strengthened glass than for the annealed glass (89 N).

The threshold loads were measured[21] using a variety of sphere materials (ferritic and stainless steels, WC–Co) and sphere sizes. The 'apparent fracture toughness' method was used to determine residual stresses in the chemically strengthened glass specimens; apparent fracture toughnesses were within the range 1.8–2.2 MPa $m^{1/2}$. The residual stress values were calculated taking the fracture toughness values for the un-strengthened glass as 0.80 MPa $m^{1/2}$. The derived values of residual compressive stress in the chemically strengthened glass are in the range 180–240 MPa. The discrepancy between these values and the 400 MPa determined by optical methods over the depths of the cracks may be because of modification of the elastic properties of the glass by the ion exchange process: very small changes in ν produce substantial changes in C (see Table 1). Interfacial friction between the elastically dissimilar test ball and test surface may also affect the results. Because of these possible errors, and also because of the largely unknown variation in stress with depth, the results should be regarded as 'estimates' rather than as accurate determinations.

Post-Grinding Stresses in Alumina–SiC 'Nanocomposites'

Most workers who have fabricated nanocomposites with Al_2O_3 grain size in the region of 2–4 μm with an incorporation of 5–10% by volume SiC of 100–300 nm particle size achieve bend strengths of about 800 MPa[24–28] compared to 350–400 MPa for polycrystalline alumina of the same grain size. These high strengths cannot be explained by a simple increase in toughness over the polycrystalline alumina; most workers have found only a modest increase in fracture toughness, from about 3 MPa$m^{1/2}$ for a similar grain size polycrystalline alumina to about 3.5 MPa $m^{1/2}$ for the nanocomposites.[25,27,28]

Figure 9a shows the Hertzian indentation fracture loads from ground or polished Al_2O_3/SiC nanocomposite and polycrystalline alumina surfaces. The minimum cracking loads measured for the polished alumina and nanocomposite specimens are results from a low surface residual stress state, and show the intrinsic fracture toughnesses of the alumina and the nanocomposite to be very similar, consistent with earlier work.[25–28] The minimum Hertzian

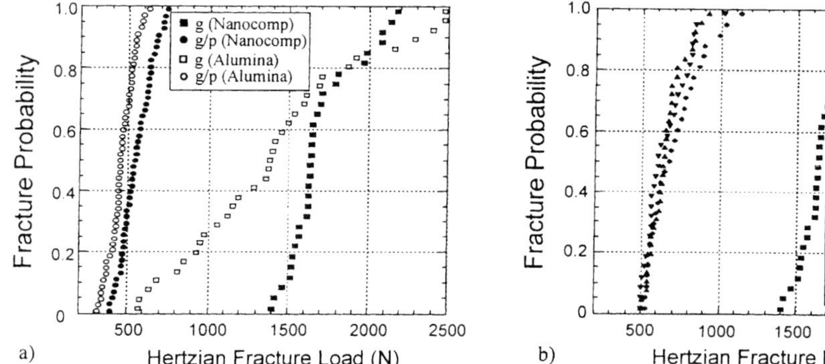

Figure 9 Hertzian indentation fracture loads, shown as a cumulative probability of failure, from: (a) nanocomposites and equivalent grain size alumina specimens after grinding, g, and polishing, p; (b) nanocomposite specimens after grinding followed by annealing at 1250 °C for 0.5, 2 and 10 hours.

fracture loads, P^*, for the as-ground surfaces are higher than for the polished surfaces for both the nanocomposite and alumina. We interpret this as the result of compressive stresses near the surface after grinding, of ~580 MPa for the nanocomposite and about 150 MPa for the alumina (these will be 'weighted average' values over the depth of the cracks, ~10 μm). Such compressive stresses provide a significant contribution to the observed high four-point bend strengths of ground nanocomposites (e.g. Refs 24, 29).

SUMMARY

Results from Hertzian fracture tests may be used to determine a wide variety of information about the surface state and mechanical properties of brittle materials. The data of interest are (a) the load initiating extension of a pre-existing surface-breaking crack, and (b) the diameter of the resulting 'ring crack'. These data can be used as follows:

1. Ring crack size and fracture load in a given test may be used to calculate the precursor flaw depth; data from a series of tests may be used to calculate densities of flaws in a given size range.
2. Fracture load/probability curves may be used to gain a rough impression of flaw density distributions.
3. In a series of tests using a given test ball size, results tend to 'focus' on particular flaw sizes; these flaw sizes are close to that for fracture at the minimum possible fracture load for the test configuration.
4. The minimum fracture load is a result of the complex elastic stress field around a Hertzian contact; it may be used in stress-free surfaces to determine fracture toughness.
5. Surface residual stresses cause shifts in the minimum load for fracture, and such shifts can be used to estimate the strength of the residual stresses and changes in residual stress with annealing, etc.

Work is continuing in developing these test methods and their applications.

ACKNOWLEDGEMENTS

My thanks to P. D. Warren and D. A. Hills for their contributions to the theoretical aspects covered in this paper, to C. W. Lawrence, Y. Bisrat, C. C. Anya and H.-Z. Wu, for provision of experimental results, to B. Derby for discussions of the nanocomposite work and to J. Bradshaw of Pilkingtons for provision of glass specimens. The research was funded in part by the European Commission, the EPSRC and the DTI; financial support was provided by Merton College, Oxford, for Y. Bisrat and the British Council for H.-Z. Wu. Thanks also to Engineering Systems (Nottingham) for their help in developing the test instrument.

REFERENCES

1. G. M. Hamilton and L. E. Goodman, *Trans. Am. Soc. Metals,* 1966, **33**, 371–6.
2. T. R. Wilshaw, *J. Phys. D.,* 1971, **4**, 1567–81.
3. F. C. Frank and B.R. Lawn, *Proc. R. Soc. Lond.,* 1967, **A229**, 291–306.
4. R. Warren, *Acta Metall.,* 1978, **26**, 1759–69.
5. R. Mouginot and D. Maugis, *J. Mater. Sci.,* 1985, **20**, 4354–76.
6. K. Zeng, K. Breder, and D. J. Rowcliffe, *Acta metall. mater.,* 1992, **40**, 2595–600.
7. K. Zeng, K. Breder, and D. J. Rowcliffe, *Acta metall. mater.,* 1992, **40**, 2601–05.
8. P. D. Warren, D. A. Hills and S. G. Roberts, *J. Mater Res.,* 1994, **9**, 3194–202.
9. D. Nowell and D. A. Hills, *J. Strain Anal.,* 1987, **22**,177-84.
10. S. Lin, P.D. Warren and D.A. Hills, *J. Euro. Ceram. Soc.,* 1998, **18**, 445–50.
11. W.E. Swindlehurst and B.R. Lawn, *J. Mater. Sci.*, 1976, **11**, 1653–60.
12. P. D. Warren, D. A. Hills and S. G. Roberts, *J. Hard Materials,* 1994, **5**, 213–28.
13. C. C. Anya and S. G. Roberts, *J. Euro. Ceram. Soc.,* 1996, **16**, 1107–14.
14. P. D. Warren, *J. Euro. Ceram Soc.,* 1995, **15**, 201–7.
15. K. L. Johnson, J. J. O'Connor and A. C. Woodward, *Proc. R. Soc. London,* 1973, **A334**, 95.
16. G. R. Anstis, P. Chantkul, B. R. Lawn and D. B. Marshall, *J. Am. Ceram. Soc.,* 1981, **64**, 533–8.
17. C. B. Ponton and R. D. Rawlings, *Mat. Sci. & Tech.,* 1989, **5**, 865–72.
18. C. B. Ponton and R. D. Rawlings, *Mat. Sci. & Tech.,* 1989, **5**, 961–76.
19. F. Auerbach, *Ann. Phys. Chem.,* 1891, **43**, 61.
20. B. R. Lawn, *J. Appl. Phys.,* 1968, **39**, 4898–36.
21. S. G. Roberts, C. W. Lawrence, P. D. Warren and D. A. Hills, *J. Am. Ceram. Soc.,* in press.
22. J. M. Bradshaw and B. Taylor, *The Physics of Non-Crystalline Materials,* L. D. Pye, W. C. La Course and H. J. Stevens eds, Taylor and Francis, London, 1992, pp 471–475.
23. T. Kishii, *Optics and Lasers in Engineering,* 1983, **4**, 25-40.
24. K. Niihara, *J. Ceram. Soc. Jpn.,* 1991, **99**, 974-82
25. L.C. Stearns. I. Zhao and M. P. Harmer, *J. Euro. Ceram. Soc.,* 1992, **10**, 473-77
26. J. Zhao, L. C. Stearns, M. P. Harmer, H. M. Chan, G. A. Miller and R. E. Cook, *J. Am. Ceram. Soc.,* 1993, **76**, 503–10.
27. C. E. Borsa, S. Jiao, R. I. Todd and R. J. Book, *J. Microscopy,* 1994, **177**, 305–12.
28. L.M. Carroll, M. Sternitzke and B. Derby, *Acta Mater.,* 1996, **44**, 4543–52.
29. H. Z. Wu, C. W. Lawrence, S. G. Roberts and B. Derby, *Acta Mater.,* 1998, **46**, 3839–48.

The Flexure Strength of Multilayer Ceramic Capacitors

R. AL-SAFFAR[*], R. FREER[*†], I. TRIBICK[¶] and P. WARD[¶]

**Materials Science Centre, University of Manchester/UMIST, Grosvenor Street, Manchester, M1 7HS, UK*

[¶] *formerly of AVX Ltd., Hiliman's Way, Coleraine, Northern Ireland, B52 2DA*

ABSTRACT

The flexure strength (MOR) of multilayer ceramic (MLC) capacitors has been investigated by three point loading tests, using a specially designed, miniaturised test rig. The starting materials were blanks and MLC capacitors of Z5U and X7R formulations containing different numbers of electrodes and ink laydown concentrations. It was found for blanks that MOR decreased as the specimen thickness increased, whilst for X7R capacitors the MOR increased with the number of electrodes. Electron microscopy showed major differences in the mode of fracture of Z5U and X7R MLC capacitors. In as-fired components the MOR varied with the ink laydown concentration, but in equivalent components which had been terminated and annealed, the flexure strength was uniformly high and independent of ink concentration.

INTRODUCTION

Multilayer ceramic (MLC) capacitors still dominate the worldwide electronic ceramic market, with total production in excess of 60 million components per year. The traditional leaded device (attached to circuit board by a pair of wires) is today frequently replaced by surface mount components, but the basic internal construction is much the same. The manufacturing processes for MLC capacitors have a number of variants, but tend to involve: tape casting of the ceramic layers; screen printing of internal electrodes; stacking (of the screen printed layers); dicing into individual components; binder burn out; sintering; mechanical tumbling to round corners; termination with external electrodes (which may involve a three layer/stage process) and a thermal anneal.

Efforts are continuing to produce still thinner dielectric layers having improved chemical and physical properties. The capacitance per unit volume depends on the square of dielectric thickness; therefore halving the thickness increases the capacity in a given component size by a factor of four. The success of the MLC capacitor as the device of choice, especially in surface mount configuration, has, however, brought its own problems. Whereas formally in the majority use of leaded devices, the number of capacitors in a unit board made parts per million failure ratesquite acceptable. This is no longer so. Modern circuits containing large numbers of capacitors in surface mount mode has resulted in parts per billion failure

[†] To whom correspondence should be addressed. e-mail: Robert.Freer@umist.ac.uk

rates being required. In addition surface mount assembly leads to the introduction of more stresses and strains into the system. The result is that manufacturers and users have now to consider much more the mechanical stresses that MLC capacitors undergo in their general use rather than just their electrical characteristics. If this is not done unsatisfactory rates of failure can occur. The mechanical performance of MLC capacitors depend upon two main factors, that of the mechanical strength of the unit and the assembly stresses put on that unit. It is the first of these which we consider in this paper.

Contamination of capacitors in the past has been caused during fabrication by hair, skin, fibres from cleaning materials and indeed from solvents/organic materials which are of lower purity than the electronic grade powders. Any gross contamination can act as a source of weakness'. Significant stresses may arise because of thermal expansion coefficient mismatch between metal, ceramic and semiconducting components in a package. Rapid thermal processing procedures, e.g. wave soldering can increase the risk of thermal shock. Other sources of stress include the evolution of gases during firing of the ceramic, and the introduction into the ceramic of impurities having different thermal expansion coefficients to that of the host composition.[1] Whilst these'external factors' may ultimately lead to the destruction of the component the mechanical reliability depends upon the type, origin and location of defects, and whether they grow to a critical level under applied mechanical, electrical or thermal stresses. Mechanical damage can occur in a variety of forms including corner chipping, chip out (flaking), delamination and major cracks perpendicular to the capacitor axis.[2,3] Bechou *et al.*[4] and Chen *et al.*[5] have developed non-destructive techniques to detect localised defects and delamination respectively.

Unfortunately, the small size and compound structure of MLC capacitors presents a serious problem for mechanical testing, and published quanitative data for MiLC chips are limited to a small number of randomly selected formulations, e.g. Refs 6, 7. Full evaluation of fracture toughness for the MLC capacitor (rather than just the dielectric) relies on a knowledge of Youngs modulus for both the ceramic and electrode metal in the capacitor environment. The necessary data for MLC capacitors are still limited although indentation-fracture toughness tests have been successfully applied to monolithic $BaTiO_3$.[8] Selected flexure strength and fracture toughness data have also been reported for COG, X7R and Z5U MLC capacitors[7] and relaxor-based components.[9] The flexure strength (modulus of rupture, MOR) is traditionally determined for ceramics by three or four point bending tests, with specimens of at least 50 mm long. Versions of such tests have been employed for MLC capacitors by Gee,[10] McKmnney *et al.*[6] and the Murata Company.[11] In the study of McKinney *et al.* large MLC capacitors of $7 \times 5 \times 2$ mm size were used. The Murata test procedure[11] strictly yields only qualitative ranking-order-data since the termination electrodes are used as the lower supports. This undoubtedly causes friction and allows stress relaxation during the test. A primary requirement for all MOR tests is that shear, friction and wedging effects must be eliminated, and it is particularly difficult to miniaturise standard MOR test rigs to meet such demands.

It is anticipated that the mechanical properties of MLC capacitors will depend upon specimen geometry, design, number of active layers, electrode laydown, post firing processes including corner-rounding, handling damage (giving rise to surface defects and plating ingress) and the application of terminations. MLC capacitors are complex systems and an understanding of the mechanical behaviour requires a knowledge of the inter-relations between the ceramic dielectric, the electrodes and terminations, and the effect of processing operations.[12]

The present study is concerned with the flexure strength of X7R and Z5U formulation MLC capacitors of size 3220. A dedicated rig has been developed for testing the MLC capacitor, and 3220 is regarded as the minimum size which can be accommodated without running the risk of friction and wedging problems. Special attention is paid to the effect of specimen thickness, the number of active electrodes and the ink laydown. Electron microscopy has been used to characterise the fracture surfaces.

In view of the importance of thermal conductivity to thermal shock resistance, a preliminary study of the thermal diffusivity of one formulation has been performed.

EXPERIMENTAL

Two sets of MLC capacitors (i) a Z5U formulation and (ii) an X7R formulation of 3220 (8.13 mm × 5.08 mm) size were prepared with different numbers of active electrodes (3, 9, 17, 25, 33 and zero, i.e. blanks) and different levels of ink laydown (9.6 to 15 mg in^2). Standard processing conditions were used for each set of specimens. Except where stated, the fired MLC capacitors were as-fired, but not terminated. In the ink laydown tests, silver-palladium terminations were used in the fully processed capacitors.

The mechanical test unit comprised a specially designed three point bend test rig. The upper 'knife edges' and lower supports were made from stainless steel rod (1.0 mm diameter). The lower rollers (separated by a span L of 4 mm) were located in grooves in a steel block (to provide near frictionless contacts) and this in turn was balanced on a steel ball bearing to encourage a uniform distribution of load on the specimens. The three point bend rig was housed in an Instron Universal Testing Machine (Model 4505) which operated at a cross head speed of 0.02 mm min^{-1}. For each specimen the modulus of rupture (MOR) was determined from

$$\text{MOR} = \frac{3P_f L}{2bd^2} \qquad (1)$$

where P_f is the load at failure, b the specimen width, and d the specimen thickness. Each data point reported is based on 20 replicate measurements under each set of conditions.

The fracture surfaces of representative samples were studied by optical microscopy and scanning electron microscopy (Philips 505 and 525 models). The thermal diffusivity of blanks of X7R formulation (in the form of discs 10 mm diameter and 1 mm thick) was determined by the laser flash technique[13] over the temperature range 70–600 °C.

RESULTS AND DISCUSSION

Table 1 summarises the results for the flexure strength of blanks of Z5U and X7R formulation for a single thickness, and also as a function of thickness for X7R blanks. In spite of the

Table 1 Flexure strength of Z5U and X7R blanks.

Formulation	Thickness, mm	MOR, MPa
Z5U	0.90	102 ± 20
X7R	0.90	136 ± 27
X7R	1.08	131 ± 29
X7R	1.33	119 ± 25

scatter in the data there is a strong suggestion that the X7R formulation is associated with higher flexure strength than the Z5U formulation, and that the strength decreases as the specimen thickness increases. This is consistent with eqn (1) which implies that MOR is proportional to l/d^2.

Examination of the fracture surfaces by scanning electron microscopy reveals the effects of processing and microstructure on the mechanical properties of the multilayer structures. Figures 1 and 2 show micrographs of the fracture surfaces of blanks of formulation Z5U and X7R respectively. In the low magnification view (for Z5U, Figure 1) two 'fracture mirrors' can be seen in the upper part of the figure. The 'fracture mirror' is the smooth, curved region which extends outwards from a flaw initiation site, as a result of a crack wave front travelling from that origin. In Fig. 1 there are clearly differences in the fracture texture between regions A (within the fracture mirror) and *B* (within the bulk of the sample). EDS analysis suggested that the fracture mirrors originated from regions where there was a high silica content, possibly as a result of reaction between the blank and silica (used in processing). The major difference between Figs 1 and 2, is the texture of the fracture surfaces for the two materials. That for Z5U (Figure 1; low magnification) is relatively smooth, reflecting the planar, brittle fracture, whilst that for the X7R blank (Fig. 2; high magnification) is comparatively rough as result of intergranular and trans-granular fracture. However, a small number of X7R blanks exhibited unusually low flexure strengths (typically 50% lower than average). SEM micrographs (e.g. Fig. 3) of the fracture surfaces revealed evidence of delamination, as a result of inadequate bonding during processing.

Figure 4 shows detail of the microstructure of a typical Z5U MLC capacitior used in the MOR tests. The dielectric layer thickness is approximately 20 μm and the internal electrodes 2 μm thick. Whilst there are occasional breaks in the electrode it should be noted that the electrode is a two dimensional network and the 'growth' of ceramic dielectric through the electrode is beneficial to the overall mechanical integrity of the device. The flexure strengths of MLC capacitors as a function of the number of active electrodes are shown in Figures 5 and 6. MOR values for the high fire Z5U capacitors are independent of the number of electrodes (Figure 5) and all the values are significantly less than values obtained for the low fire X7R MLC capacitors. The latter also show an increase in MOR with the number of electrodes (Fig. 6). The difference in trend between the two sets of data is believed to be predominantly due to the effect of the metal electrodes. Since the Z5U MLC capacitors employ a high fire dielectric ceramic formulation, whose sintering temperature is in the range 1300–1400 °C, noble metals, such as palladium, are essential for the internal electrodes. In contrast, the formulation used for the X7R MLC capacitors is of the low fire type,

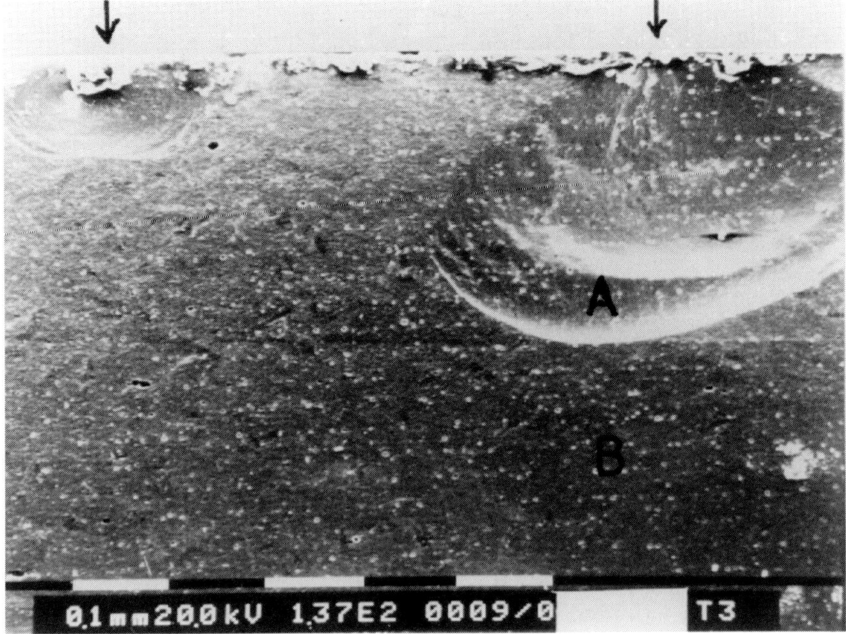

Figure 1 Fracture surface of Z5U blank after MOR test. Arrows indicate 'fracture mirror' (see text).

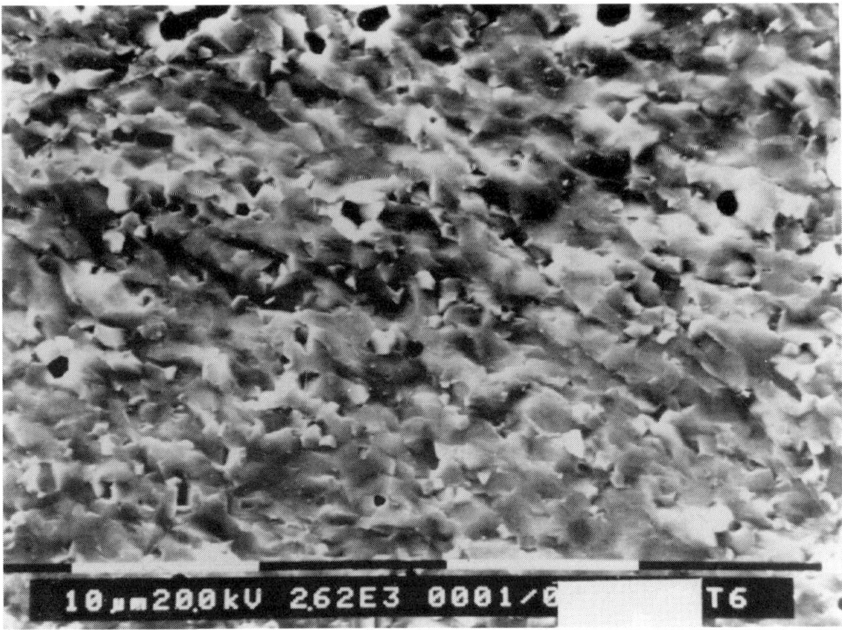

Figure 2 Fracture surface of X7R blank after MOR test.

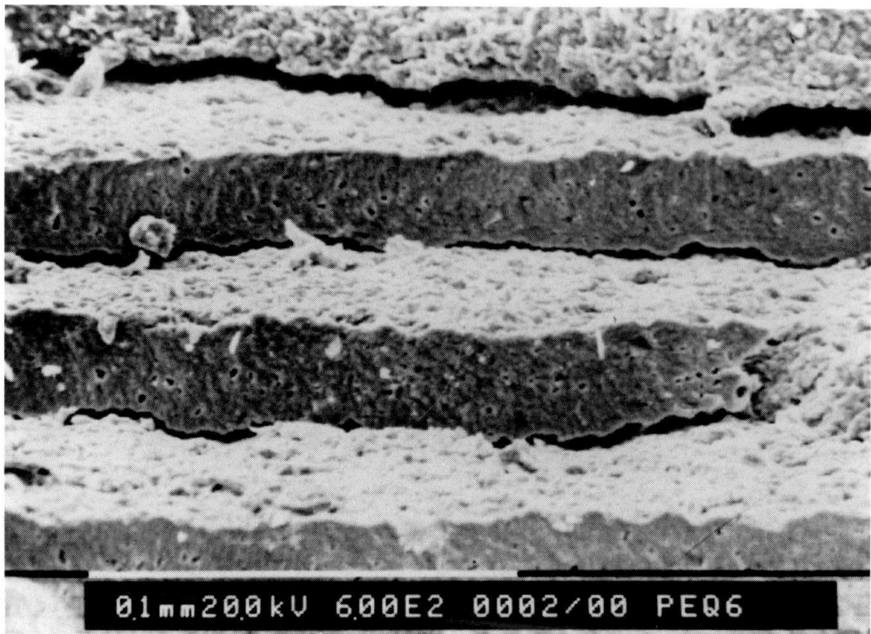

Figure 3 SEM micrograph of fracture surface of MLC blank which had suffered delamination.

where sintering temperatures are in the range 1050–1200 °C. For these, cheaper alloys, typically based on Ag–Pd with a high silver content may be used. EDS analysis confirmed that the electrodes for the Z5U capacitors were nominally pure palladium, and those for the X7R devices 67% Ag–33%Pd. Mechanical polishing of the MLC capacitors demonstrated that the Ag-rich electrodes in the X7R devices were relatively soft and tended to 'smear' leading to an irregular cross sectional profile; the Pd electrodes in the Z5U capacitors are much harder and tend to exhibit a much more uniform thickness after polishing (Fig. 4).

The modulus of elasticity (E) for the high fire electrode is estimated to be ~110 GPa, and that for the low fire electrode is estimated, by the method of mixture to be ~84 GPa.[14] The modulus of elasticity for the ceramic matrix will be of the order of 100 GPa in each case. The fact that E for the metal is lower than E for the ceramic in the case of the X7R MLC capacitors means that the electrodes should be more plastic and thus provide increasing support as the number of electrodes increase. This trend is observed in Fig. 6. In contrast, for the Z5U components E(metal) > E(ceramic) and so the metal does not provide any 'plastic support'. Even with an increasing number of electrodes, there is no increase in flexure strength (Fig. 5).

Figure 7 shows the fracture surface of a Z5U MLC capacitor, containing 25 active electrodes. The break is clean and neither the ceramic nor the metal appear to have offered much resistance to the travelling fracture wave front. This is reflected in the relatively low flexure strength (Fig. 5). Two fracture initiation sites can be seen near the tensile surface (top part of the figure). The fracture texture is the same as that observed in Fig. 1 (for blanks of the same formulation) where there were no electrodes. For selected devices, the fracture initiation

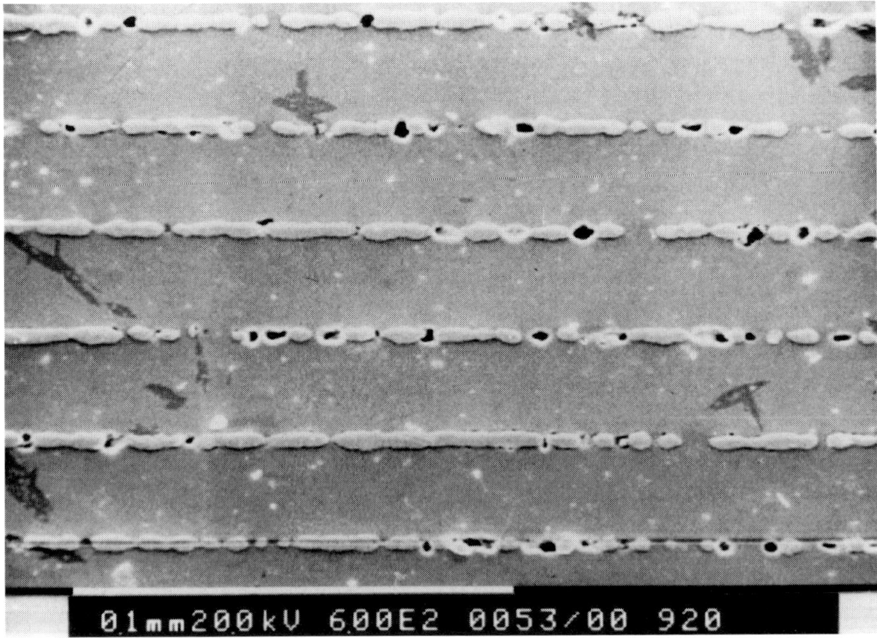

Figure 4 SEM micrograph of polished cross section of Z5U MLC capacitor used in MOR tests.

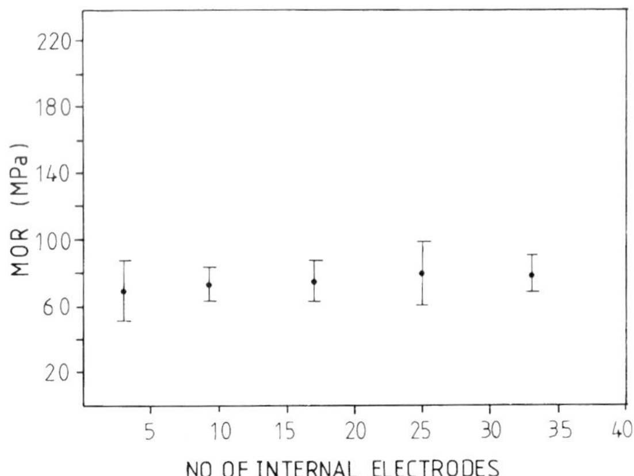

Figure 5 Flexure strength of Z5U MLC capacitors as a function of number of internal electrodes.

sites could be identified with impurities (the same as for blanks); but not all defects could be attributed to chemical sources. Even when there were 33 internal electrodes the Z5U devices exhibited the same type of relatively clean fracture and almost the same fracture strength (Fig. 5). At the electrode overlap region at the end of the electrode stack in such devices, where there is an effective transition from 33 to 16 electrodes, there was no significant difference between the fracture texture in the two regions.[15]

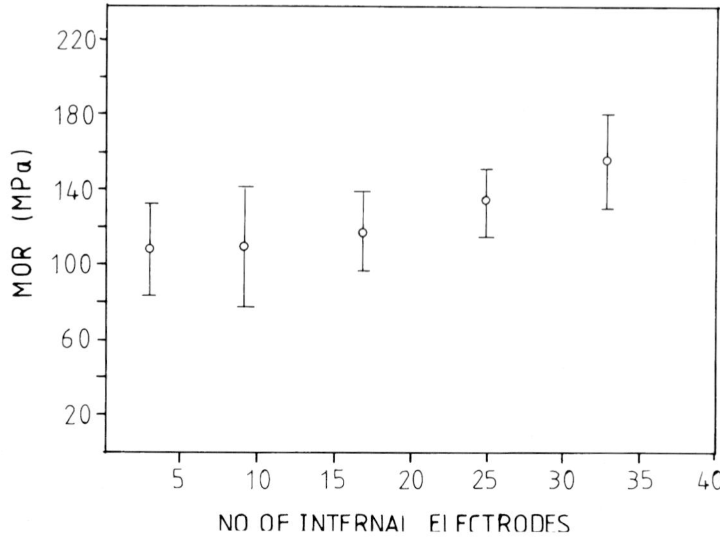

Figure 6 Flexure strength of X7R MLC capacitors as a function of number of internal electrodes.

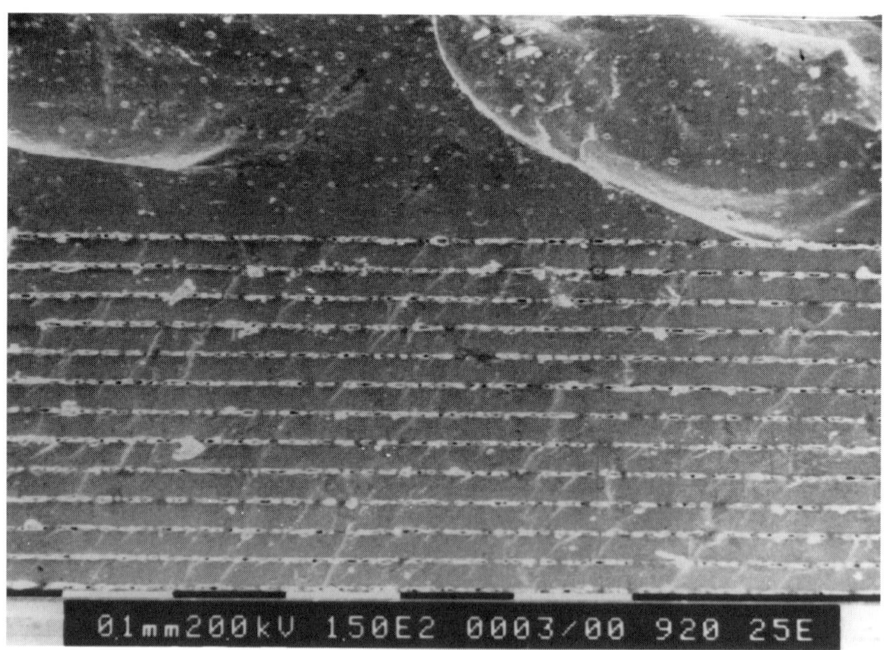

Figure 7 Fracture surface of Z5U IvILC capacitor (containing 25 active electrodes) after MOR test. The break is clean; compare with Figures 8 and 9.

In contrast, the fracture surfaces for the low fire X7R MLC capacitors are uneven and show extensive irregular cracks (Figs 8 and 9) and indicate that the fracture wave fronts

Figure 8 Fracture surface of X7R MLC capacitor (containing 17 electrodes) after MOR test. Break is irregular.

have been distorted and deflected by the electrodes. In some cases[15] there are multi-fracture initiation sites near the tensile surface, and gross fracture within the electrode region; occasionally major cracks branch into multiple structures within the electrode stack. For the X7R capacitors the more plastic internal electrodes distort and stretch, effectively absorbing some of the energy from the advancing fracture wave front. This enables the MLC capacitor to withstand higher loads before failure (Fig. 6).

Figure 9 clearly demonstrates the importance of increasing the number of electrodes. The micrograph (Fig. 9a) shows the fracture surface at the end of the electrode overlap region (Fig. 9b). There is significantly more distortion of the fracture wave front where there are more electrodes; i.e. 25 compared to 12 in the two regions shown. This was not the case for the high fire Z5U capacitors. The practical implication of the differing behaviour in the two types of devices is illustrated graphically in Figs 5 and 6.

Results for the flexure strength of X7R MLC capacitors as a function of ink laydown density are summarised in Figs 10 and 11. For the as-fired specimens there is a general increase in strength to a plateau level (~140 MPa) as the laydown density is increased (Fig. 10). Above a threshold laydown concentration the electrodes will be of uniform thickness. Under such circumstances the bonding strength between the ceramic dielectric and internal electrodes will reach its optimum value. However, the application of terminations, including the low temperature anneal, improves the flexure strength significantly. Figure 11 shows flexure strength for an equivalent set of X7R capacitors after final processing. It can be seen that there is a marked improvement in strength, especially at the lower laydown levels. The

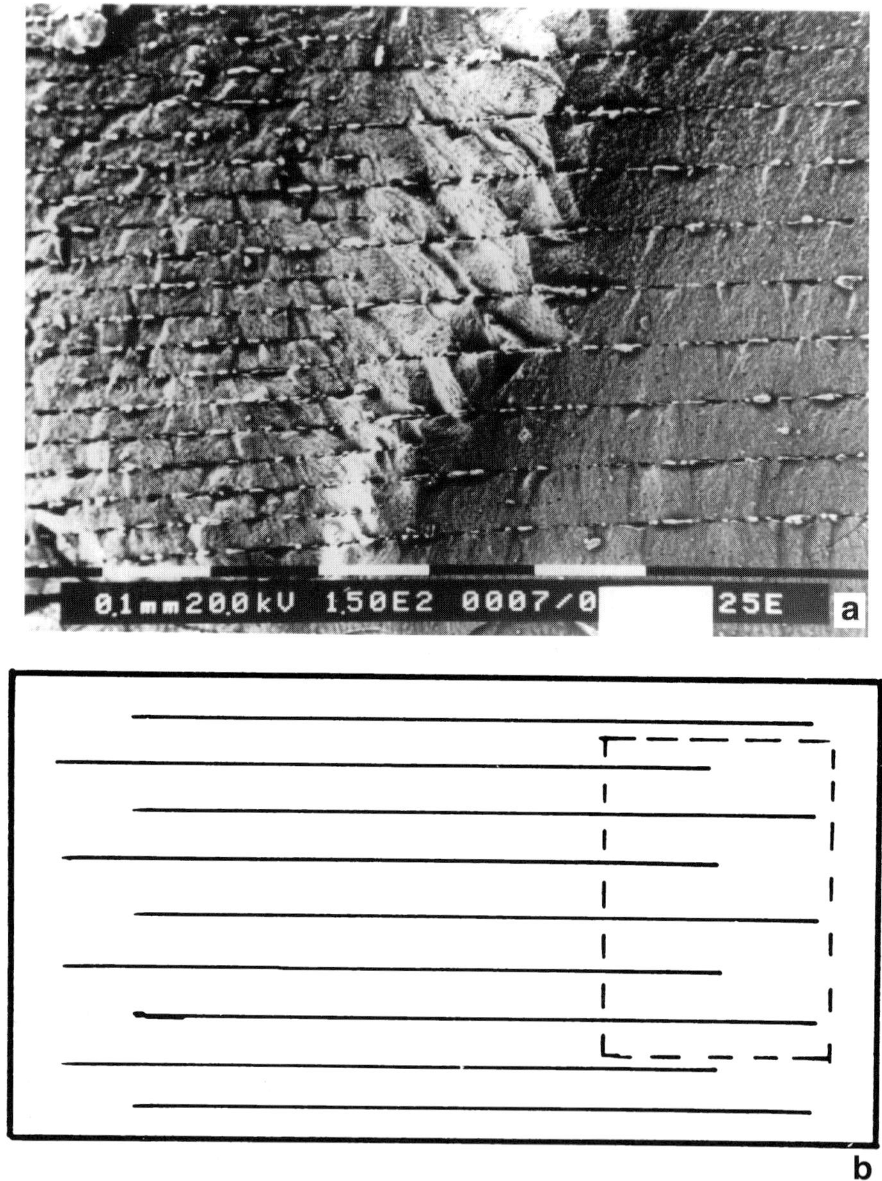

Figure 9 (a) Fracture surface of X7R MLC capacitor (containing 25 electrodes) after MOR test. The location of the fracture section on the MLC capacitor is shown in (b). Note the difference in fracture behaviour between the regions containing different numbers of electrodes.

primary reason for the change is the increased number of ceramic bonds or 'pillars' formed through the optimised electrode system at the reduced ink laydowns typically used in MLC production. A secondary beneficial effect is believed to be the effective annealing during firing of the termination paste of process-induced mechanical stresses.

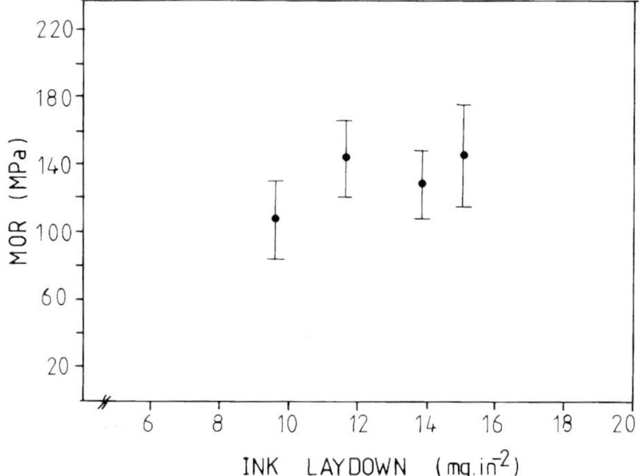

Figure 10 Flexure strength of as-fired X7R MLC capacitors (containing 17 active electrodes) as a function of ink laydown concentration.

Ideally, three-point bend tests and determination of flexure strength should be performed on monolithic ceramic specimens of known geometry. However, in the case of a MLC capacitor, which should be regarded as a complex composite system, the flexure strength determined via eqn (1) should not be taken as absolute but as a good approximation. For a full understanding, a suitable theoretical model is required, that takes into consideration the contribution of the electrode metal and the ceramic-metal interface. Various workers, e.g. Refs 16, 17, have attempted to develop composite-based models to accurately predict the stress distributions and mechanical behaviour of MLC capacitors.

The results of this study have shown that the flexure strength of MLC capacitors is dependent upon both design features and processing conditions, with the capacitors behaving as complex composite systems.

Thermal shock resistance continues to be of interest to MLC manufacturers especially in the use of non-leaded chip capacitors in surface mount form. This is because the assembly stresses of the surface mount units in some systems, notably from wave solder (where the capacitor is plunged directly into a molten solder wave in order to make the solder joint) can be very high. This is particularly true if the customer does not take notice of good practice precautions, such as necessary pre-heat and rate of heating and cooling. It is in the manufacturers interest therefore to fabricate dielectric/electrode structures which are resistant to the stresses generated by the users. Thermal diffusivity data provides at least some insight into the thermal shock resistance. Preliminary measurements of diffusivity for blanks of X7R formulation over the temperature range 70–600 °C yield results of the order of 0.006 cm^2 sec^{-1}.This is approximately a factor of 10 lower than that for alumina ceramic under similar conditions. Hence the thermal shock resistance for the bulk X7R ceramic is expected to be poorer than that for alumina, although the presence of the electrodes in the real capacitors should help to dissipate some of the heat. The next stage is to determine the thermal diffusivity and the thermal shock resistance of X7R ceramic capacitors possessing different

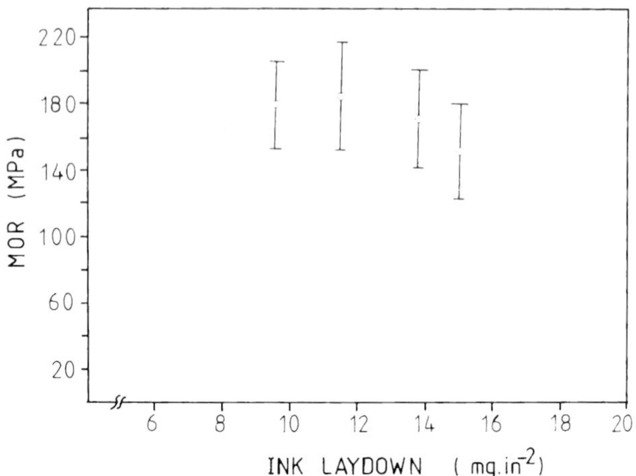

Figure 11 Flexure strength of as-fired and terminated X7R MLC capacitors (containing 17 active electrodes) as a function of ink laydown concentration.

numbers of internal electrodes. However, the relatively low thermal diffusivity of the X7R dielectric implies that the presence of the internal electrodes is beneficial to the thermal shock characteristics.

CONCLUSIONS

MLC capacitors are complex systems and it is difficult to clearly identify the effect on the mechanical properties of any one design or processing feature in isolation. The results of this study show that the flexure strength of MLC capacitors vary with composition, electrode type, number of active electrodes, ink laydown density and post -processing steps.

REFERENCES

1. S. W. Frieman and R.C. Pohanka, *J. Amer. Ceram. Soc.,* 1989, **72**, 2258–2263.
2. C. R. Koripella and H. V. DeMatos, J. Amer. Ceram Soc., 1989, **72**, 2241–2246.
3. J. G. Pepin, W. Borland, P. O'Callahan and R. J. S. Young, J. *Amer. Ceram. Soc.,* 1989, **72**, 2287–2291.
4. L. Bechou, S. Mejdi, Y. Ousten and Y. Danto, *Quality and Reliability Eng. Internat.,* 1996, 12, 43–53.
5. Y. C. Chan, X. Dai, G. E. Jin, N. K. Bao and P. S. Chung, *Microwave and Optical Technol. Letts,* 1997, **16**, 80–85.
6. K. R. McKinnery, R. W. Rice and C. Cm. Wu, *J. Amer. Ceram. Soc.,* 1986, **69**, C228–C230.
7. C. R. Koripella, *IEEE Trans. On Compts. Hybrids and Manufact. Technol.,* 1991, **14**, 718–724.
8. R. F. Cook, S. W. Frieman, B. R. Lawn and R. C. Pohanka, *Ferroelectrics,* 1983, **50**, 267–272.
9. H. Kanai, T., Fukazawa, O. Furukawo and Y. Yamishita, *J. Amer. Ceram Soc.,* 1997, **80**, 594–598.
10. M. Gee, Private communication, 1989.

11. *Chip Type Monolithic Ceramic Capacitor Technical Data,* Murata Manufacturing Company.
12. G. de With, *J. Europ. Ceram. Soc.,* 1993, **12**, 323–336.
13. R. Taylor, *J. Phys. E. Sci. Instruments,* 1980, **13**, 1193–1199.
14. R. B. Ross, *Metallic Materials Specification Handbook,* Fletcher and Sons, Norwich, U.K.
15. R. Al-Saffar, PhD Thesis, University of Manchester, 1992.
16. W. Carlson, T. Rutt and M. Wild, *Ferroelectrics. Letts,* 1996, **21**, 1–9.
17. A. A. Wereszczak, K. Breder, R. T. Bridge and L. Riestler, paper SXTII-008-98 presented at 100th American Ceramic Society Meeting, Cincinnati, May 1998.

Indentation Damage Zone in Calcium and Lead Titanate Ceramics

J. M. CALDERON-MORENO, F. GUIU, M. MEREDITH,

M. J. REECE and K. S. SOHN.

Materials Department, Queen Mary and Westfield College, London E1 4NS, UK

ABSTRACT

Indentation cracks are driven to their equilibrium length by residual stresses produced by the plastic zone beneath the indenter. In this study we investigated the shape and size of this zone for several lead-titanate based materials. These materials have a range of tetragonal distortions, or c/a ratios. For isotropic unpoled samples the size of the plastic zone can be predicted by elastoplastic theory. The shape of the plastic zone for poled materials is slightly elliptical with its long and short axes perpendicular and parallel to the poling direction respectively.

INTRODUCTION

The study of the mechanical properties of piezoelectric ceramics is becoming of increasing importance because of their increasing application under conditions that subject them to relatively large electromechanical loading. The ceramics studied in this work, lead zirconate titanate (PZT) and calcium modified lead titanate (PCT) are ferroelectrics with a wide range of applications. PZT is the most widely used electroceramic in piezoelectric devices, e.g., transducers and actuators, and PCT exhibits an unusually large ratio between thickness and planar electromechanical coupling factors, and so a high electromechanical anisotropy, which is particularly suitable for ultrasonic transducer applications. These materials transform from cubic (paraelectric) to tetragonal (ferroelectric) on cooling, and the strains generated are relieved by the formation of twin related ferroelectric domains. The domains can be switched (reoriented) by an electric field or a mechanical stress. In most applications the materials are not significantly mechanically stressed, so the study of their mechanical properties has not been a significant issue in the past. However, they are being applied in increasingly more demanding applications, requiring very long life times and large displacements, where the degradation of their electromechanical properties and mechanical failure become significant issues. The mechanical behaviour of PZT type piezoelectrics has been the subject of some recent studies: the damage produced by mechanical fatigue,[1] and the behaviour of cracks under both mechanical loads.[2,3] and electric fields.[4] The reorientation of 90° domains in the near stress field of a crack tip is considered a possible toughening mechanism .[5,6] The mechanical behavior of the PCT materials used in this study has been investigated previously.[7,8]

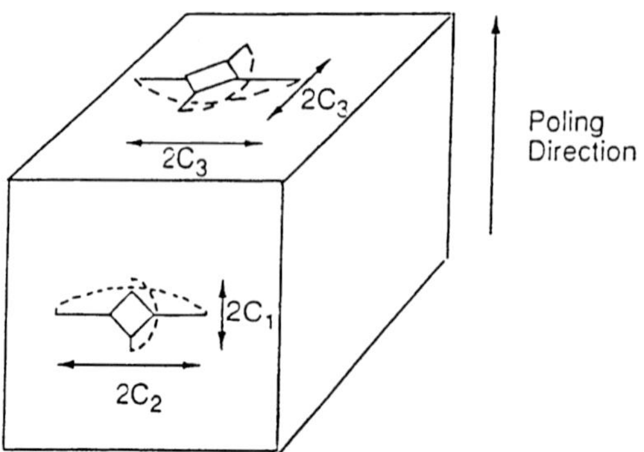

Figure 1 Different orientations of the cracks in the poled sample.

Indentation cracks are driven to their equilibrium length by residual stresses produced by the plastic zone beneath the indentation impression. In this study we investigated the shape and size of this deformation zone for several lead titanate based materials.

MATERIAL AND EXPERIMENTAL TECHNIQUES

The materials used in the present work were four different ceramics of the following compositions: PZT: 0.449$PbTiO_3$–0. 431 $PbZrO_3$–0.12$(Mn_{1/3}Sb_{2/3})O_3$, sintered at 1200 °C for 2h; PZT 4D, a commercial hard piezoelectric material; PCTl, 0.70Pb 0.30Ca $((Co_{1/2}W_{1/2})_{1/20}Ti_{19/20})$ O_3 + 1% MnO; and PCT2, 0.76Pb 0.24Ca $((Co_{1/2}W_{1/2})_{1/20}Ti_{19/20})$ O_3 + 1% MnO, both PCTs were sintered at 1050 °C for 3h., and their resulting piezoelectric properties are reported elsewhere.[3,7,9]

All the materials were single phase, with a tetragonal structure. The c/a ratios, a measure of their tetragonal distortion, were determined from XRD patterns, giving values of 1.015 for PZT, 1.016 for PZT 4D 1.025 for PCTI, and 1.038 for PCT2. Grain sizes ranged from 3 to 4 μm for both PCT's, above the minimum size needed for the formation of 90° and 180° domains inside single grains.[10] The densities were determined using the Archimedes method, giving values of 7.87, *7.57,* 6.59 and 6.73 g cm^{-3} for PZT, PZT 4D, PCT1 and PCT2 respectively. The porosity was found to be less than 6% for all the samples, based on the measured densities.

Vickers indentations, with applied loads ranging from 25 to 100 N, were made on polished surfaces of unpoled samples, and on two different surfaces in the poled samples, parallel and perpendicular to the poling axis. On the surface parallel to the poling direction, the two diagonals of the indentations were either parallel or perpendicular to the poling axis. The different possible orientations of the cracks emanating from the corners of the indentations are shown, and labeled as C_1, C_2, and C_3 in Fig. 1. The indentation cracks in the unpoled samples are referred to as C_u. The indentation impressions only give an indication of the surface crack lengths but no information about the crack profiles and plastic deformation zone. To investigate these, the indented surfaces were sequentially polished in steps of ~30 μm. Then the indentation crack lengths and the deformation zone at different depths

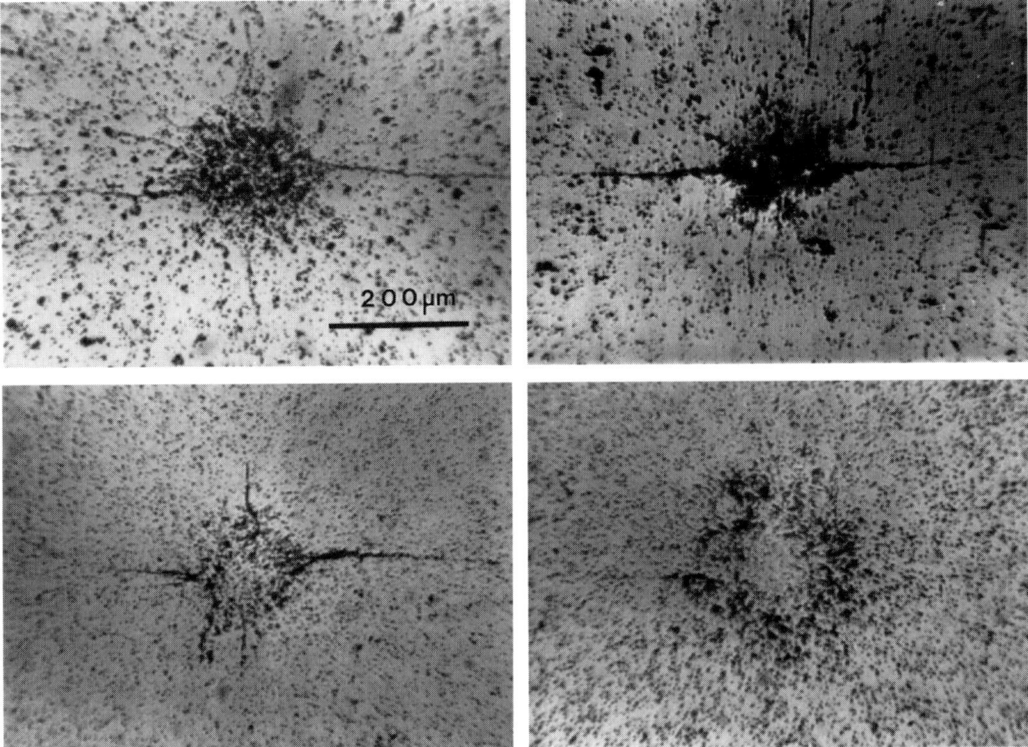

Fig. 2 Optical micrographs showing the cracks and the deformed zone at — 100 μm under the surface parallel to the poling (98 N indentation load). Clockwise from top left: PZT, PZT 4D, PCT1, PCT2.

from the initial surface were measured to determine their profiles. The plastic deformation zone could be identified by the substantial grain pull-out produced within the zone during the polishing, and by the fact that the cracks did not penetrate the zone.

Knoop indentations were also made on the different surfaces to determine the ratio, E/H, of Young's modulus tohardness from the ratio of the short and long diagonal lenghts following the method of Marshall *et al.*[11]

EXPERIMENTAL RESULTS

Figure 2 shows, for poled samples of the four materials the cross sections corresponding to a 98 N indentation on the surface parallel to the poling axis, at a depth of ~100 μm from the indented surface. In the highly plastically deformed region beneath the indenter the material was probably highly microcracked, which caused extensive grain pull-out during polishing, and offered a chance to study the anisotropy of the plastic deformation.

For the poled PZT the deformation zone below the indenter is asymmetric, with an elliptical cross section, with the elongated axis in the direction perpendicular to the poling direction. This direction also corresponds to the direction of the longest indentation cracks. In the case of the poled PCT type materials, PCT1 and PCT2, the cross section of the deformed zone in Fig. 2 has a nearly circular shape. For the PZT 4D material the damage appears more severe with large regions of connected pull-out. The cracks were again more developed in the direction perpendicular to the poling direction, showing the same type of anisotropy as the PZT. Although at the depth beneath the indenter shown in Fig. 2 the shape of the deformed zone for the PCT type materials is isotropic, it is not the case at all depths of the indentation profile. Figure 3 shows the shapes of the deformed zones at different depths. In all cases the deformed zone becomes elliptical in shape at the lowest depths.

DISCUSSION AND CONCLUSIONS

Expressions for equilibrium crack evolution in the median/radial crack system derived from elastoplastic theory can be used to evaluate the depth of the plastic zone, b, underneath the indentation impression. We shall use the equation fromRefs 12 and 13:

$$b = \alpha(E/H)^{1/2}(\text{cotg}\ \Psi/\pi)^{1/3}\ P^{1/2}/(2H)^{1/2}, \tag{2}$$

where $\Psi = 74°$ (angle between pyramidal edges of Vickers indenter), and α is a constant with a value of 0.68 131. Using the values for *E, H* and *E/H* given in the Fig. 4 we obtain values for *b*. The experimental results not only follow the dependence with $P^{1/2}$, but also quantitatively agree with the theoretical predictions from eqn (2). The material with the largest deformed zone is PZT 4D, which also has the highest $E^{1/2}/H$ ratio. The dependence of *b* with on $E^{1/2}H$ is in good agreement with the results for PZT 4D, PZT and PCT 1. The theoretical prediction for PCT2, the material with the smallest deformation zone, does not give a good fit. In this case, the calculated value for *E* from the Knoop indentation method appears unrealistically high and this could explain the discrepancy.

Table 1 Fracture toughness anisotropy and relative values of toughening.

	c/a	$K_{Ic}(C_1)$	$K_{Ic}(C_2)$	$K_{Ic}(C_2)/K_{Ic}(C_1)$
PZT	1.015	1.7 ± 0.1	0.7 ± 0.1	0.41
PZT 4D	1.016	2.0 ± 0.1	0.7 ± 0.1	0.35
PCT1	1.025	1.8 ± 0.1	1.0 ± 0.1	0.55
PCT2	1.038	2.6 ± 0.1	1.5 ± 0.1	0.58

The toughness anisotropy of these piezoelectrics has been studied in detail by the authors in previous work.[14] The values of hardness, Hv, Hk, and Young's Modulus, *E*. are the same for unpoled and poled materials, and there was no anisotropy with respect to the poling axis. K_{Ic}

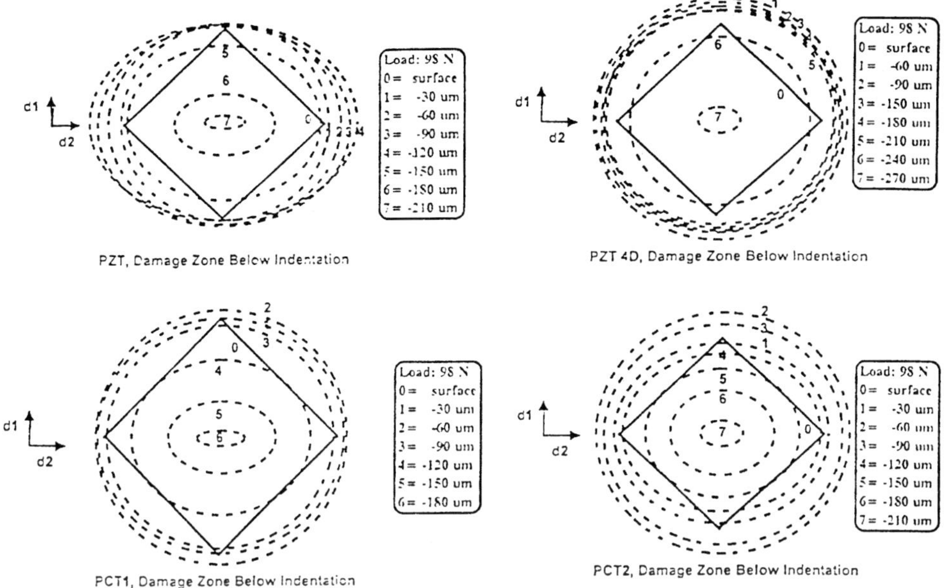

Fig. 3 Cross section of the plastically deformed zone at different depths below the indentation, in planes parallel to the poling axis, for, clockwise from top left: PZT, PZT 4D, PCT1, PCT2.

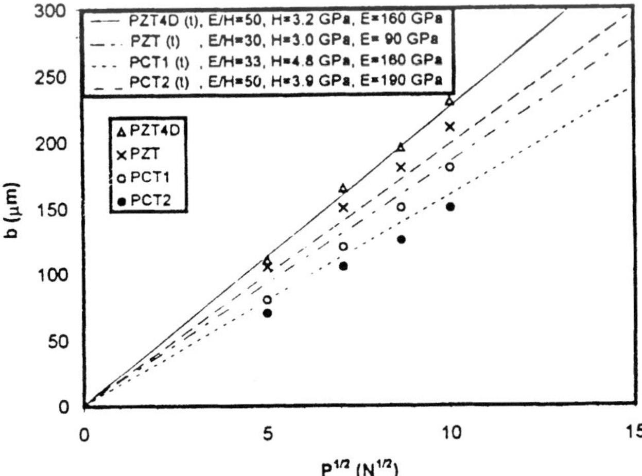

Fig. 4 Values of b (depth of the deformed zone) vs. P's, P being the indentation load, for the different piezoelectric materials, the dotted lines in the Fig. represent the theoretical values of *b* for each material.

was affected by the poling, with a significant anisotropy in the poled samples, with the highest and lowest values for crack lengths in the directions parallel and perpendicular to the poling axis, respectively. Values of K_{Ic} are shown in Table 1.

There is some uncertainty about the origin of the indentation fracture toughness anisotropy. When the fracture toughness of the same material is measured by other methods, such as the single edge notch beam (SENB), a much smaller anisotropy is obtained.[14] In previous work, this observation led the authors to suggest that the large fracture toughness anisotropy may be, at least partly, an artifact of the indentation method. Plastic anisotropy of the poled material may produce a significant anisotropy of the residual stress field that drives the indentation cracks to their equilibrium lentgh. The elastic/plastic anisotropy of piezoelectrics is easily revealed by compression testing of poled materials. They exhibit a higher pseudo elastic limit in compression in the poling direction than in compression normal to the poling direction. This effect has been observed and reported by several authors[15,16] and is particularly large for a hard PZT.

The conclusion of this study is that care should be taken when applying elastoplastic theory to anisotropic materials, such as piezoelectrics. It is possible that it may exaggerate the observed anisotropy in fracture behaviour relative to the poling direction.

ACKNOWLEDGEMENTS

This work has been supported by an Engineering and Physical Sciences Research Council Research Grant(GR/L90361).

REFERENCES

1. C. S. Lynch, L. Chen, Z. Suo, R. M. Mcmeeking and W. Yang, 'Crack Growth in Ferroelectric Ceramics Driven by Cyclic Polarization Switching', *J. Intell. Mater. Systems and Structures,* 1995, **6,** 191.
2. J. G. Bruce, W. W. Garberich and B. G. Koepke, 'Subcritical Crack Growth in PZT', D. P. H. Hasselman and F. F. Lange eds, *Fracture Mechanics of Ceramics,* Vol. 5, Plenum Publ. Co, N. Y. (1978), 687.
3. F. Guiu, B. S. Hahn, H. L. Lee and M. J. Reece, 'Growth of Indentation Cracks in Poled and Unpoled PZT', *J. Eur. Ceram. Soc.,* 1997, **17**, 505-512.
4. Hengchu Cao and A. G. Evans, 'Electric-Field Induced Fatigue Crack Growth in Piezoelectrics', *J. Am. Ceram. Soc.,* 1994, **77**, 1783–1786.
5. G. G. Pisarenko, V. M. Chushko and S. P. Kovalev, 'Anisotropy of Fracture Toughness of Piezoelectric Ceramics', *J. Am. Ceram. Soc.,* 1985, **68**, 259–265.
6. Karun Mehta and Anil V. Virkar, 'Fracture Mechanism in Ferroelectric-Ferroelastic Lead Zirconate Titanate (Zr:Ti = 0.54:0.56) Ceramics', *J. Am. Ceram. Soc.,* 1990, **73**, 567–574.
7. J. Ricote, C. Alemany and L. Pardo, 'Microstructural Effects on Dielectrics and Piezoelectric Behavior of Calcium-modified Lead Titanate Ceramics', *J. Mater. Res.,* 1995, **10**, 3194–3203.
8. J. Ricote, L. Pardo and B. Jimenez, 'Mechanical Characterisation of Calcium-modified lead titanate ceramics by Indentation Methods', *J. Mat. Sci.,* 1994, **29**, 3248–3254.
9. Catalogue of 'Piezoelectric Ceramic Products', Morgan Matroc Ltd., Unilator division, UK.
10. G. Arlt, 'The Influence of the Microstructure on the properties of Ferroelectric Ceramics', *Bol. Soc. Esp. Ceram. Vidrio,* 1995, **34** (5–6), 267–271.
11. D. B. Marshall, T. Noma and A. G. Evans, 'A Simple Method for Determining Elastic-Modulus-to-Hardness Ratios using Knoop Indentation Measurements', *J. Am. Ceram. Soc.,* 1982, **65**, C175–C176.

12. D. B. Marshall, B. R. Lawn and A. G. Evans, 'Elastic/Plastic Indentation Damage in Ceramics: The Lateral Crack System', J. Am. Ceram. Soc., 1982, **65**, 561–566.
13. M. Reece and F. Guiu, 'Repeated Indentation Method for Studying Cyclic Fatigue in Ceramics', *J. Am. Ceram. Soc.,* 1990, **73** (4), 1004–1013.
14. J. M. Calderon-Moreno, F. Guiu, M. Meredith and M. J. Reece, 'Fracture Toughness Anisotropy of PZT', *Mat. Sci. and Eng.,* 1997, **A234–236**, 1062–1066.
15. Ansgar B. Schaufele and Karl H. Hardtl, 'Ferroelastic Properties of Lead Zirconate Titanate Ceramics', *J. Am. Ceram. Soc.,* 1996, **79**, 2637–2640.
16. H. Cao and A. G. Evans, 'Nonlinear Deformation of Ferroelectric Ceramics', *J. Am. Ceram. Soc.,* 1993, **76** (4), 890–896.

In Situ Measurements of Residual Stresses in a Thermal Barrier Coating During Thermal Cycling

A. AHMED and R. I. TODD

Manchester Materials Science Centre, University of Manchester and UMIST, Grosvenor Street, Manchester, M1 7HS

L. PEJRYD and J. WIGREN

Volvo Aero Corporation, SE-461 81 Trollhättan, Sweden

ABSTRACT

The residual stress changes in the bond coat (NiCoCrAlY) and top coat (YSZ) of a heat treated, plasma sprayed thermal barrier coating during thermal cycling between 20 °C and 700 °C have been measured using Stoney's method, in which the stress changes are deduced from the change in curvature of the coated substrate. The thermomechanical properties of the bond coat and substrate materials were measured in isolation to help in the interpretation of the stress measurements. The stress cycles in the bond coat reached a steady state after three thermal cycles, with a stress range of 100MPa. The bond coat stresses could be explained in terms of elastic accommodation of its TCE mismatch with the substrate combined with inelastic deformation caused by microstructural changes. The stresses in the top coat were purely elastic in origin, with a stress range of 135MPa. There was no evidence suggesting that the stresses in the top coat were influenced by the stresses in the bond coat.

1. INTRODUCTION

The efficiency of gas turbine engines is limited by their operating temperature, and the refractory nature of many ceramics therefore makes them attractive candidates for use in such engines. The widespread use of monolithic ceramic components has been prevented, however, by the mechanical unreliability inherent in their brittleness. The beneficial properties of ceramics have been utilised to good effect nevertheless, in the form of coatings, which can reduce the temperature of the underlying cooled metallic component by more than 150 °C relative to the surroundings.

Thermal barrier coatings (TBCs) commonly comprise two components: a ceramic top coat, typically 300mm thick, and a metallic bond coat with a thickness of ~150 µm. The majority of commercial coatings are currently applied by plasma spraying, but other methods are also used, notably electron beam physical vapour deposition. The top coat normally consists of zirconia partially stabilised with yttria, and provides the thermal insulation. The role of the bond coat is largely to protect the substrate from oxidation and to provide a surface with good adherence for the top coat, but it may also play a mechanical role, as is described below. The bond coat compositions are usually based on the MCrAlX system, where M is nickel and/or cobalt, and X is an 'active element' such as yttrium which improves the adherence of the alumina scale which forms on oxidation.

Whilst TBCs already have sufficient longevity for commercial use, failure does occur eventually, usually by spallation. Failure must be avoided if TBCs are to be used to best advantage on critical components, so it is important for the mechanisms involved to be understood. Two factors thought to influence failure are: (i) the stresses caused by the volume expansion associated with oxidation of the bond coat, and (ii) residual stresses caused by thermal expansion mismatches between the top coat and the substrate. The oxidation stresses are widely considered to be the more important, but some studies comparing the lifetimes of different coatings have shown that there is no direct correlation between specific mass gain and TBC life,[2,3] so it is clear that the thermal expansion mismatch stresses also play a role in failure. The bond coat is also thought to influence the residual stresses in the top coat, and is often considered to act as a compliant layer which can allow relaxation of residual stresses by plastic deformation or creep. There is evidence that creep resistant bond coats lead to long TBC lifetimes.[4,5]

The objective of the work described here was to gain a better understanding of the residual stresses in a TBC during thermal cycling by correlating direct measurements of the stresses with modelling based on measurements of the thermomechanical properties of the individual constituents of the TBC in isolation. The work reported here focuses on the bond coat, though residual stress measurements for the top coat are also reported.

2. EXPERIMENTAL

Bond coats with a thickness of 150 μm were air plasma sprayed onto 20 mm × 100 mm rectangular Hastelloy X substrates. The bond coat alloy had the following composition: Ni-23Co, 17Cr, 12.5Al, 0.45Y. Free standing specimens of the plasma sprayed bond coat material with a thickness of 1 mm were also made by spraying onto a stainless steel substrate which was then dissolved using nitric acid. Some Hastelloy X substrates were given a complete TBC, by spraying a 300mm thick YSZ top coat onto bond coats deposited as above. All specimens were then heat treated in air, for 4 hours at 1080 °C followed by 32 hours at 870 °C.

Microstructures were examined using optical microscopy, SEM-EDX and XRD on specimens prepared using conventional metallographic techniques. Specimens for optical examination were etched using a solution of 33% conc. nitric acid, 33% acetic acid, 1% hydrofluoric acid, bal. water.

In order to measure the elastic modulus of the materials used, a four point bend testing rig capable of operating at temperatures of up to 1000 °C was constructed. The inner and outer spans were 10 and 20 mm respectively. LVDTs were used to measure accurately the displacement over the constant bending moment region between the inner pair of rollers. Tensile tests were performed in air on 20mm gauge length specimens of the free standing bond coat material at temperatures of between 20 °C and 935 °C. Vickers hardness tests were conducted on the same material at room temperature.

The thermal expansion behaviour of the bond coat and substrate materials was measured

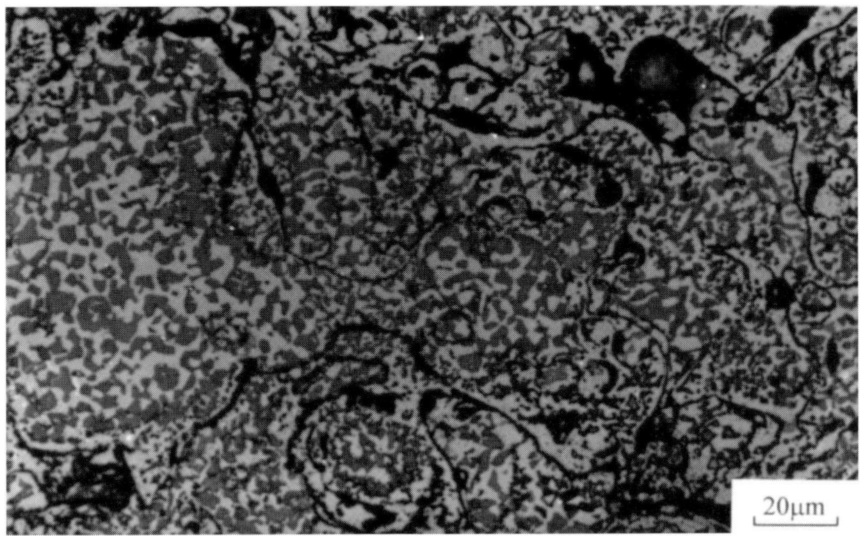

Figure 1 Optical micrograph of heat-treated bond coat microstructure.

according to methods defined in European Standard ENV 1159-1. The specimen dimensions were measured during three consecutive thermal cycles, from room temperature (cycle 1), or 100 °C (cycles 2 and 3) to 700 °C, with heating and cooling rates of 350 °C per hour and a one hour dwell at the maximum temperature.

Residual stress changes in coatings on Hastelloy X substrates were measured during the same sequence of thermal cycles as was used for the dilatometry. The measurements were made using the Stoney method,[9] in which the curvature of the substrate caused by the residual stresses is measured and used to deduce the residual stresses. The curvature changes were measured by clamping the specimens at one end such that they stood upright in a furnace equipped with a transparent window, and reflecting a laser beam from the polished, unclamped end of the substrate onto an external screen. The position of the reflected laser spot was monitored during cycling and used to deduce the specimen curvature. The tests were carried out in an argon atmosphere.

3. RESULTS AND DISCUSSION: CHARACTERISATION OF THE BOND COAT MATERIAL

3.1 Microstructure

Figure 1 shows the microstructure of the heat treated bond coat. The morphology of the material is typical of plasma sprayed coatings, consisting of an assembly of 'splats', occasional unmelted particles, and porosity and interlamellar cracks. Image analysis of micrographs such as Fig. 1 indicated around 20% porosity/cracks, but this is an overestimate as

Table 1 γ/β volume ratio and Vickers hardness as a function of thermal history. The as-treated value of γ/β volume ratio was deduced from analysis of optical micrographs, and the other values were deduced from this using XRD peak intensity measurements.

Thermal History	γ/β Ratio	H_v
as treated	1.2	230
+ 1 standard thermal cycle to 700 °C	1.6	320
+ 2 standard thermal cycles to 700 °C	1.7	290
+ 3 standard thermal cycles to 700 °C	1.9	300

Table 2 Young's modulus, E, at high and low temperature for bond coat and substrate.

Material	Temperature, °C	*E*, GPa
Bond Coat	20	118
	700	103
Ssubstrate	20	193
	700	180

some of the dark regions in the figure undoubtedly either consist of oxide, which also appears dark, or were caused by pullout during polishing.

Within the splats, the bond coat had a microduplex structure with approximately equal amounts of the two phases. The XRD and SEM-EDX results showed the two phases to be based on the γ and β-NiAl phases of the Ni-Al phase diagram, with Co substituting freely for Ni in both phases, and most of the Cr being in the γ solid solution.

The mean linear phase intercept was 5 μm in the as heat treated bond coat, and no significant change in average grain size was seen after thermal cycling. The proportions of the two phases did vary with thermal history, however. Table 1 shows how the γ/β ratio varied as a function of heat treatment. The results are consistent with the idea that the equilibrium γ/β ratio increases as the temperature is lowered. Thus, the as heat treated result corresponds to the equilibrium value at 870 °C, and subsequent cycling to the lower temperature of 700 °C increases the amount of γ at the expense of the β. The biggest change occurred during the first thermal cycle, but the further small increases were seen after the 2nd and 3rd cycles indicating that the kinetics of the phase transformation were sluggish below 700 °C.

3.2 Mechanical Properties

Table 2 shows the measured Young's modulus (*E*) for the free standing bond coat and Hastelloy X substrate at room temperature and 700 °C. Both materials showed a small reduction in stiffness at the higher temperature, this being more pronounced for the bond coat (13%) than for the substrate (7%). The room temperature value for the substrate is in good agreement with the alloy supplier's value of 200GPa, but the literature[6,7] suggests similar values hold for related bond coat alloys. The lower bond coat stiffness reported here can be explained in terms of the porosity in the plasma sprayed bond coat. A simple estimate of the effect of porosity is[8] that it reduces the Young's modulus to:

$$E = E_0 (1\text{-}2f)$$

where E_0 is the Young's modulus in the absence of porosity and f is the volume fraction of porosity. On this basis, the 20% porosity in the bond coat can be expected to reduce the Young's modulus to around 120GPa, which is in excellent agreement with the observed value of 118GPa.

Table 1 shows the hardness results for the free standing bond coat material. It is notable that the hardness of the as heat treated material increases by more than 30% on thermally cycling to 700 °C. This is attributable to increased ordering of one or both phases at lower temperatures, but the presence of ordering has yet to be investigated experimentally. After thermal cycling H_V is ~300, which corresponds to a high yield stress of ~1 GPa, despite the presence of porosity which would be expected to reduce the apparent hardness.

At temperatures of 700 °C and below, the behaviour in tension was brittle, with little or no plastic deformation preceding fracture (Table 3). (The elongations of more than 1% quoted in Table 3 are apparently at odds with this conclusion, but were measured from the crosshead displacement of the testing machine, which will include a substantial contribution from the machine compliance for such small elongations). This explains why the room temperature strength was about an order of magnitude less than the room temperature yield stress estimated from the hardness results. It is also consistent with the insensitivity of the strength at 700 °C to the strain rate.

Although the primary aim of this paper is to consider thermal cycling at temperatures of up to 700 °C, it is interesting to note the tensile properties of the bond coat at higher temperatures. At 800 °C the elongation increased sharply to more than 10%, indicating substantial plasticity, and there is also evidence of sensitivity to strain rate characteristic of creep. At 935 °C a superplastic elongation of 106% was obtained, which is remarkable in such a porous material. The low flow stress of only 4MPa is also typical of superplastic flow. Superplastic behaviour is to be expected at high temperature because of the fine grained microduplex structure described in section 3.1, which allows diffusional accommodation of grain boundary sliding because of the small grain size, while being protected against grain growth by the presence of two phases.

3.3 *Dilatometry*

Figure 2(a) shows the dilatometer trace for the bond coat during the first three thermal cycles after heat treatment. The three cycles were performed consecutively without removing the sample between cycles. Two features are noteworthy. First, it is evident that the first cycle showed a net reduction in specimen length, whereas the specimen returned to its initial length during subsequent cycles. The steady state thermal expansion coefficient averaged over the whole cycle was $15.4\ 10^{-6}\ K^{-1}$. Secondly, a small but distinct increase in thermal expansion coefficient (TCE) is apparent at temperatures exceeding 600 °C. These features will be discussed further in Section 4.1.

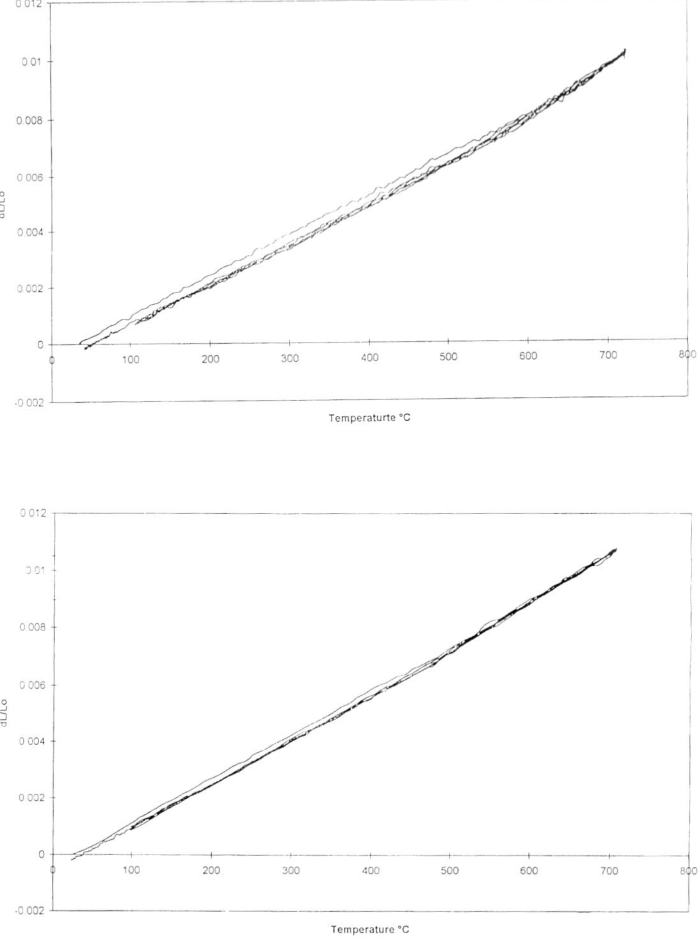

Figure 2 Dilatometer results presented as fractional increase in length versus temperature: (a) freestanding bond coat material; (b) Hastelloy X substrate.

Table 3 Tensile test results for free-standing bond coat specimens.

Temperature, °C	Strain rate, s^{-1}	Tensile strength, MPa	Elongation, %
20	4×10^{-5}	118	1.5
700	4×10^{-5}	100	2.3
	4×10^{-4}	95	2.9
800	4×10^{-5}	68	14
	4×10^{-4}	74	11
935	4×10^{-4}	4	106

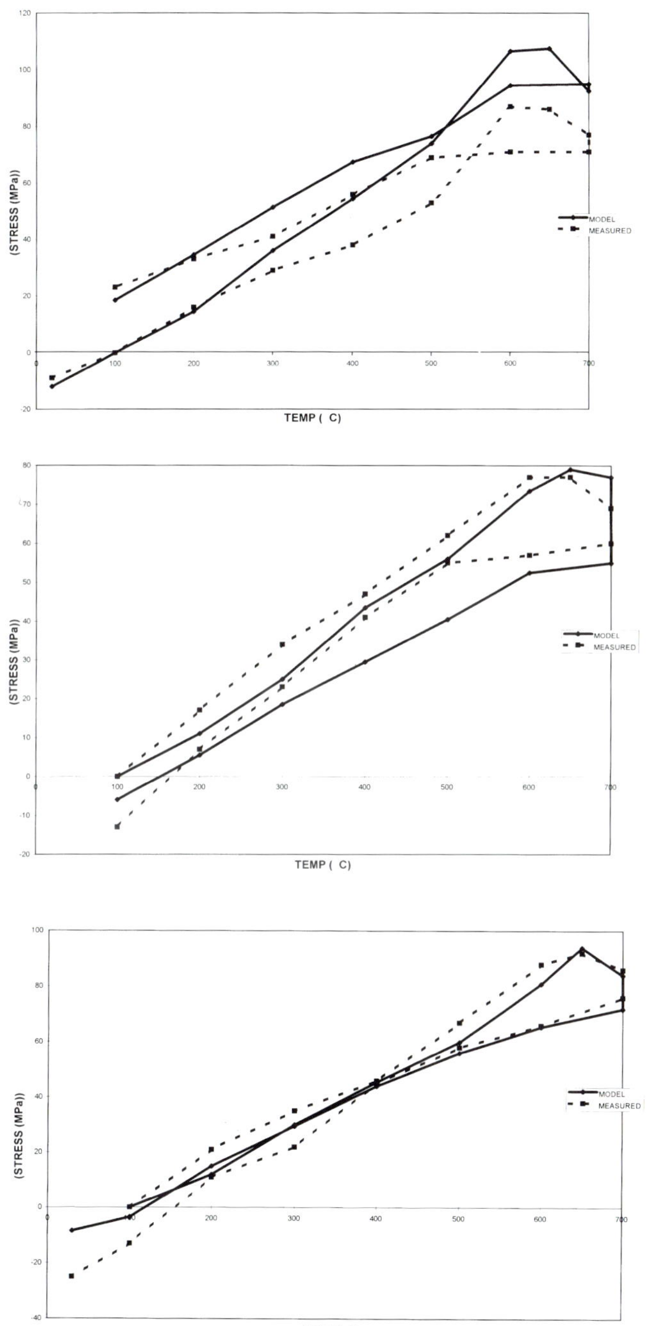

Figure 3 Residual stress changes in the bond coat during thermal cycling. Bold line = direct experimental result, dotted line = predicted using thermomechanical property results: (a) cycle 1; (b) cycle 2; (c) cycle 3.

Figure 2(b) shows the dilatometer results for the substrate material. There is much less hysteresis during the first cycle than in the bond coat, and although the TCE increases slightly with temperature (as is the case with all normal materials) there is no marked discontinuity as for the bond coat. The average TCE was 16.3 10^{-6} K^{-1}, i.e. greater than the average TCE of the bond coat.

4. RESULTS AND DISCUSSION: RESIDUAL STRESSES DURING THERMAL CYCLING

4.1 Bond Coat

The residual stress changes measured in the bond coat during the first three thermal cycles after heat treatment are shown in Figure 3. The stress change for each cycle is arbitrarily defined to be zero at 100 °C during heating. The results are the average of the results obtained from two specimens. At temperatures below 500 °C the stress change is linear with temperature, with the same gradient for all three cycles during both heating and cooling. There is a net positive stress change over the first cycle as a whole, but by the third cycle a steady state has been reached so that the stresses at the beginning and end of the cycle are the same. All three cycles show that the rate of increase of residual stress with temperature first diminishes, and then becomes negative above 600 °C, and there is considerable hysteresis in this temperature range. Hysteresis is observed during the third cycle despite the fact that there is no net stress change of the cycle as a whole. The steady state (cycle 3) stress range is ~100 MPa.

Figure 3 also show the predictions of a model for the stresses. The model assumes a uniform biaxial stress in the coating, and is based on simple beam theory. Only elastic deformation is accounted for explicitly in the model, and the experimental measurements of the relevant thermomechanical properties (TCE, *E* – assumed constant with the room temperature value) of the bond coat described in section 3 have been used to obtain the predictions of residual stress change shown. The predictions agree with the experimental results to within about 20MPa in absolute terms, which is within the combined error range of the techniques used. More significantly, the distinctive features of the experimental stress measurements described above are reproduced in great detail, including the net increase in stress during the first cycle, the linear portion of the curve, on heating and cooling, and the peak in stress and hysteresis above 500 °C.

The good agreement between model and experiment allows the processes which give rise to the residual stress changes to be identified. It is clear that the linear portion of the stress change versus stress plots below 500 °C is a straightforward elastic response to the thermal expansion mismatch between two materials with approximately constant (but differing) TCE. The hysteresis at higher temperature, however, indicates inelastic behaviour, and it is tempting to ascribe this to conventional plastic deformation or creep. These processes cannot be responsible for the hysteresis, however, because they are correctly predicted by the model, which assumes only elastic behaviour. The hysteresis predicted by the model is a consequence of hysteresis in the dilatometer results for the same thermal cycle, and the reversal of the stress increase on heating

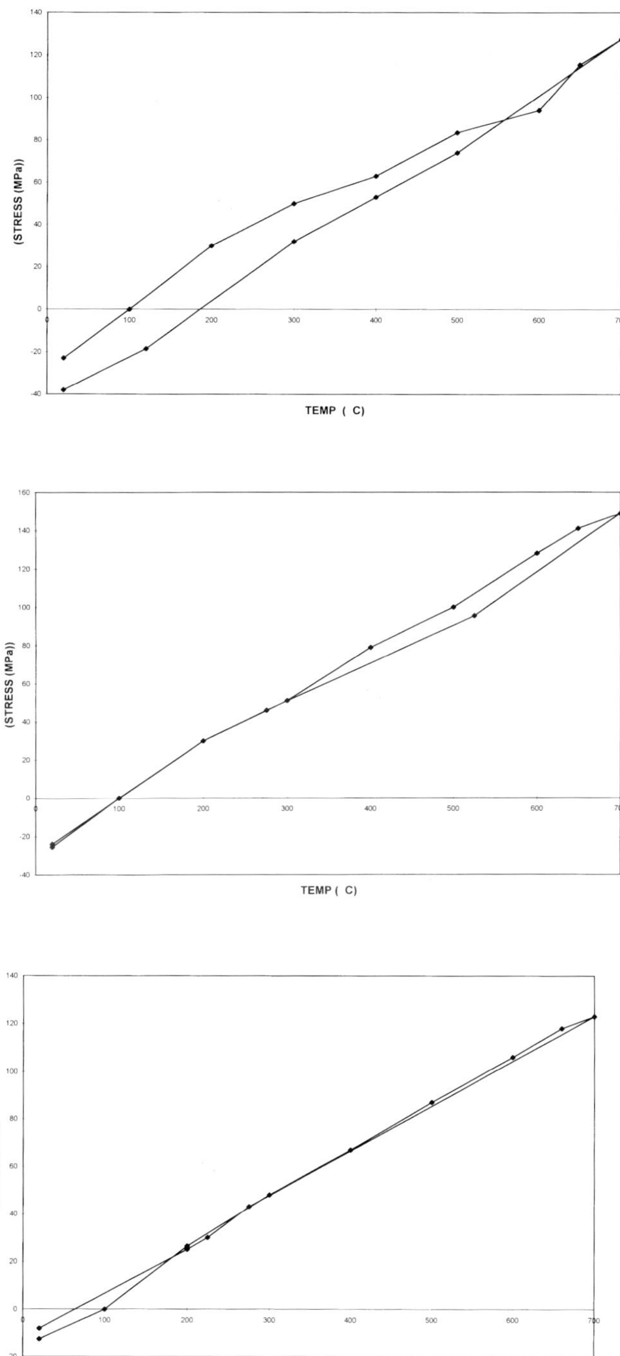

Figure 4 Measured stress changes in the top coat during thermal cycling: (a) cycle 1; (b) cycle 2; (c) cycle 3.

above 500 °C is caused by the change in the apparent TCE of the bond coat above this temperature described in Section 3.3. Since there is negligible applied stress in the material when in the dilatometer, these effects are not the result of stress-dependent processes such as creep or plastic deformation. This conclusion is supported by the tensile mechanical properties reported in Section 3.2. Although we cannot determine the absolute value of the stress from these results, it is reasonable to assume that the tensile stress in the bond coat cannot exceed its strength, and no plastic deformation was observed in the tensile tests for temperatures of 700 °C and below.

The results can be explained, however, by considering the microstructure of the bond coat as described in Section 3.1, and in particular the dependence of the equilibrium γ/β volume ratio on temperature. As the temperature rises, the amount of β phase increases at the expense of the fcc γ. It is reasonable to suppose that this causes a net expansion of the material, because the bcc-based β has less efficient atomic packing than the close packed γ, although the specific volume change also depends on the atomic radii of the atoms transferred between the two phases. Nevertheless, this offers a convincing explanation for the results, in that the apparent TCE of the bond coat would be increased at higher temperatures, where the diffusion rate would be sufficiently rapid to allow the phase ratio to change in the time scale of the thermal cycle, but would still be sufficiently slow to cause hysteresis.

4.2 Top Coat

By assuming that the residual stress during thermal cycling is unchanged in the presence of a ceramic top coat, the stress changes in the top coat can be estimated experimentally by measuring the average stress in the combined top coat and bond coat, and subtracting the contribution of the bond coat. The stress change in the YSZ top coat for the first three cycles after heat treatment are shown in Fig. 4. There is a small net stress change of about 15 MPa during the first cycle, which may be simply experimental error, but by the third cycle a steady state has been attained in which the stress against temperature plot approximates closely to a straight line, with no hysteresis. This suggests that the top coat is deforming in a purely elastic manner. The stress range is 135 MPa. For a thin coating on a thick substrate, the biaxial stress in the coating is given by $E\Delta\alpha\Delta T/(1-\nu)$, where E and ν are the Young's modulus and Poisson's ratio of the coating, and $\Delta\alpha\Delta T$ is the thermal expansion mismatch with the substrate. Using the following values for the top coat properties: $E = 17$GPa (see Ref. 10), $\nu = 0.25$, $\alpha = 10.0 \times 10^{-6}$ K^{-1}(see Ref. 10), the measured value of the substrate TCE of $16.3\ 10^{-6}$ K^{-1}, and the appropriate value for ΔT of 680 °C, gives a predicted stress of 97MPa, which is in very good agreement with the measurements given the uncertainties in the thermomechanical properties of the top coat used.

5. RELATION TO TBC LIFETIME

The main conclusion to be drawn from the present results is that if the TBC used here is cycled between 20 °C and 700 °C, a steady state will eventually be established in which both the top coat and the bond coat will deform in a predominantly elastic manner, the only exception being a small amount of inelastic deformation in the bond coat caused by reversible microstructural changes. There is little evidence for relaxation of the residual stresses in the top coat as a result of the presence of the bond coat. In the present case, this may be attributed to the high stiffness, yield stress and creep resistance of the bond coat in the temperature range considered, although the very large lateral dimensions of the coating compared to its thickness make it doubtful that simple relaxation of the in-plane stresses by macroscopic bond coat deformation would occur no matter how compliant its nature. Local relaxation at cracks in the top coat or ridges at its interface with the bond coat may be possible under conditions in which the bond coat is more compliant, and the extremely low flow stress of the bond coat found at temperatures in excess of 900 °C certainly make this a possibility during thermal cycling to higher temperatures. Future work will investigate this possibility.

Another question raised by the results concerns the events occurring during heating of an as-fabricated coating for the first time (for heat treatment in the present case). It is known that both the bond coat and the top coat are in tension in the present system immediately after spraying, and if the elastic stresses measured here occurred during initial heating the strengths of both coatings would be exceeded. Either the coatings must fragment by the formation of cracks normal to the plane of the coating, or the stresses must be lower because the thermomechanical properties of the coating materials are different to those measured here, because the microstructure of the as-sprayed material is different, for example. This is certainly true of the bond coat used here in that the as sprayed microstructure is in a highly strained metastable state with limited crystallinity. On crystallisation, the grain size can be anticipated to be much smaller than in the annealed structure reported here, and the material may, therefore, undergo superplastic deformation at lower temperatures and faster strain rates. Work is also in progress to determine which of these possibilities is correct.

6. CONCLUSIONS

1. The residual stress changes in a heat treated thermal barrier coating have been measured during thermal cycling between 20 °C and 700 °C.
2. The stress range in the bond coat was 100 MPa, and could be explained in terms of a combination of elastic accommodation of its TCE mismatch with the substrate, and inelastic deformation caused by microstructural changes.
3. The stresses in the top coat were elastic with a range of 135 MPa.
4. There was no evidence suggesting that the stresses in the top coat were influenced by the stresses in the bond coat.

ACKNOWLEDGEMENTS

This work was funded by Volvo Aero Corporation. We are grateful to Mr A. Wallwork for assistance with the dilatometry, and to Mr I. Easdon for assistance with the mechanical testing.

REFERENCES

1. S. Stecura, *Thin Solid Films,* 1980, **73**, 481.
2. S. Stecura, *Thin Solid Films,* 1989, **182**, 121.
3. L. S. Cook, A. Wolfenden and W. J. Brindley, *J. Mat. Sci.,* 1994, **29**, 5104.
4. W. J. Brindley and J. D. Whittenberger, *Materials Science and Engineering,* 1993, **A163**, 33.
5. D. J. Wortman, E. C. Duderstadt and W. A. Nelson, *J. Eng. for Gas Turbines and Power,* 1990, **112**, 527.
6. J. M. Veys, A. Riviere and R. Mevrel, *Proc. 1st Plasma Technik Symposium (1988)*, Wohlen, Switzerland, Plasma Technik AG, 1988, vol. 2, 115.
7. P.A. Siemers and R. L. Mehan, *Ceramic Sci. Eng.,* 1983, **4**, 9.
8. R. W. Davidge, *Mechanical Behaviour of Ceramics,* Cambridge University Press, 1979, 27.
9. G.G. Stoney, *Proc. R. Soc. London,* 1909, **A 82**, 172.
10. Woraphat Phucharden, D. Eng. Thesis, Cleveland State University, 1990.

Towards More Predictable Deformation in Sanitaryware Products

H.J. GLASS and A.M. KOOL

Delft University of Technology, Subfaculty of Applied Earth Sciences, Mijnbouwstraat 120, 2628 RX, Delft, The Netherlands

F. BLÖMER

Koninklijke Sphinx Sanitaryware, P.O. Box 1019, 6201 BA, Maastricht, the Netherlands

ABSTRACT

During the casting and drying of green sanitaryware products, stresses develop which lead to diferential shrinkage. By subjecting a series of green products to three-point bending, the stress-strain relation has been measured as a function of the moisture content. The experimental results have been characterized with strain hardening and mechanical models. The model parameters can be used for simulations serving to speed up development and to optimize the processing of green sanitaryware products. Furthermore, a qualitative explanation for the trends in the parameters is developed.

INTRODUCTION

The production of sanitaryware consists of three main stages: (1) preparation of the powder, (2) consolidation of the powder into the desired product shape and (3) sintering of the green product. Slip-casting is often used as a consolidation technique, leading to homogeneous but moist green products. Drying is necessary before the green product can be sintered. Different types of stress are exerted on the green product dunng the later stages of slip casting, and throughout the drying and sintering stages. As a result, the product deforms and assumes a different shape. In an optimised process, the deformation is such that the desired product shape is obtained at the end of the process. In practice, designing a process for a new product shape is costly and time-consuming. Important aspects are the shape of the casting mould and the selection of additives. The experimental effort can be significantly reduced by performing computer simulations where the product is subdivided into elements.[2] Finite element simulations require that suitable models describing the deformation behaviour are available. Therefore, the first step would be characterization of the deformation behaviour of the material. A setback of this approach is that subdivision of the product is quite complex. Consequently, a simplified but reliable alternative method is sought for predicting product deformation.

Deformation of a product is due to differential shrinkage and/or local failure. Shrinkage occurs during sintering, when a liquid phase is formed, and during drying of the product. Shrinkage is not likely to be isotropic[1] on account of gravity. Although differential shrinkage is largest during during the sintering stage, this contribution will focus on the deformation during the later stages of slip-casting and drying. Deformation is intimately related to the mechanical properties of the material, which indicate how a material responds to stresses created during drying or handling. Consequently, the material will be characterized in terms of the mechanical properties.

MECHANICAL PROPERTIES

The mechanical properties will be established as a function of the moisture content, which is a good measure to describe the progress of the drying process. For this purpose, a variety of techniques, which are generally destructive in nature,[3,4] can be used. This makes it difficult to determine the mechanical properties during the drying process itself. Instead, a series of samples with varying moisture contents are prepared by double-sided slip-casting. The latter should ensure that gradients in the moisture content within the sample are small. Techniques for measuring mechanical behaviour can be classified according to the type of stress which is created in the material. The resistance of a material to deformation increases with the type of stress in the following order:[5] tension, shear, compression. In practice, the effect of each type of stress will be influenced by the shape and the orientation of the product during drying. Therefore, the strongest deformation will be observed in those sections of the product where tensile stresses are present. This makes measurement of the material behaviour under influence of tensile stresses relevant. A simple tensile test is not suitable because the fastening of the product at either end affects the data. As this problem is avoided with bending tests, we have used the three-point bending test in this study.

EXPERIMENTAL

A slip is typically a water-based slurry containing a mixture of clay (mainly kaolinite), silica sand, fluxes (feldspar, nepheline), recycled scrap and deflocculants. Three different compositions are used during this study. These will be referenced as mass *A, B,* and *C.* Mass *A* and *B* both consist of roughly 40 % hard materials, 30 % ball clays and 30 % china clays. However, the amount and type of additives differ, leading to larger agglomerates in mass *B* than in mass *A*. Mass *C* contains approximately 50 % hard materials, 35 % ball clays and 15 % china clays. The agglomerate size is more or less equal to that of mass *A*.

The rheological properties of a slip are characterized in terms of the density, viscosity and thixotropy (thickening tendency). The density of the slip is conveniently expressed in terms of the moisture content, *M*. According to the British Standard:

$$M = 100\,(\rho - \rho_0)/\,\rho_0 \tag{1}$$

where ρ is the mass of the slip and ρ_0 is the density of the solids. Using a slip with $M = 36$ %, a series of rectangular products (length 150 mm, width 40 mm, height 12 mm) are obtained by double-sided casting. The residence time of the slip in the mould is maintained at 1.5 hours. After release of the products from the mould, *M* is approximately equal to 21 %. Prior

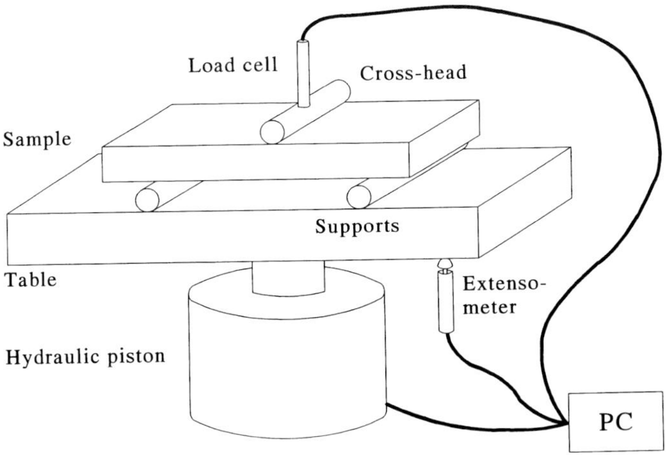

Figure 1 Schematic representation of the three-point bending test.

to testing, the product is dried in an oven (temperature 50 °C) for a certain time before the moisture content is determined. The testing equipment is shown in Fig. 1. During a test, the force is increased at a prescribed rate while the resulting deflection is measured as a function of time (rate: 1 measurement per second). Relatively high loading rates are used during this investigation. This loading rate is applicable when an external mechanical stress is applied. This could be case when the excess slip is removed from the mould, when the green product is released from the mould or when the product is handled. During drying, evaporation of moisture produces a so-called drying stress which also produces deformation. Deformation as a result of the drying stress is considered to be a creep process. However, when the ambient conditions during drying are carefully controlled, the drying stress is rather small compared to the aforementioned stresses. Figure 2 shows the relation between the moisture content and the (expected) deformation. The test is terminated when failure occurs, i.e. the product breaks. After each test, the moisture content is re-determined. Results show that the moisture content drops by, on average, 0.2 % during a test.

INTERPRETATION OF FORCE-DEFLECTION DATA

During bending, stresses are created within the product. By using narrow, rectangular products, the longitudinal stress is significantly larger than the lateral stresses. Consequently, the force applied by the cross-head at midspan can be directly related to the longitudinal or axial stress. It should be noted that the stress at midspan varies with the vertical position: the stress is compressive (+) near the upper surface, while the stress is tensile (–) near the lower surface.

When the deflection is sufficiently small, the deformation of the product is likely to be elastic. In that case, the axial stress, σ, is related to the axial strain, ε, by the modulus of elasticity modulus, E:

$$\sigma = E\varepsilon \qquad (2)$$

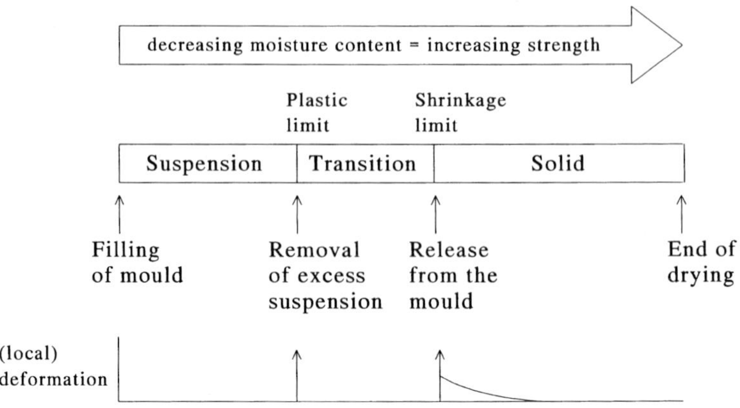

Fig. 2 Deformation of the green product prior to sintering.

For simply-supported beams, the force, F, applied by the cross-head is related to the axial stress, σ, as follows:[6]

$$\sigma = Mz/I \tag{3}$$

where M is the bending moment, z the vertical coordinate and I the moment of inertia. At midspan, the bending moment M equals $FL/4$, where L is the distance between the supports. The vertical coordinate z and the moment of inertia I are defined with respect to the position of the neutral axis, which marks the transition between compression and tension. When defining the vertical coordinate of the neutral axis as $z=0$, the upper and lower surfaces are initially located at $z = h/2$ and $z = -h/2$ respectively. By integrating between these boundaries, the moment of inertia I equals $bh^3/12$. Eqation (2) becomes:

$$\sigma = 3FLz/bh^3 \tag{4}$$

We are interested in the stress at the bottom surface, $\sigma_{-h/2}$, given by:

$$\sigma_{-h/2} = 3FLz/2bh^2 \tag{5}$$

During bending, the neutral axis will shift slightly because the deformation of the material subjected to tension or compression is probably different. Because deflection is rather small, this effect has not been considered.

Using eqn (5), the force can be converted to the axial stress as long as the deformation is reversible or elastic. When a certain yield stress, σ_y, is exceeded, any further increase of the stress will be reduced by plastic deformation. It is estimated that plasticity reduces the incremental stress by on average 25 %. However, this reduction could be compensated by shifting of the neutral axis: for large strains, the neutral axis shifts upwards. The latter effectively increases the stress at the bottom surface. For example, the 25 % reduction of the incremental stress is neutralised when the neutral axis has moved 0.08h upwards. In practice, the effects of plasticity and of motion of the neutral axis are not measured. We could assume that the two effects approximately cancel out or that the net effect is within the limits of random variations. In that case, eqn (5) is also used to convert the force when the deformation is (partly) plastic. This approach is adopted when mechanical elements are used to model the stress-strain relation.

We could also convert the force to a stress assuming that the deformation is strain-hardening plastic. Strain-hardening can be characterised by a strain hardening exponent, n, and a modulus of plasticity, H. This leads to the following stress-strain relation:

$$\sigma = H\varepsilon^{n} \tag{6}$$

When n has a value between zero and unity, the stress-strain curve has a progressively decreasing slope, which is typical of strain hardening behaviour. The conversion of the force into the stress is then given by:

$$\sigma = FL(n + 2)/2bh^{2} \tag{7}$$

The dependency of the stress on the hardening exponent, n, implies that the stress cannot be determined as such. Instead, the hardening exponent, n, as well as the modulus of plasticity, H, is determined from a graph of the force, F, versus the strain, ε. In this case, no other model is employed.

The strain, ε, can be related to the deflection, ϕ, by assuming that the bent product forms part of the perimeter of a circle. The axial strain, ε, is then given by ϕ/R, where R is the circle radius. The circle radius, R, can be expressed in terms of the span, L, and the deflection, ϕ:

$$R = (L^{2}/4 + \sim\phi^{2})/2\phi \tag{8}$$

Consequently, the axial strain at the bottom surface $\varepsilon_{-h/2}$ is given by:

$$\varepsilon_{-h/2} = 2\phi^{2}/(L^{2}/4 + \phi^{2}) \tag{9}$$

Although the axial strain at the top surface is actually measured, the variation as a function of the vertical coordinate, z, is small. For the conversion of the deflection, ϕ, to the strain, ε, it is assumed that the bent beam has the shape of a semi-circle. Equation (9) will be used to convert the deflection to the strain irrespective of the nature of the deformation. This is acceptable because the deflections are small.

STRESS–STRAIN MODELLING

The measured force–deflection curve, shown in Fig. 3, is typically non-linear over the entire strain range. In theory, finite-element modelling of these curve is necessary in order to extract the parameters characterising the material properties. However, we will model the

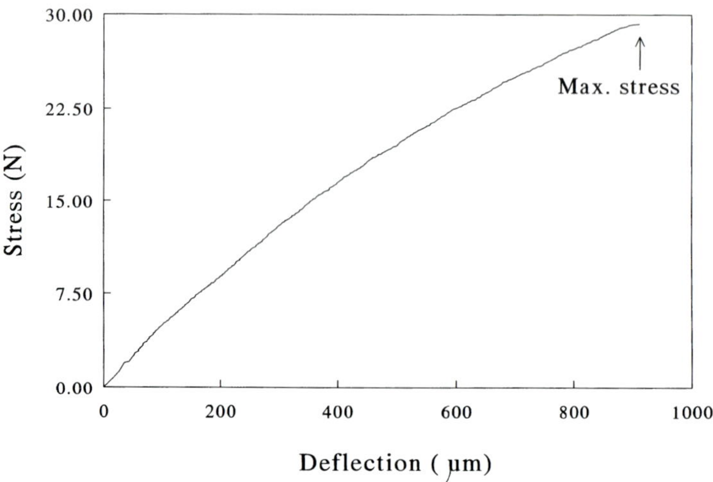

Fig. 3 Typical force-deflection curve.

measured relations directly, neglecting any deviation arising from deformation in any other plane than the bottom surface. Two types of models are considered: (1) a model consisting of mechanical elements and (2) a strain-hardening model. In addition, the stress at failure, which constitutes a measure for the green strength, can be determined.

For developing a model consisting of mechanical elements, the measured data is first converted into stresses and strains. This produces a non-linear relation which is best fitted with a model which consists of two imaginary dashpots and a friction element. A dashpot consists of a plunger sealing a fluid in a container. When displacing the plunger, the axial stress σ depends on the axial strain rate, $d\varepsilon/dt$, rather than the axial strain, ε. Given a linear correlation of the stress and the strain rate, the deformation behaviour is termed viscous and characterised by the viscosity, η. A friction element, on the other hand, can be seen as buckle which snaps open when a so-called yield stress, σ_0, is exceeded.

For a constant loading rate, as applied during the bending tests, the following relation applies for a single dashpot:

$$d\varepsilon/dt = \sigma = rt \tag{10}$$

where r is the loading rate and t is the duration of loading. With the boundary condition, $t = 0$, $\varepsilon = 0$, integration leads to:

$$\varepsilon = rt^2/2\eta = \sigma^2/(2r\eta_1) \tag{11}$$

Use of eqn (11) is not sufficient to obtain an adequate fit of the stress-strain relation. Especially the progress at higher strains is not adequately represented. A solution is to introduce a yield stress, σ_0, and to add a second dashpot to the model. The second dashpot is placed in series and only deforms when the yield stress, σ_0, is exceeded (Fig. 4). The deformation of the second dashpot is given by:

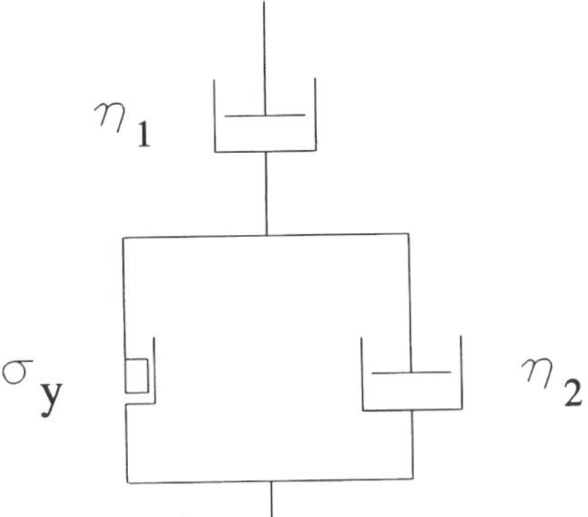

Fig. 4 Mechanical model esribing the measured stress-strain curve.

$$\varepsilon = (\sigma - \sigma_0)^2/(2r\eta_2) \tag{12}$$

For two dashpots, labelled 1 and 2, in series, the total strain is obtained by summing the strains of the individual dashpots. The model contains three parameters: the viscosities, η_1 and η_2, and the yield stress, σ_0. The values of these parameters are determined by regression.

A second type of model is the strain-hardening model described earlier. The strain-hardening model has two independent parameters, the modulus of plasticity, H, and the strain-hardening exponent, n. Unfortunately, the model does not provide a good fit over the entire range of strains. Therefore, different strain-hardening models are applied to different strain ranges. The transition stress, σ_t, constitutes an additional parameter. With this model, even better correspondence with the measured data was obtained than with the model consisting of mechanical elements.

INTERPRETATION

All the model parameters vary with the moisture content. Plotting the parameters as a function of the moisture content reveals details of the factors which determine the mechanical properties of the material. Figure 5 shows the development of strength, which corresponds to the stress required to produce failure, during drying. three stages can be distinguished: a rapid increase both at relatively high and low moisture contents and a slowly increasing strength at intermediate moisture contents. An explanation is sought by considering the interaction between moisture and the particles. The strength increases when the particles are in direct contact with each other and decreases when moisture is present. Moisture will reduce the interparticle friction by lubricating the sliding motion of particles with respect to each other. We postulate that the effect of moisture depends on the state of the moisture. The lubricating effect of moisture increases in the following sequence of states: chemisorbed moisture, physisorbed moisture, free water. Between moisture contents of 21 and 16 %, the free water evaporates from the product. This reduction causes the appreciable increase of the strength. The assumption that the free water has evaporated at a moisture content of

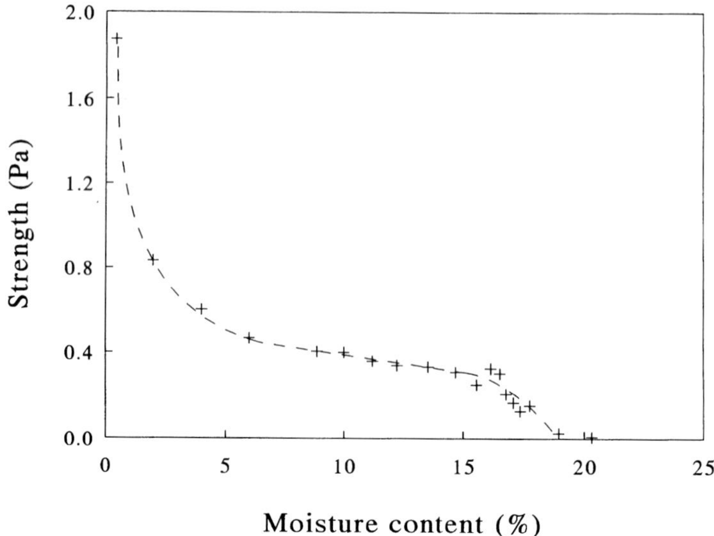

Fig. 5 General tend in the strength during drying.

roughly 16 %, is supported by the observation that the limit to shrinkage also occurs at 16 %. This implies that the deformation potential of the product greatly decreases. The slow increase in the strength with decreasing moisture content reflects the still relatively high mobility of physisorbed moisture. Once the physisorbed moisture is evaporated, only the chemisorbed moisture remains. This moisture is less mobile, so that the strength increases. In addition, the coverage of the particles decreases, increasing the likelyhood of high interparticle friction.

The development of the model parameters as a function of the moisture content is remarkably similar for the compositions tested. Fig. 6 shows the trend which is generally observed for all parameters with a single exception (the strain hardening exponent, n_1). The value of the parameter is clearly lower for mass *C* at high moisture contents but surpasses the parameters of mass *A* and *B* for lower moisture contents. This could be explained by considering the weaker bonding of moisture by hard materials in comparison to clay. For a high moisture content, the fraction of free water is relatively high, leading to a low strength of mass *C*. Once the coverage becomes thinner, the friction between the hard material particles rapidly gains significance. This increase is much larger than in the case of clay.

The difference between mass *A* and mass *B* is less pronounced. The main effect of the larger agglomerates in mass *B* is to reduce the initial deformation at higher moisture contents. After the yield or transition stress, the difference between mass *A* and mass *B* becomes smaller. This indicates that the distinct nature of agglomerates effectively disappears at lower moisture contents.

APPLICATION TO PRODUCTS

The parameters of the models described above offer insight into the deformation behaviour

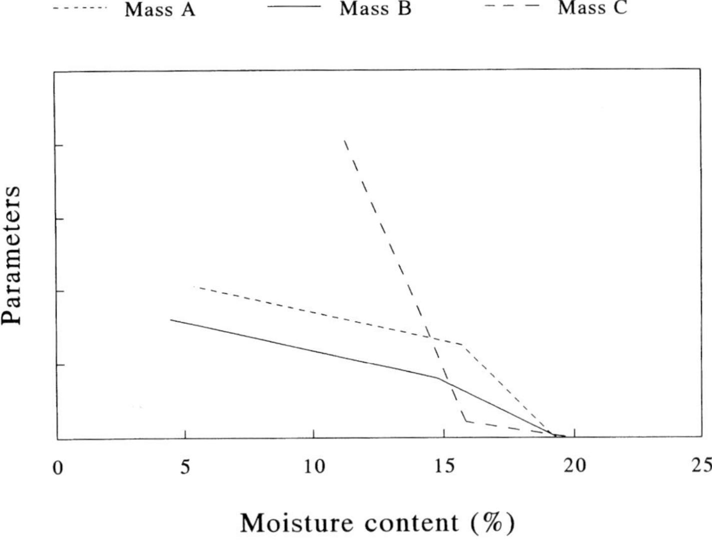

Fig. 6 General trend in the model parameters during drying.

of green sanitaryware products. The application of these models to real products is not always straightforward. First of all, a relatively high loading rate is used.

In practice, this occurs only during removal of the excess slip and during handling of the product. The effect of removing excess slip can be traced to certain features like sills. At these positions, removal of the excess slip produces a vacuum, leading to large deformation. Deformation of the product during handling could be tackled; however, the stresses exerted on a product are difficult to determine. Finally, the deformation as a result of the drying stress could be considered. This deformation is relatively large at moisture contents above the shrinkage limit. However, the product is only released from the mould at or directly below the shrinkage limit. In other words, the significance of supporting the product at moisture contents above the shrinkage limit is well-recognised. The remaining deformation during drying is very small. This deformation could be calculated with a finite-element approach. Given the arduous exercise of discretising modern sanitaryware products, we propose another approach. First, the product is mapped in three dimensions with a grid which coincides with the skeleton of the product. Stresses acting on the intersections of the grid lines while the deformation of grid lines is given by either the model containing mechanical elements or the strain-hardening model. By specifying the drying rate, the deformation for a given time interval is calculated. This approach, although similar to a finite element method (FEM), is computationally much less demanding than its well-known cousin.

FINAL REMARK

The major attraction of measuring the mechanical properties is to provide a measure for the deformation behaviour of different compositions. Comparison of the parameters for different compositions allows the optimisation of the composition of the masses. It also generates an initimate understanding of the material under study.

REFERENCES

1. J. Clausen and R. Fish, 'Successfully drying sanitary ware', *Amer. Ceram. Soc. Bul.,* 1995, **74**, 40.
2. A. Broese van Groenou, R. Lissenburg, 'Inhomogeneous density in die compaction: experiments and finite element calculations', *Comm. Amer. Ceram. Soc.,* 1983, **62**, C156.
3. G. D. Quinn and R. Morrell, 'Design data for engineering ceramics: a review of the flexure test', *J. Am. Ceram. Soc.,* 1991, **74**, 2037.
4. L.-Y. Chao and D. K. Shetty, 'Reliability analysis of structural ceramics subjected to biaxial flexure', *J. Am. Ceram. Soc.,* 1991, **74**, 333.
5. H. J. Glass, *Compaction behaviour of (Mn,Zn)–ferrite granulate,* Letru Press, Helmond, 1994.
6. S. Timoshenko and J. Goodier, *Theory of Elasticity,* 3rd edn, McGraw-Hill, 1989.

FRACTURE, EROSION AND WEAR

Influence of Environment on Delayed Failure of Alumina Ceramics

S. M. BARINOV, N. V. IVANOV, S. Y. ORLOV and V. Ya. SHEVCHENKO

High Tech Ceramics Research Centre, Russian Academy of Sciences, Ozernaya 48, Moscow, 119361 Russia

ABSTRACT

Dynamic fatigue testing of various alumina ceramics in humid air, water, acidic and basic environments was performed. It is revealed that the fatigue resistance of glass-bonded alumina in air and in water is significantly reduced as compared to that for high-purity magnesia-doped alumina. The crack-growth velocity exponent is independent of the composition of the glassy grain-boundary phase, as well as of the water content in environment. Hydrochloric acid slightly lowers the crack-growth velocity exponent for the yttria-alumosilicate glass bonded ceramics, whereas the exponent decreases drastically for the magnesia-doped material in an acidic environment. Possible mechanisms controlling the slow crack growth in the ceramics are outlined.

INTRODUCTION

Alumina-based ceramics are intended for application in many engineering devices, often being subjected to the action of corrosive-active environment whilst under the load. Like many inorganic solids, alumina ceramics exhibit delayed failure due to the subcritical crack growth. Delayed failure in aluminium oxide has been demonstrated for both the polycrystalline ceramics and sapphire single crystals.[1] This phenomenon is believed to result primarely from a stress-enhanced chemical reaction between the environment and the bonds at the tips of various microstructurural stress concentrators. The slow crack growth in alumina occurs even in a relatively mild environment such as water.

It is supposed that the susceptibility to delayed failure is an inherent feature of a certain solid/environment system. Two main mechanisms are used to explain stress corrosion in solids are under consideration: the dissociative chemisorbtion of environmental species at strained crack tip bonds and ion solvation. In the former, the essential conditions for fatigue are the capacity of reactive species in environment to donate electron and proton, and a polarizability of the crack-tip bonds in solid.[2] The slow crack growth mechanism in alumina is considered to be similar to that in fused silica. The latter has been proposed by Mechalske and Freiman.[3] It was envisaged that incoming molecule of liquid environment, e.g. water, interacts with the critically stretched Si–O–Si bond followed by production of the two new bonds and formation of Si–OH. The fracture surface is saturated with hydroxyl groups. Both silica and alumina undergo similar surface reactions with adsorbed species of environment, particularly with the water molecules. In the solvation mechanism, the incoming species promote crack growth without dissociating. This case is inherent for highly ionic solids, e.g. for magnesium halides.

In the alumina ceramics processing, sintering additives are used which form specific grain-boundary phases that can influence or control the subcritical crack growth process. The main component of this phase is a silicate glass for the glass-bonded aluminas, or magnesia – for high-purity alumina ceramics. The responce of the grain boundaries to the stress-enhanced chemical attack of environment depends on the composition of both the additive and environment, and seems to be complicated. In this context, the present work is aimed at the investigation of some aspects of the subcritical crack growth in various alumina-base ceramics.

EXPERIMENTAL DETAILS

1. *Methods*

In the fracture mechanics approach, it is supposed that the crack tip stress causing the crack growth is directly proportional to the stress intensity factor, K. Therefore, the crack-growth velocity, v, can be directly related to the stress intensity factor. Generally, there is a trimodal dependence of v on K, each of three regions results from a different mechanism of crack growth. It is considered that the region 1 where the crack growth velocity is a power function of the applied stress intensity factor

$$v = AK^n \tag{1}$$

gives the major contribution to the time-to-failure of ceramics. In Equation (1) A is a constant and n is the crack velocity exponent.

To estimate the crack-growth velocity exponent, a method of dynamic fatigue was utilised, where the strength is measured as a function of strain rate[1,4–6] In this method, the measured strength, σ, is related to the strain rate, $\dot{\varepsilon}$, as

$$\log\sigma = C + \left[1/(1+n)\right]\ \log\dot{\varepsilon} \tag{2}$$

and the crack velocity exponent can be evaluated by the slope of the log σ versus log $\dot{\varepsilon}$ plot, where C is a constant.

10 to 20 specimens were used for each experiment. The specimens were $7 \times 8 \times 35$ mm^3 in size. A UTS-100 screw-drived testing machine (UTS Testsysteme GmbH, Germany) was used. The cross-head speeds used range was from 0.03 to 15 mm min^{-1}. Data from the dynamic fatigue test were least-squares fitted to the Eqn 2.

Fracture toughness K_{Ic} was measured by three-point bending of single-edge notched beam (SENB) configuration fracture mechanics specimens. A thin notch of 0.5 specimen's width was machined with a diamond saw-wheel. The measured radius of curvature at the notch tip was about 50 μm. The specimens were loaded at a span of 32 mm. To calculate fracture toughness, a common calibration polynomal for wide range stress intensity factor was applied.[7]

2. Materials

Experiments were performed with the specimens of four series of alumina ceramics prepared by cold uniaxial pressing followed by sintering. The ceramics were:

A1 – 99.5 wt% alumina with sintering additive of 0.4 wt% MgO, other impurities content (silica, alkali oxides) totalled up to 0.1 wt%. The specimens were pore-free with the equiaxial alpha-alumina grains of average size of about 11 μm.

A2 – 95 wt% pore-free alumina bonded with a Y_2O_3–Al_2O_3–SiO_2 glass; the composition of the sintering additive is (in mol.%): 8% Y_2O_3, 15% Al_2O_3, and 77% SiO_2. There are elongated grains of alumina 5 to 35 μm in length. The glassy phase occupies about 13 vol.%.

A3 – 95 wt.% pore-free materials containing about 12 vol.% of the grain boundary alkali-alkali earth-alumo-borosilicate glassy phase of the composition (in mol.%): 30% alkali earth oxides, 30% B_2O_3, 25% SiO_2, 14.5% Al_2O_3, 0.5% alkali oxides. Alumina grains are 15–40 μm in length and 7–15 μm thick. These grains are surrounded by a glassy grain boundary which is about 2 μm thick.

A4 – commercial glass-bonded ceramics GB-7, their composition is as follows (in wt%): 97.1% Al_2O_3, 0.9% SiO_2, 0.9% CGaO, 0.92% B_2O_3, 0.09% Na_2O. Ceramics contain 91–92 wt% crystalline and 8–9 wt% glassy phase, the alumina grains are up to 15 μm length.

The tests were performed in an ambient air atmosphere at relative humidity RH 59–62% and a temperature 20-22°C, in water, in 0.1 N HCl solution (pH = 1), in a mixture of 0.1 N HCl with a buffer solution which has been prepared by mixing of citric acid and NaOH (pH = 2), and in NaOH solution (pH 12).

RESULTS AND DISCUSSION

Shown in Fig. 1 is the dynamic fatigue data for A4 ceramics series tested in air, as an example. The strength increases with an increase of the straining rate. A fit of the data to eqn 2 gives the crack-growth velocity exponent equals to 30.3, the correlation coefficient being 0.82. Fig. 2 is a Weibull plot for these specimens tested in air at a cross-head velocity 1 mm/min. Cumulative failure probability, $P(\sigma)$, was calculated using an estimator $P(\sigma) = (i + 0.5)/N$ where i is the number of specimens that failured under the stress σ or less; N is the total number of the tested specimens. A two-parameter Weibull function was used. The modulus of the function was estimated to be equal to 8.7. Similar estimates were obtained for other ceramics under investigation.

Shown in Fig. 3 are the results of the regression analysis of the experimental data for the specimens of **A1** series ceramics tested in air, water and acidic environments. It can be supposed from these data, that the dymanic strength values at the straining rate of 10^{-2} s^{-1} do not generally depend on environment, the logarithm of the strength being invariable within the error of measurements.

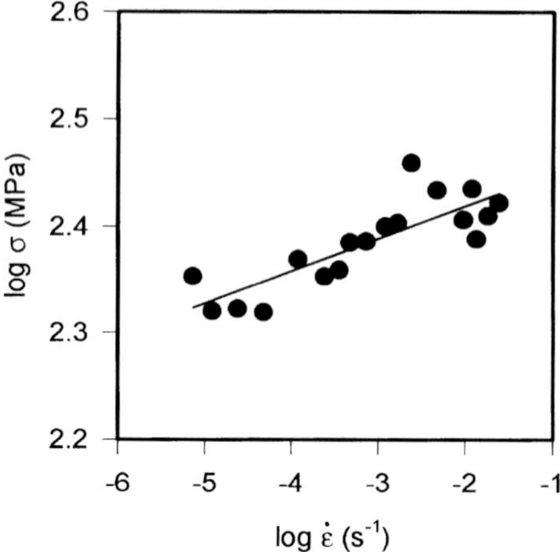

Figure 1 Strength versus straining rate for A4 ceramics tested in air.

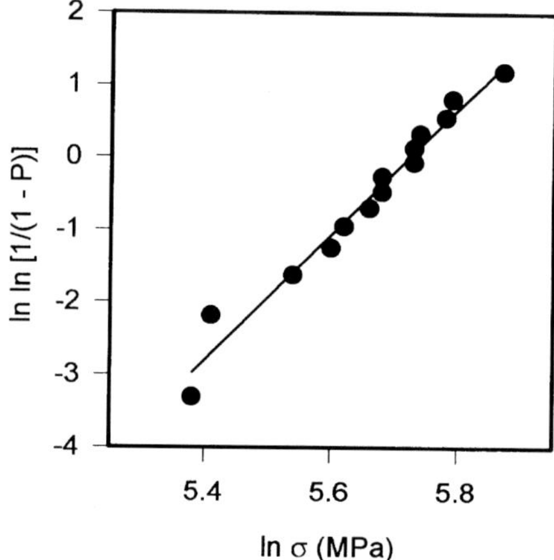

Figure 2 Weibull cumulative probability plot for A4 ceramics tested in air.

Table 1 summarises the results of the crack-growth velocity exponent estimation for the ceramics under investigation. Generally, the crack-growth velocity exponent n for all glass-bonded aluminas is lower than that for 99.5% alumina ceramics when tested in air, being approximately invariable for all the glass-bonded materials. Somewhat more fatigue resistant are the specimens of the **A2** series as compared to the specimens of **A3** and **A4** series.

The water content in the environment decreases slightly the value of n for the specimens of both **A1** and **A2** series. These data are in accordance with known estimations of n-value in the range 35 to 68 for aluminas in an aqueous environment.[5]

Hydrochloric acid reduces drastically the resistance of the 99.5% alumina–0.4% magnesia ceramics to delayed failure although it has influence on the crack-growth velocity exponent for

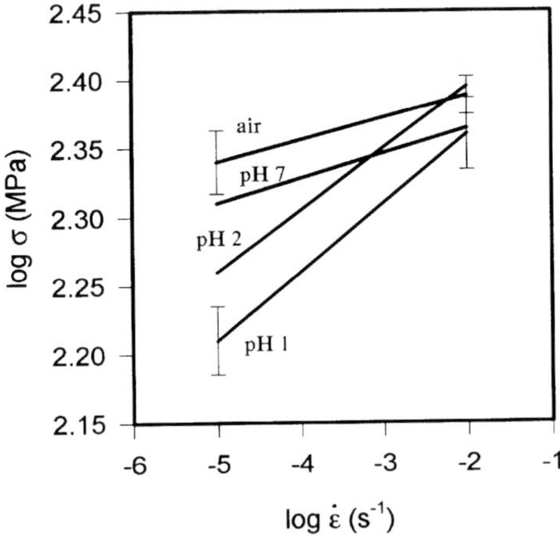

Figure 3 Extrapolated strength *versus* straining rate plots for Al ceramics tested in air, water, and acidic solutions with pH = 1 and pH = 2.

Table 1 Crack-growth velocity exponent for the ceramics tested.

Series	Environment				
	air	Ph = 1	Ph = 2	Ph = 7	Ph = 12
A1	60±4	19±2	20±3	54±3	49±4
A2	36±3	34±4	34±5	33±2	–
A3	32±4	–	–	33±3	–
A4	30±1	–	–	–	–

the glass-bonded ceramics of **A2** series. The difference in behaviour of these ceramics is obvious, as it can be seen in Fig. 4 summarising these data. It should be noted that the fatigue resistance of **A1** ceramics in an alkaline environment is approximately the same as in the water.

Scanning electron microscopy of fracture surfaces revealed that the mode of crack propagation changes from mixed trans/intergranular in air to the predominantly intergamular when ceramics of Al series are tested in water and acids (Fig. 5). Therefore, the behaviour of grain boundaries with respect to environment seems to be the major factor influencing both the slow subcritical crack growth and fast fracture.

The environment influences the fracture toughness, too. In particular, the critical stress intensity factor, K_{Ic}, for the Al series ceramics as measured at a moderate cross-head speed of 0.2 mm min^{-1} decreases from 3.8 MPa $m^{1/2}$ for the specimens tested in air down to 3.4 MPa $m^{1/2}$ for these tested in water, and further to 2.9 MPa $m^{1/2}$ when the tests were performed in acidic solution with pH = 1. In a basic solution, the critical stress intensity factor is estimeted to be 3.1 MPa $m^{1/2}$. Above tendency is similar to that for the variation in the crack-growth velocity exponent.

Measured in acidic environment at pH = 1, the critical stress intensity factor for **A1** ceramics depends significantly on straining rate (Fig. 6). A least-squares fit of the data to the

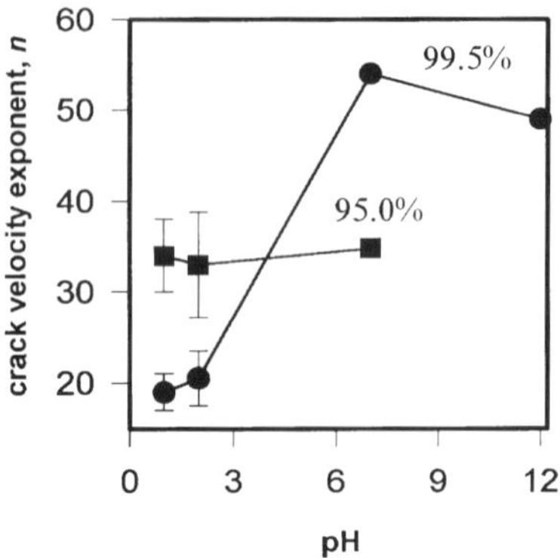

Figure 4 Effect of pH on the crack-growth velocity exponent for Al and A2 ceramics.

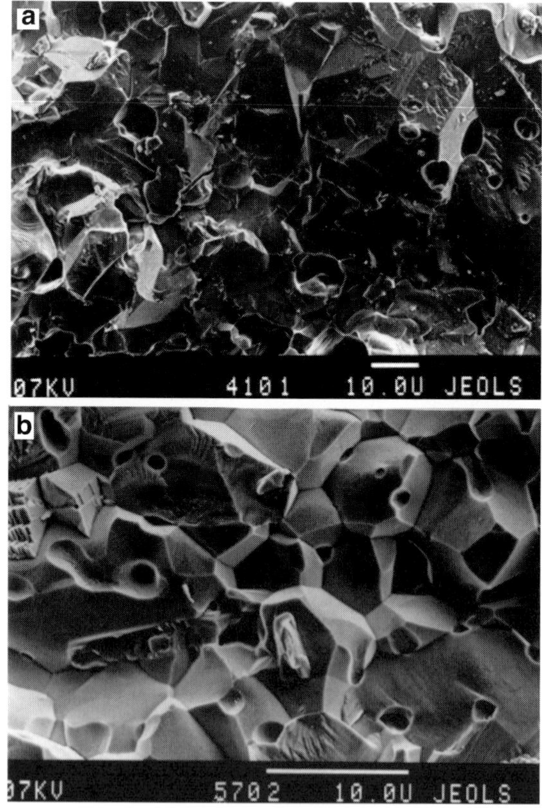

Figure 5 SEM micrographs of fracture surfaces of Al ceramics tested in air (a) and in acid at pH = l (b).

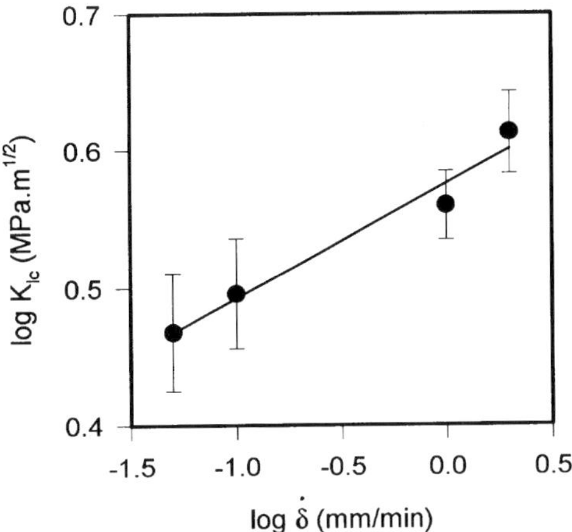

Figure 6 Fracture toughnes *versus* cross-head speed for Al ceramics tested in acid with pHl.

linear-regression equation $\log K_{Ic} = A + m \log \dot{\delta}$, where $\dot{\delta}$ is the cross-head speed of testing machine results in $m = 0.083$ and $A = 0.576$. It can be pointed out that the value of m is very close to the value of $[1/(n+1)] = 0.05$, the latter being the proportionality coefficient in Eqn 2. Therefore, it can be considered that the subcritical crack growth occurs during the fracture toughness testing of the SENB-type long-crack containing specimens. This is in agreement with the data of Okabe *et al.*[8] Therefore, there is certain similarity between the short- and long-crack propagation mechanisms.

It is well known that alumina ceramics undergo stress-corrosion cracking, as does glass. Assuming the subcritical crack growth in ceramics is reaction-rate limited, it becomes clear why water content in environment (tests performed in air and in water) affects only slightly the crack-growth velocity exponent: the chemical potential of the water does not depend on state of stress at the crack tip.[9]

The difference in the behaviour of **A1** and glass-bonded ceramics when tested in air and in water is a result of the hydration reaction response from the grain-boundary phase. The principal component of the glassy phase in **A2** series ceramics is silica which interacts actively with water under the applied stress resulting in stress-corrosion cracking.[1] Partial substitution of boron oxide for silica in the **A3** and **A4** ceramics does not result in decrease of the susceptibility to stress-corrosion cracking, probably because the boron oxide is highly acidic oxide too. The stress-enhanced dissolution of glassy phase in **A2** ceramics is not increased in acidic testing media as compared to the neutral environment (water). This is supposed to be the reason why the crack-growth velocity exponent for **A2** ceramics is approximately invariable when tested in water and in acids. This conclusion is in agreement with the experimental data indicating that the crack-growth velocity exponent for silica-base glasses does not practically depend on the pH value in the range 1 to 4.[10] Besides, these results are not contrary to the data,[11] where it has been shown that the acidic environment does not decrease the crack-growth velocity exponent in a silica-base glass.

Contrary to the glass-bonded aluminas, the delayed failure behaviour of the 99.5% alumina ceramics is rather susceptible to the acidic environment. It is well known that magnesium ions segregate to the grain boundaries in alumina ceramics sintered with small additions of magnesia.[12] The exact form of existence of magnesia additions within the grain boundaries is not clear yet, because the small amount of MgO at the boundary is small enough as to be difficult to investigate. It has been suggested that unstable spinel clusters can be formed at the grain boundaries when the amount of magnesia additive is small.[13] The magnesium ions in these clusters are forming the ionic bonds with oxygen within the interfaces. Ionically-bond solids are highly susceptible to stress-corrosion cracking,[2] unlike covalent solids like SiC or Si_3N_4.[1,14] For MgO of different grades this has particularly been demonstrated by delayed-failure tests in water.[15] The crack propagation mode in the Al series ceramics tested in water and in acid was predominantly intergranular, while it has been mixed trans- and intergranular in air. Thus, it can be supposed that the inherently low stress-corrosion resistance of the ionic Mg–O bonds at the grain boundaries is the reason for the enhanced susceptibility of the Al ceramics to the delayed failure in water with respect to that in air. An alternative explanation can be proposed on the basis of the ion solvation model.[2] In this model, the electrostatic attraction between oppositely charged magnesium and oxygen ion pairs at the crack tip is reduced as the result of dielectric screening, the latter being dependent on the dielectric constant of the environment.

The acidic media enhances the chemical attack. on the magnesium-containing grain boundaries resulting in a decrease of the crack-growth velocity exponent, i.e. making the ceramics less resistant to stress corrosion. Because magnesia is a highly basic oxide, the influence of the basic solution on the crack-growth velocity exponent is significantly less than that of acidic environment.

CONCLUSION

Dynamic fatigue tests of four alumina-base ceramics was performed in various environments. It is revealed that the subcritical crack-growth velocity exponent for the glass-bonded ceramics tested in air is significantly less than that for the 99.5% alumina–0.4% magnesia ceramics. The crack-growth velocity exponent is practically invariant with respect to the composition of the glassy grain-boundary silicate phase. Influence of the water and acidic solutions on the crack-growth velocity exponent depends strongly on the composition of ceramics. The exponent for high-alumina magnesia-bonded ceramics decreases drastically in hydrochloric acid solutions with pH 1 and 2, while basic solution with pH 12 affects the crack-growth velocity exponent unsignificantly. The changes in the fracture toughness have the same tendency as for the crack-growth velocity exponent. There is a staining-rate dependence of the fracture toughness indicating the slow long-crack growth in the fracture mechanics specimens. Contrary to that behaviour, the acidic environment had only slight effect on the crack-growth velocity exponent for the yttria–alumosilicate glass bonded alumina ceramics. This material seems to be more reliable for the acidic-media applications than high-alumina magnesia-doped ceramics.

ACKNOWLEDGEMENT

The work is supported by RFBR, grant N 97-03-33599.

REFERENCES

1. S. M. Wiederhorn, in *Fracture Mechanics of Ceramics,* Vol. 2, R. C. Bradt *et al.* eds, Plenum, 1974, 613–646.
2. T. A. Mechalske, B.C. Bunker and S. W. Freiman, *J. Am. Ceram. Soc.,* 1986, **69** (10), 721–724.
3 T. A. Mechalske and S.W. Freiman, *Nature,* 1982, **295** (5849), 511–512.
4. R. W. Davidge, J. R. McLaren and G. Tappin, *J. Mater. Sci.,* 1973, **8** (11), 1699–1705.
5. J. E. Ritter, jr, N. Bandyophyay and K. Jakus, *J. Am. Ceram. Soc.,* 1979, **62** (9–10), 542–543.
6. S. Lauf and R. F. Pabst, in Ceramic Materials and Components for Engines, W.Bunk *et al.* eds, DKG, 1986, 961–978.
7. J. E. Srawley, *Int. J. Fract.,* 1976, **12** (5), 475–476.
8. T. Okabe, M. Kido and T. Miyhara, *Eng. Fract. Mech.,* 1994, **48** (1), 137–146.
9. S. M. Wiederhorn, S. W. Freiman, E. R. Fuller and C.J. Simmons, *J. Mater. Sci.,* 1982, **17** (12), 3460–3478.
10. S. Takeda and I. Tari, in *Fracture Mechanics of Ceramics,* Vol. 9, R. C. Bradt *et al.* eds, Plenum, 1992, 575–588.
11. S. M. Wiederhorn and H. Johnson, *J. Am. Ceram. Soc.,* 1973, **56** (2), 192–196.
12. E. S. Lukin, *Refractories and Techn. Ceram.* (Russia), 1996, **4,** 7-13.
13. V. A. Lotov and A.T. Dobrolubov, *Glass and Ceramics* (Russia), 1997, **11,** 10–12.
14. K. Breder, *J. Am. Ceram. Soc.,* 1995, **78** (10), 2680–2684.
15. W. H. Rhodes, R. M. Cannon and T. Vasilos, in *Fracture Mechanics of Ceramics,* Vol. 2, R.C.Bradt *et al.,* Plenum, 1974, 709–733.

An Experimental Study of the Fracture Behaviour of Magnesium Zinc Ferrites

R. M. POOL, T. J. DAVIES and F. R. SALE

Manchester Materials Science Centre, Manchester University, UK

W. M. DAWSON,

Philips Components Ltd, Southport, UK

ABSTRACT

Magnesium zinc ferrite yoke rings are manufactured by a traditional mixed oxide route involving powder preparation, dry pressing and conventional sintering in air. The final stage of production entails splitting the yoke ring into identical halves for subsequent use in the final assembly. This splitting operation is effected by controlled fracture along grooves which are introduced into the product during compaction.

This study utilises techniques such as fractography, SEM and optical microscopy to assess how process variables affect the cracking behaviour, particularly the crack features on the fracture faces, the fragmentation of granules during compaction and development of microstructure.

The investigations have shown that the fracture process is critically dependent on product geometry, with other variables having a secondary influence.

1. INTRODUCTION

Soft ferrites are widely used in many fields of telecommunications and electronic engineering, particularly as high frequency inductors and transformers.[1,2] MgZn ferrites in particular are commonly manufactured as yoke rings for magnetic deflection systems. Such ferrites have a high resistivity enabling the yoke ring to be directly wound prior to mounting in a cathode ray tube; this requires, as a final stage in production a means of splitting the yoke rings into identical halves, to allow direct winding with copper wire. For a yoke ring to be of a standard acceptable by the customer, the fracture plane has to occur within specific boundaries on the product, i.e. within the cracking grooves introduced during pressing. If, during this fracture stage the crack propagates beyond these grooves then a phenomenon called 'off line cracking' has occurred and the yoke ring is unacceptable for final assembly. Off line cracking is a major source of yield loss during yoke ring production. Since it occurs intermittently for reasons which are not completely understood, a greater level of understanding and control of this stage of the process is required.

2. EXPERIMENTAL METHODS

All yoke rings used in the following experimental studies were produced at Philips Components, Southport, UK and consisted of iron deficient grades with a composition of

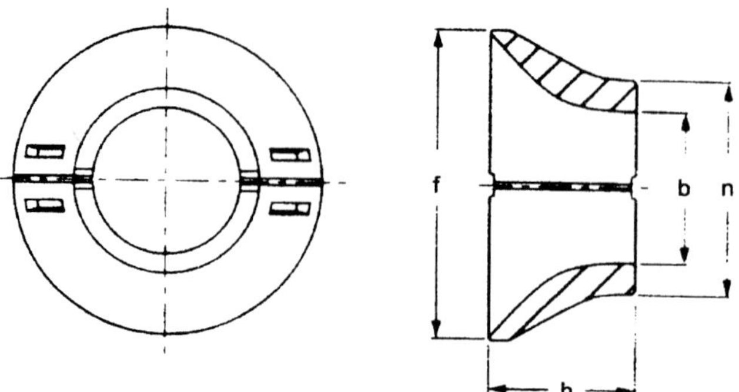

Figure 1 Engineering drawing showing the dimensions that are altered to give different yoke ring geometires: **f** = flare; **n** = neck; **b** = bore; **h** = height.

Table 1 Geometries of the yoke rings studied.

Yoke Ring	dimensions (mm)			
	flare (f)	neck (n)	bore (b)	height (h)
X	89.00	62.00	51.00	33.00
Y	100.00	62.00	48.00	44.50
Z	102.30	64.26	51.50	42.00

$(Mg_{0.63}Zn_{0.377})(Mn_{0.1}Fe_{1.8})O_{3.85}$ that were produced by a traditional mixed oxide route. Details of this production route and small variations that were introduced to meet special requirements of this experimental programme are described below.

The starting materials, MgO, ZnO, Mn_3O_4 and Fe_2O_3, were dry mixed with sinter aids (CaO and SiO_2), granulated, prefired (1050 °C for 2 hours), ball milled and then spray dried with 0.4 wt% PVA binder. Granule sizes were selected in the range 150 to 550 μm with a view to assessing their effect on compaction. Yoke ring compacts were isostatically pressed (at 50 MPa) under identical conditions to densities of approximately 2.55 g cm^{-3}. This pressing mechanism approximated a double ended operation wherein an upper ram moved downwards pressing powder against the die wall, the upper ram and die wall then moved as a single assembly onto the lower ram, thus compacting the base of the compact. The resultant compacts were subsequently sintered at 1250°C with a heating/cooling rate of 155 K h^{-1} and 1 hour at peak temperature. Sintering atmospheres were also varied to study their effect on fracture; compacts were sintered either in air followed by air or N_2 cooling or entirely N_2. The yoke rings were ultimately split into halves by a controlled fracture procedure.

A third variable investigated was product geometry. In practice all yoke rings are manufactured to specific customer requirements which include the precise size and shape of the yoke ring; the geometry is varied by altering the yoke ring dimensions shown in Fig. 1. Details of the different yoke ring geometries studied are shown in Table 1, and Table 2 summarises all variations in production.

The techniques used to assess how the production variables affected fracture and/or generated yield losses (through off-line cracking) are now described.

Table 2 Production variations (w.r.t. yoke ring geometry).

Yoke Ring Geometry	Sintering Atmosphere	Spray Dried Granule Size (μm)
X	Air	>350
		<350
Y	Air	>350
		<350
Z	Air	~350
	Air/Nitrogen	
	Nitrogen	

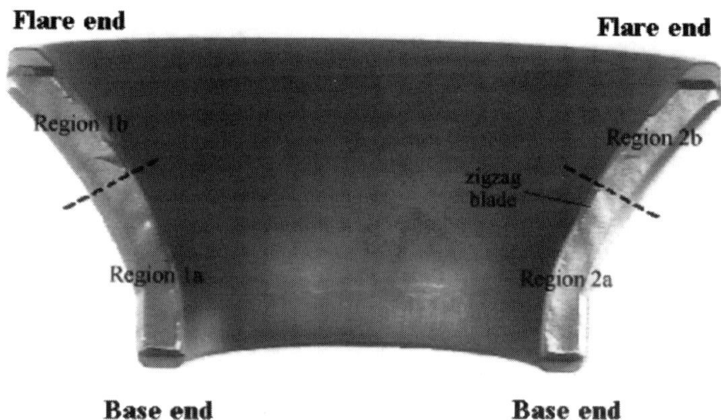

Figure 2 Split yoke ring with notation used for defect characterisation.

2.1. Fractography

Fractography was used to study the variation in type and distribution of defects on the fracture faces of yoke rings. Figure 2 shows a yoke ring (half) after splitting. Features of interest on the fracture faces shown here, are the 'zigzag' blades (also shown in Fig. 3), these are depressions in the cracking line introduced during compaction, (often requested by customers to aid re-assembly of the yoke ring before winding). Also shown are the regions (1a, 1b, 2a, 2b) on the fracture face used for locating the cracks during fractography studies; base end and flare end refer to either end of the yoke ring.

A statistical assessment of the type, location and frequency of occurrence of these defects was carried out, Table 3 provides examples of the classification used to identify cracks during data collection and examples of such cracks are shown in Fig. 3. After data collection, the frequency at which the various defects occurred was recorded with particular emphasis being placed on their location on a fracture face. A total of 38 yoke rings that were within customer specifications and some off-line cracked rings were selected at random from plant trials; this number (38) allowed a statistical study to be carried out such that the mean of the normal distribution would be no wider than 0.1 1σ at a 50 % confidence level.[3]

Table 3 Description of cracks.

Type of Defect	Orientation on Fracture Face
A: Crack	**A**: Touches both edges **B**: Touches inner edge **C**: Touches outer edge **D**: Central and transverse **E**: Central and longitudinal

Type of Defect	General Appearance
A: Crack	**F**: Single line **G**: Feathers

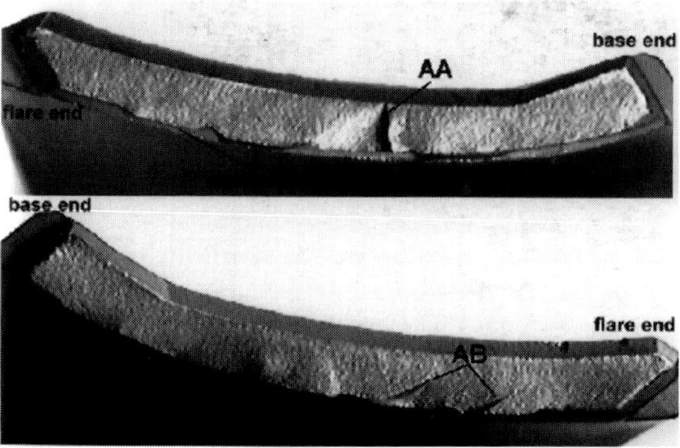

Figure 3 Examples of typical cracks on fracture faces.

2.2 Scanning Electron Microscopy of Green Yoke Ring Compacts

The study concentrated on variations in fracture characteristics in green yoke ring compacts made from spray dried granulate.

Two regions from the compacts were studied, firstly the green fracture faces and secondly parallel central sections, which were taken 90° from the fracture face as shown in Fig. 4. In the first case, compacts were fractured into halves along the cracking line and the central sections were cut using a laboratory scalpel; the sections were then mounted onto aluminium stubs and conductive silver paint applied prior to examinations in a Philips 505 S.E.M. at an accelerating voltage of 20 kV and spot size of 50 nm. Digitised SEM images of every field of view on the fracture faces and central section were obtained and subsequently analysed for percentage area of intergranular fracture (using computer image analyses).

2.3 Optical Microscopy of Sintered Yoke Rings

Optical microscopy was used to study the microstructure of sintered yoke rings, specifically to assess how porosity, residual granule history and grain size varied with the processing variations described in Table 2.

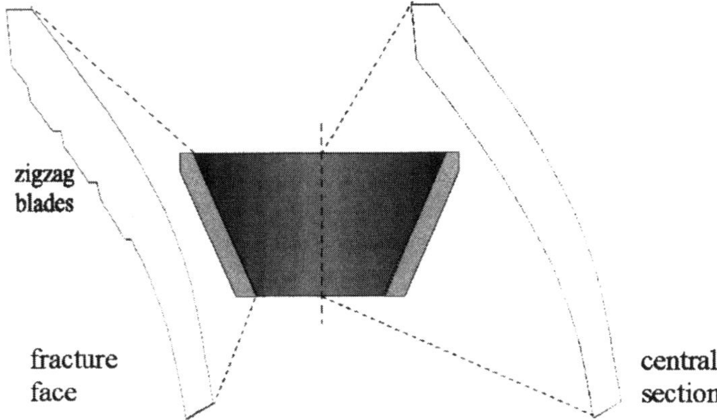

Figure 4 Schematic of green body showing sections studied to assess differences in compaction.

To enable such a study, transverse sections were cut from the entire length of fracture faces using a mechanical Isomet precision saw with a diamond blade. Sections were hot mounted in diallyl phthalate powder under a pressure of *1.5* bar. Microscopy preparation involved grinding samples with fine grade SiC paper (600 grit), to minimise grain pull-out. Samples were then polished with colloidal silica for approximately 6 hours and later etched in a solution of 30 ml HF, 10 ml HNO_3 and 130 ml H_2O, specifically for grain size observations.

To gain quantitative knowledge of the microstructure, computer image analysis was used to measure the % area porosity, % area residual granule history and mean grain size.

2.4 Density Measurements

Density measurements by Archimedes' method using mercury were performed on green compacts and sintered yoke rings to support the SEM and optical microscopy studies.

3. RESULTS AND DISCUSSION

3.1. Cracking Behaviour

The results of the fractography studies are shown in Figs. 5–10. In the figures, the frequency of a type of defect is shown with respect to its location on the fracture face. Each crack-type was classified as shown in Table 3, for example AC describes a major crack (one which extends beyond half the width of a fracture face) which touches the outer edge of a fracture face. The results indicate that cracking behaviour is affected by a number of factors. Firstly, a consistent cracking pattern is observed in all yoke rings (standard and off-line) **of the same geometry**, where a specific type of defect exists at the same position on the fracture face. When the geometries are different (as shown by X, Y or Z), the cracking patterns also differ as shown through the quantifiable variations in the different histograms in Figs 5 and 6.

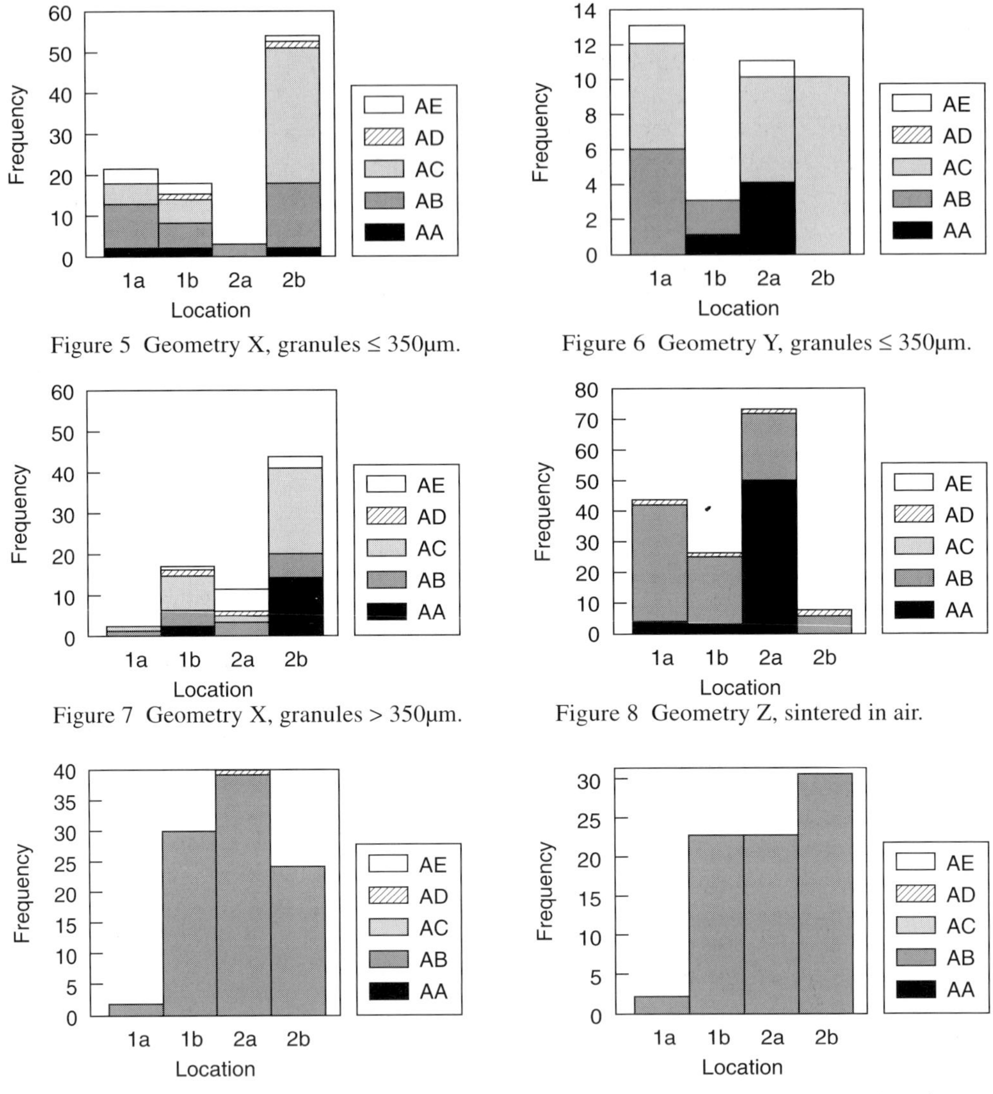

Figure 5 Geometry X, granules ≤ 350μm.

Figure 6 Geometry Y, granules ≤ 350μm.

Figure 7 Geometry X, granules > 350μm.

Figure 8 Geometry Z, sintered in air.

Figure 9 Geometry Z, sintered in air/nitrogen.

Figure 10 Geometry Z, sintered in nitrogen.

The histograms also show that the size **of spray dried granule agglomerates** affect cracking behaviour. The data in Figs 5 and 7 shows that for yoke rings with identical geometries the respective histograms are different as a result of a different spray dried granule agglomerate size. Green bodies with the smaller granulate size (≤ 350 μm) had a lower yield loss (through off-line cracking) during fracture (after sintering). Previous literature[5,6] has described how agglomerate size affects granule re-arrangement and packing during compaction; the current results suggest that the yoke rings with the smaller spray dried granule size compact to a higher green density. Glass *et al.*[7] showed that a higher green density led to a reduction in overall % shrinkage during sintering in manganese zinc ferrite compacts; the existence of a similar effect in magnesium zinc ferrites may affect yoke ring fracture and off-line cracking.

Sintering conditions also alter cracking behaviour. Yoke rings of geometry 'Z' (see Table 1) when sintered in different atmospheres (as described in Table 2) had different cracking patterns. Specifically, the introduction of nitrogen into the sintering atmosphere either during heating to temperature or on cooling changed the cracking behaviour, for example unlike yoke rings sintered in air, those sintered in a nitrogen atmosphere do not contain any 'AA' type defects on the fracture faces. The data recorded in Figs 8–10 shows such an event.

To summarise, the geometry of a yoke ring is a major factor determining cracking behaviour. Data also show that variations in spray dried granule agglomerate size and sintering conditions contribute to differences in cracking behaviour.

3.2 Compaction Characteristics

3.2.1 Qualitative Observations

The fragmentation during compaction of aggregate powders in green bodies (of different geometries) was studied by S.E.M. and the following variations in spray dried granule fracture observed. Firstly, differences in compaction were detected between fracture face sections and central sections; these variations were dependent on yoke ring geometry. Observations showed that 77 % of the green bodies had a greater proportion of transgranular fracture at the fracture faces whereas intergranular fracture was more frequent at the central sections. In contrast, the remainder (23 %) showed the reverse situation where transgranular fracture occurred predominantly along central sections.

In both locations (the fracture face and central section as shown in Fig. 4), variations in granule fracture were observed along the lengths of these sections. Granule fracture at the base and flare ends is predominantly transgranular whereas intergranular fracture predominates at locations away from the ends of the sections. Here, individual granules have deformed in a mode approximating to the classic dodecahedron condition.[8]

Variations in granule fracture at the zigzag blade regions were also observed. Typical examples are shown in Figs 11a and 11b. At the corners of zigzag blades, transgranular fracture is dominant. In contrast, at the middle of the blades the fracture of granules is a combination of transgranular and intergranular, showing both the dodecahedron form and granule fragmentation.

3.2.2. Quantitative Analysis

To quantify the observed differences in compaction, the degree of granule fragmentation on the fracture faces, zigzag blade locations and in the central section (see Fig. 4) was assessed by measuring the % area of intergranular fracture of granules; examples are given in Fig. 12. The data labels c and m in Fig. 12c refer to corners and middle of zigzag blades respectively. A 6th order polynomial line of best fit highlights how the general trend in variations changes across the fracture face and central section. Variations in the width of the fracture face are also shown.

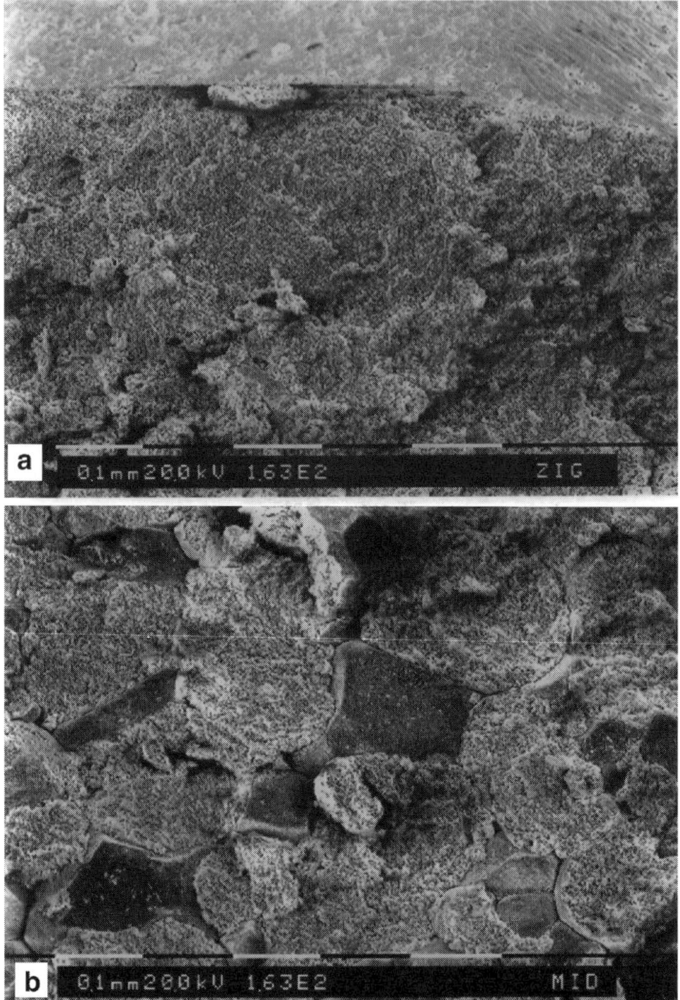

Figure 11 (a) Spray-dried granule fracture at corner of zigzag blade region; (b) spray-dried granule fracture at middle of zigzag blade region.

The data accumulated confirm the initial observations made on the green bodies. Firstly, better compaction occurs at the extreme ends, i.e. base and flare ends, and secondly, there is a greater proportion of intergranular fracture on the fracture face rather than the central section (Figs. 12a and b). Density measurements in these regions supported these results showing an average green density of 2.67 ± 0.02 g cm^{-3} at regions of transgranular fracture and one of 2.56 ± 0.02 g cm^{-3} at regions of intergranular fracture.

The data suggests that granule fracture may be linked to forces acting during compaction. Typically, during compaction friction and shear forces build up between the powder and the adjacent die walls; small differences in powder flow and aggregate size can cause variations in local pressures and therefore compaction. In a yoke ring geometry this effect is particularly exaggerated at the base and flare regions, as shown by the predominance of transgranular

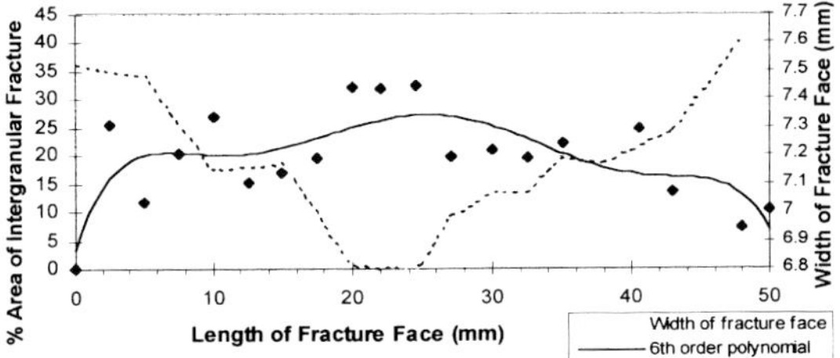

Fig. 12a : % Intergranular fracture along the length of a fracture face

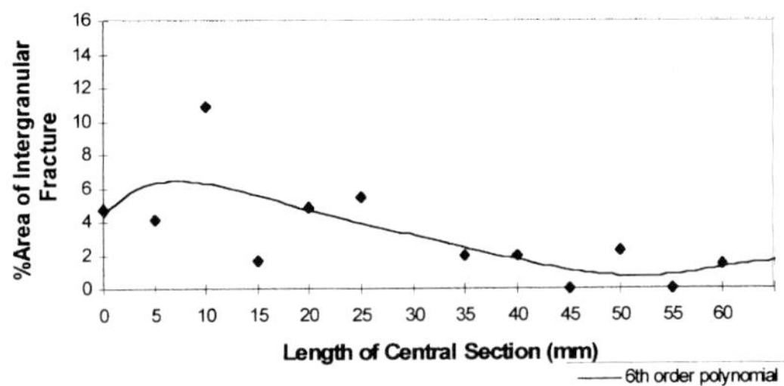

Fig. 12b : % Intergranular fracture along the length of a central section

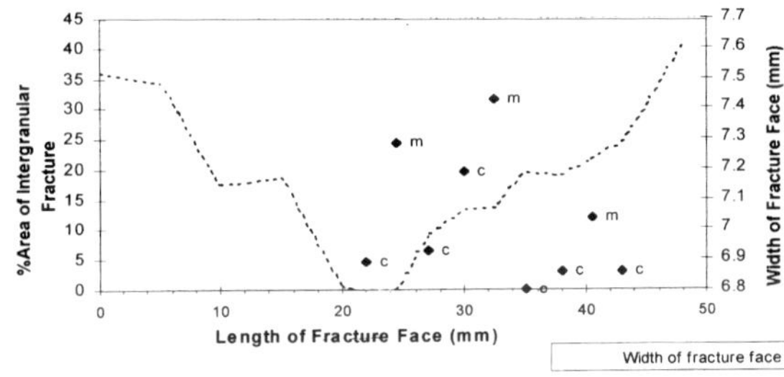

Fig. 12c: % Intergranular fracture at the zigzag blades

Figure 12 a–c % Intergranular fracture of spray-dried agglomerates.

fracture in these regions. The variations detected at the zigzag blade locations (see Fig. 12c) are a result of the local geometry which probably gives rise to localised shear stresses; these will cause particle-particle interactions and particle-die wall interactions to be greater at the corners of the zigzag contact; fracture was found to be ~ 95 % transgranular in these regions. It is likely that variations in compaction at such localised levels can result in differential shrinkage during sintering, which could affect fracture as already discussed in Section 3.1.

It is also interesting to note that the % area of intergranular fracture of granules varies (approximately) inversely with the width of the fracture face.

3.3 Microstructural Features

Preliminary optical microscopy results have shown differences in the microstructural features studied. Porosity levels varied randomly between 4 % and 16 % along the length of the fracture faces of the standard and off-line yoke rings studied. Measurement of the % area residual granule history content showed random variations of 0 % to 1 % along the length of the fracture faces of standard and off-line yoke rings studied; the mean grain diameter varied from 10 μm to 18 μm in all yoke rings, standard and off-line.

Results suggest that the microstructural features in sintered yoke rings develop independently of spray dried granule agglomerate size. Differences in compaction and green densities (as discussed in Section 3.2) that occur along the length of these fracture faces (prior to sintering) do not appear to cause any microstructural or macrostructural variations. Furthermore, any qualitative differences (in microstructural and macrostructural features) between standard and off-line yoke rings indicate that the effect of porosity, residual granule history and mean grain diameter (at the fracture face) on cracking behaviour are minimal. However, density measurements carried out over the length of fracture faces and central sections on sintered yoke rings showed similar trends to those detected in the green bodies; densities were consistently higher across the fracture face compared to the central section and were at a minimum away from the base and flare ends. Hence differences in green densities at fracture face regions and central regions lead to differences in the final sintered densities; these are indicative of microstructural variations. The fact that the results from optical microscopy are inconclusive is probably due to the small area of the total yoke ring studied. Work will continue with further microstructural and macrostructural studies on other areas of the yoke rings. In addition, yoke rings of different geometries and sintering conditions will be studied.

CONCLUSIONS

1. Fracture is primarily influenced by the geometry of a yoke ring although spray dried granule agglomerate size and sintering atmosphere also alter cracking behaviour.
2. Variations in spray dried granule agglomerate fracture occurring throughout green bodies are linked to friction and shear forces acting during compaction. This effect is exaggerated at the base and flare ends, and zigzag blade locations.
3. Variations in the green density lead to differences in the sintered density of a yoke ring. These are indicative of microstructural variations and are likely to affect fracture of a yoke ring.

REFERENCES

1. E. C. Snelling, *Soft Ferrites: properties and applications,* 2nd edn, Butterworths, 1988.
2. R. J. Brook (ed.), *Concise Encyclopaedia of Advanced Ceramic Materials,* Pergamon Press, 1991.
3. G. J. Hahn and W. Q. Mecker, *Statistical Intervals, A Guide for Practitioners,* Probability and Mathematical Statistics series, Wiley, 1991.
4. A. S. James, W. M. Dawson and T. J. Davies, *Powder Metallurgy,* 1987, **30** (4), 267–271.
5. S. T. Rasmussen, W. Ngaji-Okumu, K. Boenke and W. J. O'Brien, *Dental Materials,* 1997, **13**, 43–50.
6. B. I. Briscoe and N.Ozkan, *Powder Technology,* 1997, **90**, 195–203.
7. H. J. Glass, G. De With, M. J. M. Graaf and R. J. A. Van Der Drift, J. Mat. Sci. 1995, **30**, 3162–3170.
8. P. J. James, *Powder Metallurgy,* 1977, **20** (4), 199–200.

Erosive Wear of Alumina/SiC Nanocomposite Ceramics

C. W. LAWRENCE, S. G. ROBERTS and B. DERBY*

Department of Materials, University of Oxford, Parks Road, Oxford OX1 3PH, UK

ABSTRACT

The wet erosive wear behaviour of alumina/silicon carbide nanocomposites has been examined using a purpose built Rotating Jet Slurry Wear Tester (RJSWT). The wear rates of nanocomposites fabricated by pressureless sintering and hot pressing and incorporating varying volume fractions of silicon carbide nanoparticles (up to 20% by volume) has been measured. Scanning electron microscopy has been used to examine the worn surfaces of the nanocomposites Results indicate that nanocomposites have a lower wear rate than equivalent alumina specimens and that the volume fraction of nanoparticles does not strongly affect the wear properties of the nanocomposites.

1. INTRODUCTION

Alumina based nanocomposites have received considerable interest due to their potentially superior strength and creep resistance[1,2] compared to monolithic alumina. With the notable exception of Davidge *et al.*,[3] little attention has been paid to the possible benefits in wear resistance that this new class of materials offers. This study will present initial results from an investigation of the wet erosive wear behaviour of nanocomposites consisting of silicon carbide (SiC) nanoparticles (~ 200nm mean particle size) in alumina ($A1_2O_3$). The following aspects of the wear behaviour are of particular interest and will be addressed in the current study:

1. Does the inclusion of silicon carbide particles in alumina significantly lower its erosive wear rate?
2. Is the erosive wear mechanism for alumina based nanocomposites the same as for monolithic alumina?
3. Does the volume fraction of silicon carbide nanoparticles control the erosive wear resistance?
4. Does the nanocomposite fabrication process affect the erosive wear rate?

2. EXPERIMENTAL DETAILS

2.1 Specimen Preparation

Specimens of alumina/silicon carbide nanocomposites were fabricated by both pressureless sintering (PS) and hot pressing (HP) using a conventional powder processing route (see

* Now at Manchester Material Science Centre, Grosvenor Street, Manchester M1 7HS

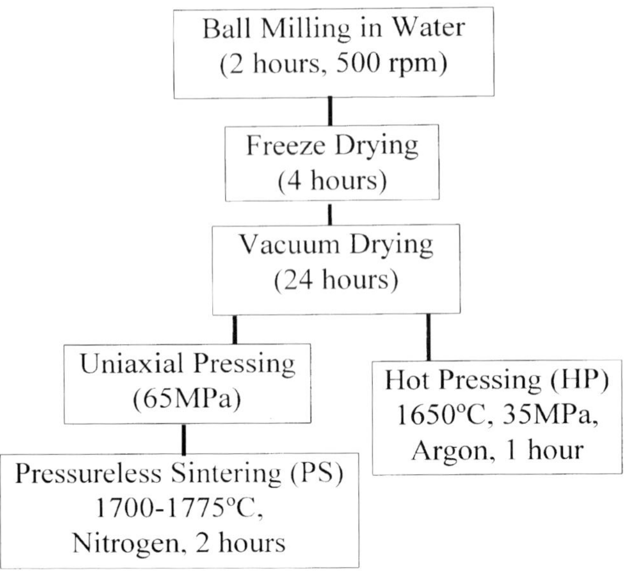

Figure 1 Nanocomposite processing route.

Fig. 1) For the pressureless sintered nanocomposites Sumitomo AES-l IC alumina powder (mean particle size ~400 nm) was used for the matrix, for the hot pressed nanocomposites Sumitomo AKP-53 alumina powder was used (mean particle size 500nm). In both cases H. C. Starck UF-45 silicon carbide particles (~200 nm mean particle size) was added as the reinforcing phase. The powders were mixed by ball milling in water (2 hours at 500rpm with zirconia media) followed by freeze drying (4 hours) and vacuum drying (24 hours). The resulting nanocomposite powders were then densified by either pressureless sintering in flowing nitrogen (after uniaxial pressing at 65MPa) or hot pressing in argon. Table 1 summarises the various nanocomposite and alumina specimens produced and the relevant processing conditions. All specimens had densities (as measured by the water immersion method) of greater than 99% theoretical.

Specimens for erosive wear testing were cut from the as-produced nanocomposites using a high-speed diamond saw. The specimen dimensions were thickness 3.5 mm, width I 2mm and breadth 10 mm. One face of each specimen was then ground (removing ~ 0.3 mm of material) and then polished using successively finer grades of diamond slurry (25 μm, 8 μm, 3 μm and 1 μm). It was noted that the nanocomposite materials exhibited less grain pullout during the polishing process.

2.2 Erosive Wear Tests

A purpose-built Rotating Jet Slurry Wear Tester (RJSWT) was used to perform erosive wear tests on the nanocomposite and alumina specimens. The RJSWT essentially consists of four parts. (I) a slurry pot containing the erosive slurry in which (2) a specimen holder, fixed in a et body is rotated by (3) an electric motor and is supported on (4) a frame The particular advantage of this equipment over previous ones is that the jet body allows control of the

Table 1 Nanocomposite and alumina processing conditions.

Matrix	% SiC	Process	Temperature (°C)
AES-l 1C	0	PS	1600
AES-l 1C	1	PS	1700
AES-l 1C	2	PS	1700
AES-l 1C	5	PS	1775
AKP-53	0	HP	1500
AKP-53	5	HP	1650
AKP-53	10	HP	1700
AKP-53	20	HP	1700

slurry media impinging on the specimen during testing allowing constant and repeatable test conditions to be used. Full details of the RJSWT can be found in Ref. 4.

A slurry consisting of 1.5 kg of silicon carbide abrasive (mean particle size of ~780 μm) in 8 litres of water was used for all wear tests with a motor speed of 180 rpm. Specimens of each nanocomposite and alumina were tested as follows. The dry weight of the specimen was measured and recorded and the specimen inserted in the sample holder. The equipment was then run for one hour and the motor then stopped. The specimen was then removed, washed, dried and weighed before being re-inserted in the specimen holder This piocedure was repeated until each specimen had experienced a total of 15 hours erosion. Fresh slurry was used for each specimen as there is a small change in the particle size distribution of the silicon carbide abrasive during the course of the erosion measurements.

3. RESULTS

The erosive wear rate of each specimen was calculated by plotting the weight loss of each specimen as a function of erosion time and calculating the gradient of the plot. Dividing this value by the wear area (47.2×10^{-5} m^2 as measured from photographs of worn specimen surfaces) yields the erosive wear rate. Figure 2 shows a plot of some of the measured weight loss data and the erosive wear rates measured for each specimen are tabulated in Table 2. Also listed in Table 2 are grain size measurements for some of the specimens, determined from SEM images using the linear intercept method The measurement of strain sizes for the remaining specimens is on going, though SEM images of the worn surfaces of the nanocomnposite specimens indicate that they all have a similar grain size in the range 1–5 μm.

It is apparent from the data in Table 2 that with regards the 'pure' alumina specimens, the erosive wear rate is dependent on grain size. This result agrees with the findings of other workers.[3,5] The data presented in Table 2 reveals that all the nanocomposites possess a significantly lower wear rate than either of the two aluminas. Further, the data indicates that there is very little difference in wear rate between the different nanocomposites despite the wide range in volume fraction of silicon carbide reinforcement (1–20%). However it can be

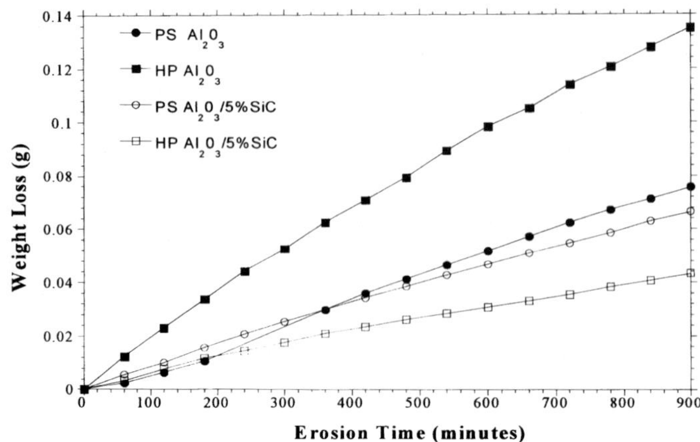

Figure 2 Weight loss as a function of erosion time for selected specimens.

Table 2 Nanocomposite and alumina erosive wear rates.

Material	Wear Rate (10^{-5} kg m^{-2} s^{-1})	Grain Size (μm)
Al_2O_3 – PS	3.1	3.4
Al_2O_3 – HP	4.9	14.9
Al_2O_3/5% SiC	1.5	
Al_2O_3/10% SiC	1.8	
Al_2O_3/20% SiC	1.3	3.0
Al_2O_3/1% SiC	2.1	
Al_2O_3/2% SiC	2.6	
Al_2O_3/5% SiC	2.4	

seen that in general the nanocomposites produced by hot pressing have lower erosive wear rates than those produced by pressureless sintering.

Figures 3 and 4 show SEM micrographs of the wear surfaces of a pressureless sintered alumina and an Al_2O_3/5 vol.% SiC nanocomposite respectively. The alumina surface exhibits more grain 'pull-out' (i.e. completely removed alumina grains) than the nanocomposite and a greater degree of grain boundary fracture. Nanocomposite wear surfaces were also noticeably smoother than the alumina and suggestive of plastic deformation There was no appreciable difference between the wear surfaces of the different nanocomposites or between the alumina specimens.

4. CONCLUSIONS

The erosive wear rates measured for alumina and alumina based nanocomposites indicate that the inclusion of nanosize silicon carbide particles reduces the wear rate compared to equivalent grain size polycrystalline alumina. We find that for equivalent grain sized alumina and nano-

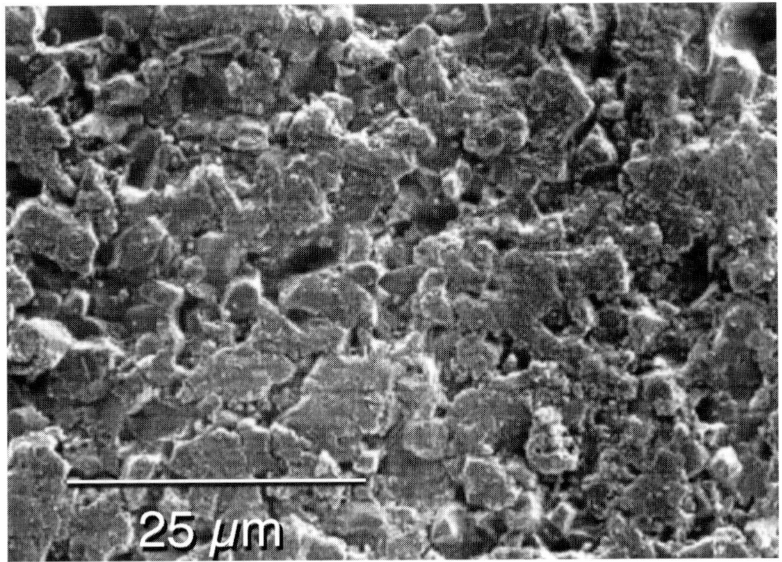

Figure 3 Wear surface of pressureless sintered alumina.

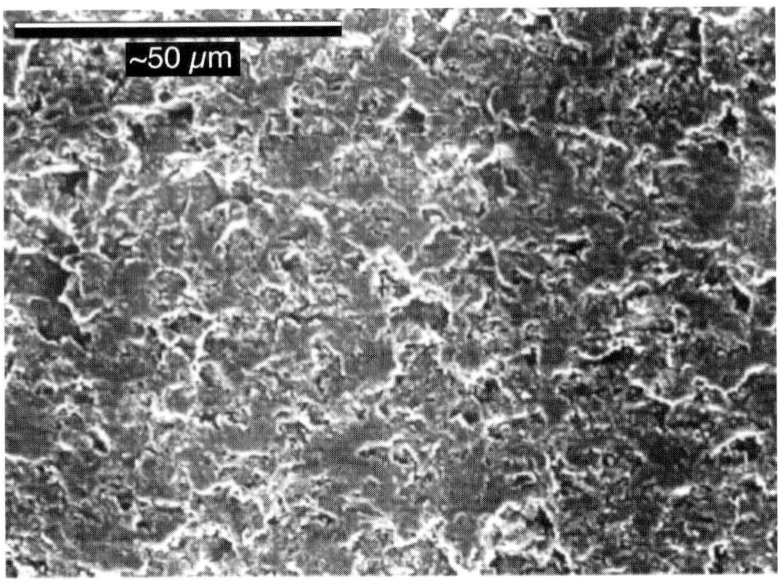

Figure 4 Wear surface of pressureless sintered alumina/5 vol.% SiC.

composites (compare the data for the pressureless sintered alumina and hot pressed Al_2O_3 20 vol.% SiC in Table 2) the wear rate is reduced by a factor of' 2–3 in agreement with the results of Davidge *et al.*[3] The experimental data also supports the conclusion of Davidge *et al.*[3] that an important factor in the excellent wear resistance of alumina based nanocomposites is the role of the silicon carbide nano-phase in inhibiting alumina grain growth.

As can be seen from Table 2, the improvement in wear rate appears to be insensitive to the volume traction of silicon carbide particles over the range 1–20 vol.%. This indicates that only a small amount of the comparatively expensive silicon carbide particles are needed to achieve a significant reduction in erosive wear rate compared to 'pure' alumina. Further commercial benefits arise in that the processing conditions (notably sintering temperature) needed to produce high quality nanocomposites decrease with the volume fraction of silicon carbide Hence low volume fractions (<5 vol.%) of silicon carbide particles are preferable.

Comparing the wear data for the hot pressed nanocomposites with that for the pressureless sintered nanocomposites we find that the latter have slightly higher wear rates (by a factor of up to 2). Direct comparison of the two is at the moment difficult as (1) further work is needed to measure the grain sizes of the materials, and (2) the two nanocomposite types were made using differing alumina powders. Further work is in progress in order to optimise the various nanocomposites by altering the processing conditions

SEM images of the wear surfaces indicate that the inclusion of nano-size silicon carbide particles in the alumina results in a change in the wear mechanism. The degree of grain pullout is reduced as is the degree of grain boundary fracture and the degree of plastic deformation increases. Work on characterising the wear mechanism operating in alumina based nanocomposites is on going though these initial results agree with other researchers' work[3,4,6] This is that as the grain size of alumina/nanocomposite approaches 1 μm the dominant wear mechanism changes from brittle-fracture to polishing/plastic deformation controlled.

5. REFERENCES

1. K. Niihara, *J. Ceram. Soc. Jpn,* 1991, **99,** 974–982.
2. A. M. Thompson, J. Chang, H. M. Chan and M. P. Harmer, *Ceramic Transactions,* 1995, **51**, 671-678
3. R W Davidge, P C Twigg and FL Riley, *J. Euro. Ceram. Soc.,* 1996, **16** 799–802.
4. A. Franco Jr, 'Erosive Wear of Alumina', D.Phil Thesis, University of Oxford, 1996.
5. A. Franco Jr and S. G. Roberts, *J. Euro. Ceram. Soc.,* 1996 **16** 1365–1375.
6. S. J. Choun, B. J. Hockley, B. R. Lawn and J. Bennison, *J. Am. Ceram. Soc.,* 1989, **72**, 1249–52.

Sialon/TiB_2 Ceramic Composites: Synthesis, Properties and Tribology

A. H. JONES, R. S. DOBEDOE, M. H. LEWIS and R. J. LUMBY

Centre for Advanced Materials, Department of Physics, University of Warwick, Coventry, UK

ABSTRACT

The addition of titanium diboride particulates to a sialon matrix offers the possibility of ultrahard ceramic materials suitable for tribological applications (e.g. ball bearings, cutting tools etc.). Sialon matrix composites have been fabricated by hot pressing and pressureless sintering with TiB_2 added by both direct addition and in-situ reaction sintering. Both processes require a dual atmosphere sintering cycle of argon followed by nitrogen at the sintering temperature. Hot pressing has been used to produce materials which approach full density independent of the amount of TiB_2 present. However, the densification of pressureless sintered materials is inhibited by the presence of TiB_2.

The study of hardness, indentation fracture toughness, microstructure and tribology has shown significant improvements in properties. At >40vol.% TiB_2 the low resistivity of these materials allows the possibility of electro-discharge machining (EDM).

A mini pin-on-disc apparatus was constructed to measure the coefficient of friction (μ) and the wear rate (k) of the composites. Worn surfaces were studied by electron microscopy. The wear behaviour is dominated by tribo-chemical reactions, forming oxygen rich tribofilms adhered to the discs surface. Wear then occurs by degradation of the tribofilm and the polishing action of wear debris. Inherent problems with this type of wear test are discussed.

1. INTRODUCTION

The excellent engineering properties of Si_3N_4 and the related sialon materials are well known and varying grades of sialon are used in applications such as cutting tools and high performance bearings. The inclusion of secondary phases into the ceramic microstructure is a widely used approach to further improve or tailor the key properties of ceramic materials.[1,2,3] The secondary phases used must be compatible with both the matrix and with the environment in which the materials may be used, i.e. high temperatures, chemically hostile environment or tribological situations. The properties of TiB_2 listed in Table 1 make it a candidate for use as a reinforcement in sialon materials.[4]

The research presented here has investigated the production and properties of β-sialon matrix materials containing particulate TiB_2. Sialon/TiB_2 composites offer the possibility of higher hardness, improved fracture toughness and other benefits such as ceramics that can be machined by electro-discharge machining, an alternative and more versatile technique than diamond machining. Silicon nitride and sialon materials are already used in many tribological applications[5] and the property enhancement brought about by the addition of TiB_2 has the potential to produce a superior wear material.

Table 1 Some properties of TiB_2.[4]

Property			
Hardness, Hv, (GPa) [Load]	33 [0.3N]	25.5 [1N]	20.8 [5N]
E Modulus, (GPa)	500–545		
Resistivity, (Ω cm)	9×10^{-6}		
Thermal Expansion Coefficient (10^{-6} K^{-1})	6.63 (a & b)	8.6S (c)	
Oxidation resistance (mg cm^{-2})	30 @1000°C	(100 hours)	

2. EXPERIMENTAL PROCEDURE

The sialon/TiB_2 composites were produced by several methods and each method was evaluated primarily in terms of its success at producing the required composite but also in terms of its ease of use and potential for economic production of components. The resulting materials were characterised with respect to some of their key engineering properties, their microstructure and their tribological performance. The results of such tests were compared to identical tests carried out on some commercially available grades of sialon in order to determine whether there was any significant change in materials properties with TiB_2 content.

2.1 Materials Production

The introduction of TiB_2 particles into the sialon matrix was achieved by the direct addition (DA) of TiB_2 or by the in-situ reaction sintering (ISR).

For direct addition, fine grades of TiB_2 powders were mixed directly with those used to form the matrix. A fine grade of TiB_2 powder (H. C. Stark, grade F) was mixed with a commercial, β-sialon forming powder containing ~6wt% yttria as the densification aid. This grade has a small, even particle size (D10=0.9-1.6 μm, D50 = 2.0–3.5μm, D90 = 3.6–5.5 μm) and is of a high purity,(O<2%, N<0.25%, C<0.25%, Fe<0.25%). Direct addition was used to produce 40vol.% TiB_2 composites.

An alternative approach to the direct addition method is the synthesis of TiB_2 in-situ during the sintering process. The reaction of TiN with BN (eqn 1) was used, with the excess nitrogen produced being reacted with silicon to simultaneously produce the matrix phase. The oxide impurities present on all constituents were accounted for and also used to produce the matrix and particle phase (eqn 2).

$$4TiN + 8BN + AlN + 0.5Si_3N_4 + 9Si + 0.5SiO_2 = 4TiB_2 + Si_{11}AlON_{15} \quad (1)$$

$$2\,TiO_2 + 2B_2O_3 + 10AlN + 5Si = 2TiB_2 + Si_5Al_{10}O_{10}N_{10} \quad (2)$$

Powders used were α Si_3N_4 (UBE, Japan, SN E-10), TiN (Tioxide Chemicals, UK, TTN70),

BN (BC&C, UK, HQ), Si (Kemanord, Sweden, 4E) and AlN (H.C. Starck GmBH, Germany, C). The densification aid was a glass powder with Y_2O_3:Al_2O_3:SiO_2 in the ratio 40:25:35wt%.

Powders were wet milled with iso-2-propanol in high density polyethylene bottles with sialon milling media for 24-72hrs. The resulting slurries were passed through a 38μm sieve, dried, ground and dry sieved through 500 and 250 μm meshes to beak up drying agglomerates. Powder mixtures containing 10, 20, 30 & 40vol.% TiB_2 were produced in this way.

From both powder production methods green bodies were produced by both uniaxially pressing discs (~8 × 25mm dia.) at 40 to 60MPa or by isostatically pressing cylinders (~30 × 23mm dia.) at ~120MPa.

Composites were densified using both hot pressing and pressureless sintering techniques. Hot pressing was carried out in BN lined graphite dies with a maximum applied pressure of 30MPa. Pressureless sintering was carried out in sialon lined, graphite crucibles. The atmosphere in either furnace could be nitrogen (max. 1 atm.), argon (max. 1 atm.) or vacuum ($<10^{-5}$ mbar).

The sintering of these composites, which can contain silicon or TiB_2 in the green state, is not as straight forward as the sintering of the matrix alone. Reaction and densification temperatures along with sintering times and atmosphere changes were determined experimentally and are presented in the results section.

2.2 Characterisation

The resulting composites were tested to determine some key engineering properties. Hardness (Hv) was measured using Vickers diamond indentation at 5,10 and 50 N applied loads with indents sizes measured optically. The indentation fracture toughness (K_{Ic}) was determined from cracked Vickers indents made at 50 N load. The crack size $2c'$ was measured optically and K_{Ic} was calculated after the method of Anstis.[6] Determination of Young's modulus (E) for K_{Ic} calculation was carried out using pulse-echo ultrasonic techniques.

Examination of the crack path, determination of porosity and TiB_2 volume fractions and imaging of the microstructure was carried out using scanning electron microscopy and image analysis software. Phase identification was carried out using X-ray diffraction on the surface of polished, solid polycrystalline specimens.

2.3 Tribological Testing

The tribological behaviour of the composites was examined using a mini pin-on-disc apparatus specially designed to test small specimens (~20–25 mm diameter disc, 3 × 3 × 10 mm pins). The apparatus is shown schematically in Fig. 1 and was enclosed so as to allow control over ambient temperature and humidity. The pin end was hemispherical and finished using 25, 6 and 1 μm diamond polishing to a radius (R) of ~3 mm. The disc was machined flat and parallel and finished using the same polishing method as for the pin. Testing was carried out at various loads and sliding speeds. The wear volume of the pin was recorded by optical measurement of the pin wear scar diameter. The volume removed (V_p) is calculated from $V_p = \pi r^4/4r$, where r is the wear scar radius and R is the radius of the hemispherical ended pin. The wear volume of the disc was not easily measurable and the reasons for this are explained in the results section. From the worn volume it is possible to calculate the

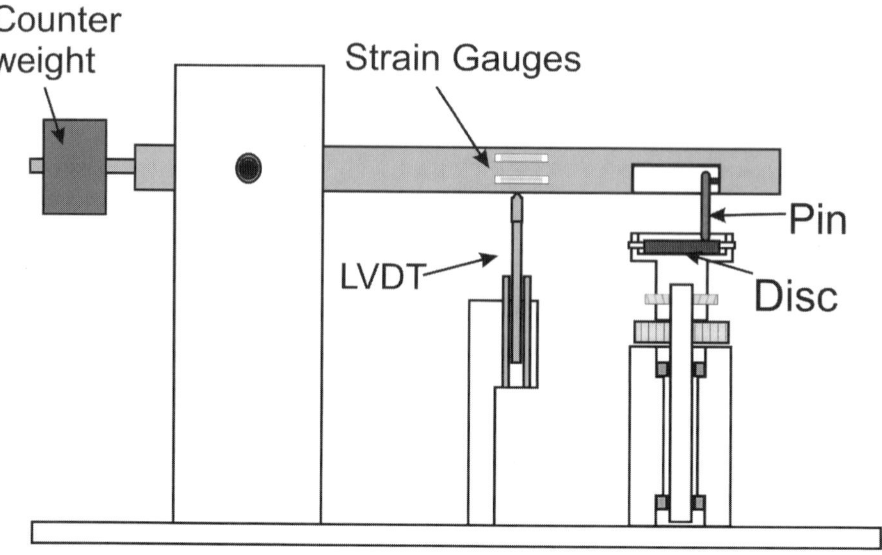

Figure 1 The pin-on-disc wear test apparatus. The disc specimen is usually 20–25mm in diameter.

wear coefficient, k, which has units of $mm^3\ N^{-1}\ m^{-1}$ and is a more useful measure of wear than weight loss or volume loss as it is independent of sliding distance and load (providing the test is carried out under constant conditions). The frictional force experienced by the load arm (F) was recorded via a strain gauge set up and continuously monitored. The coefficient of friction (μ) was calculated from $F = \mu W$, where W is the applied load in N. Wear debris were collected and these, along with wear surfaces of both the pin and the disc, were examined using SEM.

To investigate the effect of varying test conditions, loads (W) were varied from 1.96 to 9.8 N, sliding speeds (V) from 0.05 m s^{-1} to 1.25 m s^{-1} and a relative humidity (RH) from 5 to 95% RH. The pin radius was 3mm. Typically tests were run over sliding distances (d) of 500m. The resulting Hertzian stresses(σ_r) were calculated from the following equations. For a stationary spherically ended pin on a flat surface,

$$\sigma_r = \frac{1-2\nu}{3} P_0 \tag{3}$$

where ν is Poisson's ratio and P_0 is the maximum pressure at the axis of contact given by

$$P_0 = \frac{3W}{2\pi r^2} \tag{4}$$

where W is the load and r is the radius of contact given by

$$r = \left(\frac{3WR}{4E^*}\right)^{1/3} \tag{5}$$

E^* is given by

$$E^* = \left[\left(\frac{1-v_1^2}{E_1}\right)+\left(\frac{1-v_2^2}{E_2}\right)\right]^{-1} \quad (6)$$

where E is the elastic modulus and ν is Poisson's ratio, the subscripts 1 and 2 represent the 2 materials in contact. In sliding contact the stress is greater and is concentrated at the trailing edge and is dependent upon μ. This trailing edge stress is given by

$$\sigma_s = \left\{\frac{(1-2v)}{3}+\left[\frac{(4+v)}{8}\right]\pi\mu\right\}P_0 \quad (7)$$

In sliding contact it is this magnified stress that can cause damage to the materials surface. This magnified stress is illustrated in Fig. 2.

3. RESULTS

The materials were characterised with regards to density, phase content, microstructure, mechanical properties and tribological performance.

3.1 Materials Production

Normal densification of sialon uses an atmosphere of nitrogen throughout the sintering process. However, in the presence of TiB_2 (in the case of direct addition materials) or silicon (in the case of in-situ reacted materials) it was found necessary to use an argon atmosphere up to a point where the structure had achieved closed porosity. This was due to the potential reactions between the specimen and the nitrogen atmosphere potentially used for sintering sialon:

$$2TiB_2 + 3N \Rightarrow 2TiN + 4BN\ (\Delta G_{1900\,^{\circ}C} = -831\ \mathrm{kJ\ mol^{-1}}) \quad (8)$$

forming TiN and BN throughout the composite and

$$3Si(s) + 2N_2(g) \Rightarrow Si_3N_4(s) \quad (9)$$

consuming all the silicon required for the *in-situ* reaction process. Nitrogen was found to be necessary at the sintering temperature to avoid decomposition of the matrix but by introducing the nitrogen atmosphere after closed porosity had been achieved then any reactions were limited to the surface of the specimen.

The optimum temperature/time profile for densification of direct addition materials was found to be a ramp rate of 15°C min^{-1} to a maximum of 1700 °C for ~1 to 4 hours, depending on whether hot pressing or pressureless sintering was being used. Temperatures in excess of 1700 °C resulted in decomposition of the sample. For hot pressing a maximum pressure of 30 MPa was found to be sufficient to achieve maximum densification.

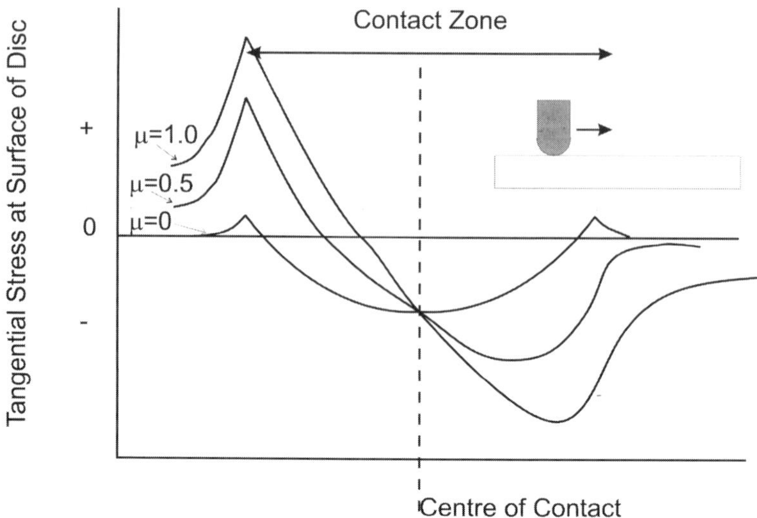

Figure 2 The effect of sliding with a coefficient of friction μ on the tangential Hertzian stress (σ_s) induced in the surface of the disc.

For *in-situ* reacted materials an additional temperature hold was necessary to allow the reaction between TiN and BN to proceed. The optimum temperature for this reaction was found to 1380 °C. Lower temperatures resulted in unreacted TiN and BN while higher temperatures resulted in the melting of the silicon which was then unable to react with N_2 to form the matrix phase. The maximum densification temperature for *in-situ* reacted materials was found to be 1650 °C, above which the specimen decomposed. The optimum sintering profile for a pressureless sintered, *in-situ* reacted material is shown in Fig. 3.

The resulting composites consist of a low substitution β-sialon and TiB_2 as shown by XRD in Fig. 4. The starting compounds of a Si_3N_4, AlN, TiN, BN and glassy sintering aid result in a sintered composition of β-sialon, TiB_2 and a yttrium aluminium garnet phase, YAG ($Y_3Al_5O_{12}$), formed from the crystallisation of the residual sintering aid.
Hot pressing achieved the best result in terms of densification. Direct addition materials could be easily hot pressed to >98% of the theoretical at up to 40vol.% TiB_2 while in-situ reacted materials reached ~95–98% of theoretical but were limited to less than 20 vol.% TiB_2. For pressureless sintered materials only those with low vol.% of TiB_2 reached >90% of the theoretical density. The results of sintering trials are summarised in Table 2.

From these sintering trials it was found that only hot pressed materials could be densified to a level where the composites were both dense enough and contained enough TiB_2 to make any improvements in properties possible.

3.2 *Microstructure*

The microstructure of polished specimens were examined to determine the homogeneity, particle size and phase content of materials produced by all methods. Energy dispersive x-ray spectroscopy (EDXS) was used to identify the composition of phases in the microstructure.

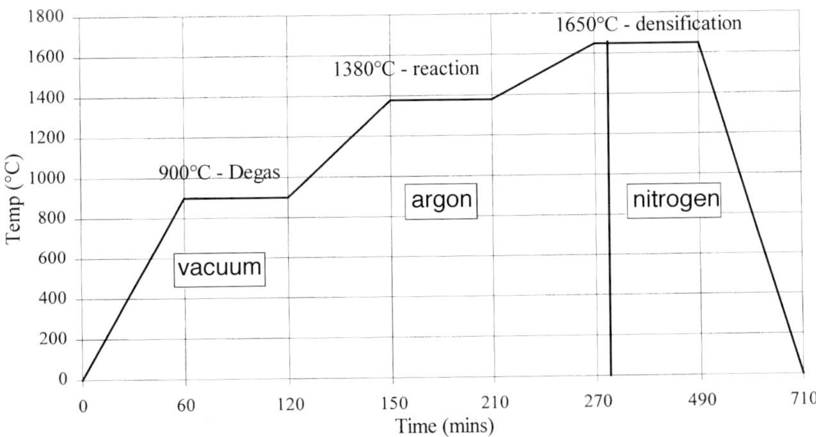

Figure 3 The optimised sintering cycle for pressureless sintered materials produced by *in-situ* reaction sintering

Table 2 The densities achieved by materials produced by different methods (IS = in-situ reaction, DA= direct addition, PS = pressurless sintered, HP = hot pressed).

Material ID	Addition/Production Method	Vol.% TiB_2	Density, g cm^{-3}, (% of theoretical)
HP40DA	DA/HP	40	3.656 (99)
HP20IS	IS/HP	20	3.408 (98)
PS10 IS	IS/PS	10	3.317 (97)
PS10 DA	DA/PS	10	3.177 (95)
PS20DA	DA/PS	20	3.095 (89)
PS30DA	DA/PS	30	3.150 (88)
PS40DA	DA/PS	40	3.260 (87)

3.2.1 In-situ reacted materials

In-situ reacted materials consist of a homogeneous distribution of TiB_2 grains in a β-sialon matrix. Using back scattered electron imaging on the SEM confirms that all the TiN and BN has been converted to TiB_2 particles (confirmed by EDX of particles and XRD) and that the α-Si_3N_4 to β-sialon phase transformation has taken place (see Fig. 5). The dark phase in the micrographs is the sialon matrix and the grey particles have been identified by EDX as TiB_2.

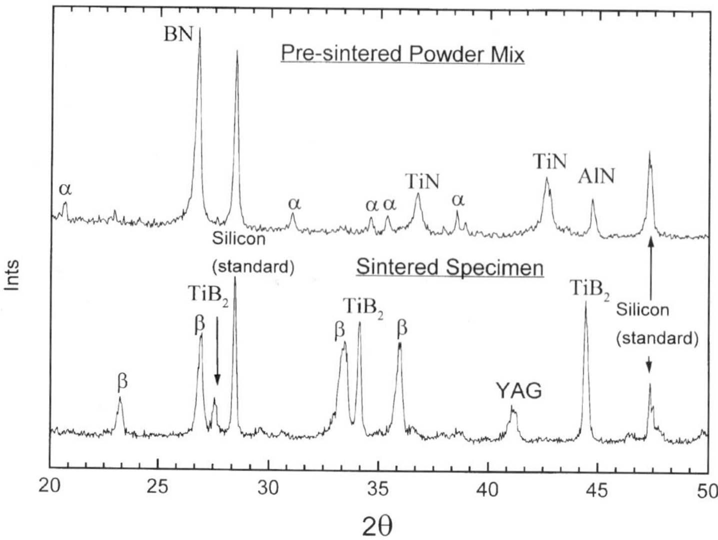

Figure 4 XRD trace from the powder mix for in-situ reacted material and a sintered specimen. The a Si_3N_4, TiN and BN have been transformed to β-sialon and TiB_2.

The TiB_2 grain size is usually below 3 μm while the sialon retains a very fine grain size of slightly elongated β-sialon grains. In the case of pressureless sintered materials the residual sintering aid remains as an intergranular, yttria rich glass phase or as YAG which shows up bright in back scattered electron imaging (see Fig. 6). In hot pressed materials where <1wt% yttria was used there is no detectable intergranular phase and hence it is not possible to observe the sialon grain size or morphology.

3.2.2 Direct Addition Materials

Direct addition material contained ~6wt% yttria whether hot pressed or pressureless sintered and were fabricated with a higher vol.% of TiB_2. Despite sintering times in excess of 60mins in the hot press the α-Si_3N_4 to β-sialon transformation has not been completed and there appears to be both α-Si_3N_4 and β-sialon present in the structure (confirmed by XRD). At 40 vol.% some of the TiB_2 particles are in contact and coalesce to form grains larger than the original particle size. The intergranular phase contains yttria but has not crystallised on cooling as in the pressureless sintered case.

3.3 Hardness and Fracture Toughness

Only materials with less than 2–3% porosity were used for property measurement as above this level the porosity dominates the results for hardness, toughness and elastic modulus. Table 3 presents the porosity level, hardness (Hv) at differing loads and indentation fracture toughness (K_{Ic}) values for a hot pressed material containing 40 vol.% TiB_2 composite and values obtained from comparable commercial materials.

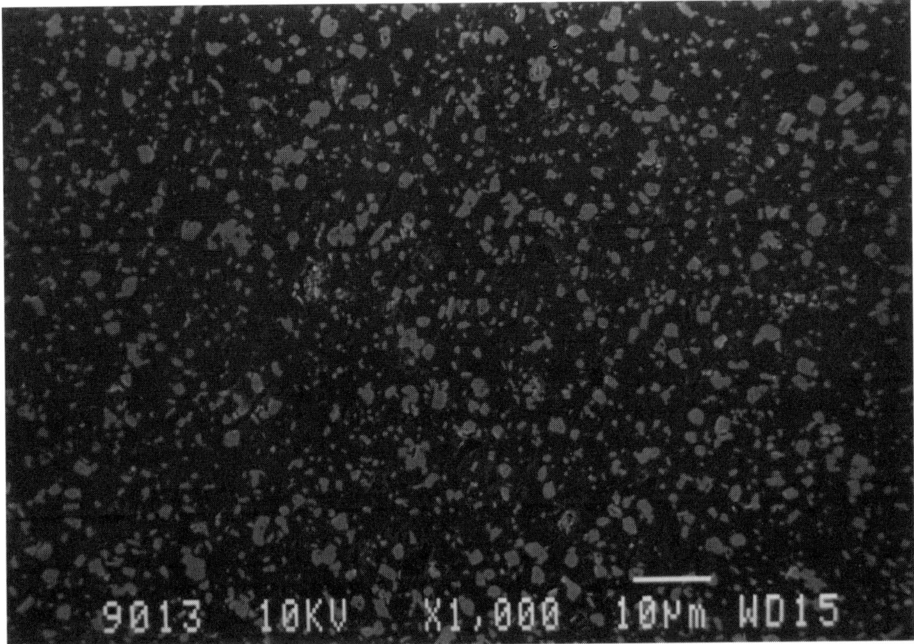

Figure 5 SEM image (back scattered electron mode) of a hot pressed 20 vol.% TiB_2 sialon composite formed by *in-situ* rection sintering.

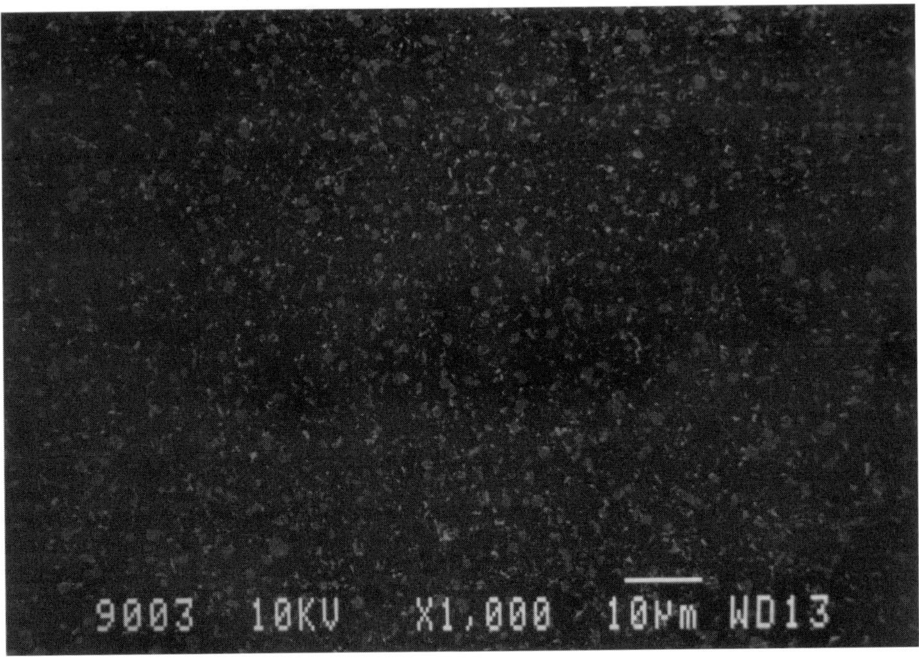

Figure 6 SEM image (back scattered electron mode) of a pressureless sintered 10 vol.% TiB_2 sialon formed by *in-situ* reaction sintering.

Figure 7 SEM images showing the crack path through a 10 vol.% pressureless sintered material.

Table 3 Hardness and indentation fracture toughness for a hot pressed 40vol% TiB_2 material compared with several comparable commercial materials.

Material	Hv [50N], GPa	K_{Ic} $MPam^{1/2}$ (Anstis)	K_{Ic} (adjusted)	Conducts (Y/N)
HP40DA	**16.8**	**6.4**	**9.0**	**Y**
Sialon (cutting tool grade)	13.9	4.4	6.2	N
Si_3N_4 (bearing grade)	15.7	4.7	6.6	N
Si_3N_4 + 30wt%TiN[11]	15.3	5	–	Y

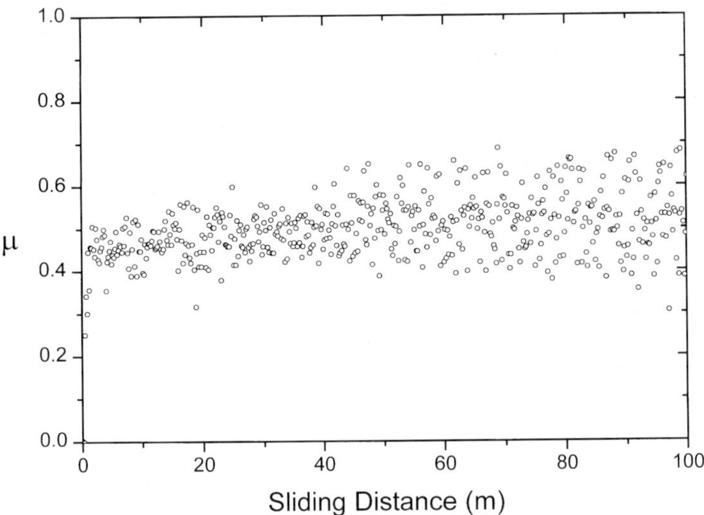

Figure 8 The noise in the recorded coefficient of friction (μ). The noise is inherent to the pin-on-disc test The true value of μ was taken as the top of the 'noise envelope' and not the mean.

Table 4 The value of μ for several grades of sialon/TiB_2 composites under dry sliding and lubricated conditions (* = cutting tool grade sialon).

Pin	Disc	Conditions W(N), V(ms^{-1}), RH(%)	μ
HP20IS	HP20IS	W = 9.88, V = 0.05, RH = 40	0.78
PS10IS	HP20IS	W = 9.88, V = 0.05, RH=45	0.56
HP20IS	PS10IS	W = 9.88, V = 0.05, RH = 43	0.4
HP20IS	HP20IS	W = 9.88, V = 0.05, water lubricated	0.67
HP20IS	H20IS	W = 9.88, V = 0.05, paraffin lubricated	0.15
sialon*	sialon*	W = 9.88, V = 0.05, RH = 45	0.69

The TiB_2 containing material is harder than both the commercial grades and an alternative conductive sialon composite. The fracture toughness as measured by indentation is significantly higher than for the commercial materials. The column showing K_{Ic} (adjusted) is calculated from equations routinely used in industry to compensate for the fact that indentation toughness of Si_3N_4 materials is invariably lower than toughness measured by SENB or indent and break techniques. The hot pressed material containing 40 vol.% TiB_2 has been found to be conductive to the point where it was easily machined using EDM.

Examination of the crack path from Vickers indents (see Fig. 7) reveals that the path includes both intra and inter granular paths whilst being deflected by both sialon and TiB_2 grains. Some TiB_2 particles are observed to bridge the crack front.

Table 5 The variation of μ with relative humidity for 20 vol.% TiB2, hot pressed material sliding on itself.

Relative Humidity (%)	μ
<10	0.8
40	0.75
>95	0.7

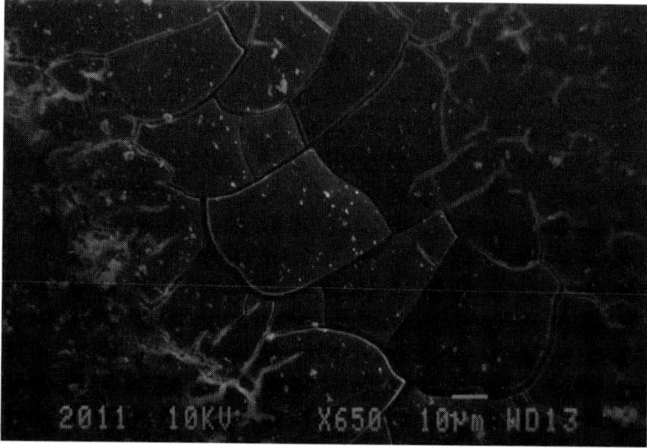

Figure 9 The tribofilm adhered to the surface of the disc after dry sliding wear of sialon/TiB_2 on itself.

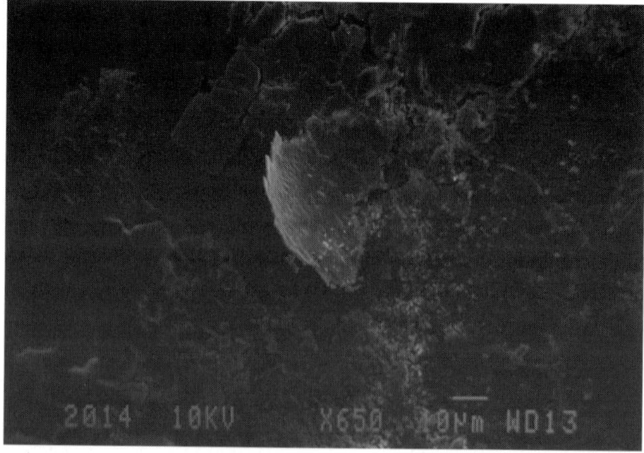

Figure 10 An area of tribofilm which has been partially removed demonstrating it to be of ~1μm thickness.

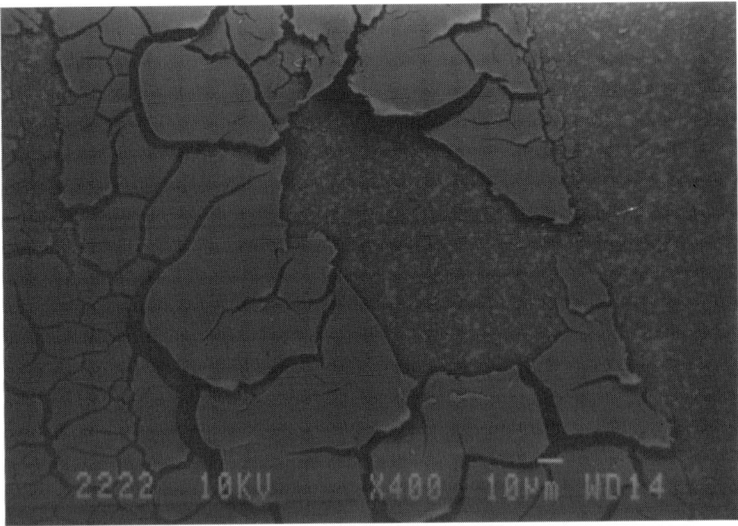

Figure 11 The removal of pieces of tribofilm during wear reveals that the underlying surface of the disc is undamaged by the passage of the pin.

3.4 Tribological Behaviour

Several grades of the sialon/TiB_2 composite were tested under dry sliding conditions. An example of the dynamically recorded value of μ is shown in Fig. 8. The variation in the measured frictional force could be large due to the nature of the pin-on-disc interactions. The frictional force was taken as an average of the maximum values at the top of the spread of values. Table 4 shows the values of *m* recorded for both dry sliding and some limited lubricated tests.

The variation of test conditions did not reveal any significant variation of μ. The variation of μ with relative humidity is shown in Table 5. The variation of load (1.96 N, 4.98 N, 9.9 N) did not change the value of μ significantly for any material pairing.

While measurement of the wear volume removed from the pin was straight forward the material removed form the disc was not easily measurable for several reasons discussed later. Hence all wear coefficients (*k*) are for the pin material sliding on the stated disc and are summarised in Table 6.

An SEM image of the disc surface after dry sliding of a sialon/TiB_2 composite is shown in Fig. 9. A layer has formed between the pin and the disc and is adhered to the disc surface. The layer has been extensively cracked after repeated passage of the pin. This 'tribofilm' has been demonstrated to be very thin (<1µm) as shown in Fig. 10. The tribofilm appears to protect the underlying disc from wear as demonstrated in Fig. 11 where the removal of the tribofilm reveals a surface that has very little damage. The tribofilm and the wear debris were removed and analysed using EDXS on the SEM. Both the wear debris and the tribofilm have an increased oxygen content (see Fig. 12) and also contain traces of all the elements present in the specimen (i.e. Si, Al, O, N, Y, Ti).

Table 6 Wear coefficients (k) measured for pin material for several tribo pairs of sialon/TiB_2 materials.

Tribo Pair		Test Conditions W(N), V(ms^{-1}), RH(%)	Wear Coefficient (k) $\times 10^{-6}$ ($mm^3 N^{-1} m^{-1}$)
Pin	Disc		
HP2OIS	HP2OIS	W = 9.88, V = 0.05, RH = 41	4.6
HP2OIS	PS10IS	W = 9.88, V = 0.05, RH = 40	2.5
PS10IS	HP20IS	W = 9.88, V = 0.05, RH = 45	1.1
sialon (cutting tool grade)	sialon (cutting tool grade)	W = 9.88, V = 0.05, RH =4 0	6.0

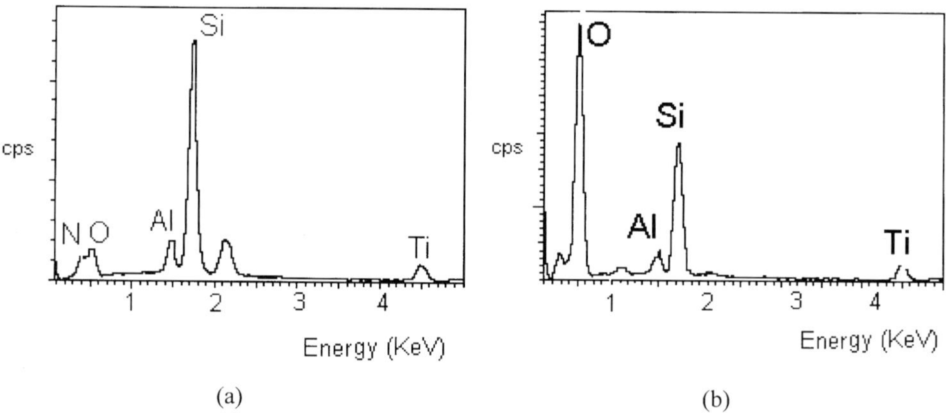

(a) (b)

Figure 12 EDX spectra from, (a) particles of tribofilm and (b) the unworn surface. Tribofilms contain large amounts of oxygen.

4. DISCUSSION

4.1 Materials Synthesis

To densify sialon/TiB_2 composites a dual atmosphere sintering route has had to be adopted to prevent the sintering atmosphere from reacting with materials being fired. The presence of TiB_2 in pressureless sintered materials inhibits the densification process. The densification of sialon takes place primarily by a liquid phase sintering route in which the a Si_3N_4 starting materials is dissolved in the liquid phase and precipitates out as β-sialon on existing grains, shrinking pores. The TiB_2 does not take part in this process and much of the porosity associated with a TiB_2 particle is not eliminated. In the presence of TiB_2 only hot pressing was successful in producing materials that can compete with existing commercial grades. The *in-situ* reaction process has potential advantages in terms of cost of production and the formation of near net shape components but currently is limited by its inability to produce dense, high vol.% TiB_2 materials.

4.2 Hardness and Toughness

Fully dense materials exhibit an increase in hardness of ~20% over a cutting tool grade sialon and ~7% over a bearing grade Si_3N_4. This was accompanied by an increase in fracture toughness of up to 45%. It should be noted that even in hot pressed materials there is some residual porosity and residual sintering aid. The optimum material would minimise both and it is expected that the hardness would then be considerably higher. However, this may also be accompanied by a drop in the fracture toughness.

Possible toughening mechanisms include crack deflection,[7] crack bridging,[8] cracks encountering residual stresses in the matrix[9] and micro-cracking. The two latter mechanisms can be caused by the mismatch in the coefficient of thermal expansion between sialon and TiB_2 (α_{TiB2}= 6.6 to 8.6x10^{-6}/K, α_{Si3N4}~3×10^{-6} K^{-1}) but are difficult to demonstrate in practice.

4.3 Tribology

The complicated nature of the wear process makes interpretation of results of wear tests difficult and sometimes ambiguous. Many researcher working with similar tests and materials often report contradictory results.

4.3.1 Friction

The noise in the frictional force signal is inherent to this type of wear test and can be reduced but not eliminated. Measured values of μ compare well with values measured in similar ways in the literature and it appears that TiB_2 has no major effect on the value of μ. However, lower values of μ were generally seen when one or more of the tribo-pair contained significant amounts of residual sintering aid (Y_2O_3) and it may be contributing to a solid lubrication effect. The reverse situation was seen for steel pins on ceramic discs where materials containing Y_2O_3 (pressureless sintered) had a higher value of μ than those with smaller amounts (hot pressed). Water was not capable of providing hydrodynamic lubrication under the test conditions used but the reduction of μ with paraffin suggests that the materials do respond to conventional boundary lubrication.

4.3.2 Wear

The materials exhibit comparable wear to commercial materials ($k \approx 1$–5×10^{-6} $mm^3 N^{-1} m^{-1}$) but quantification of wear is complicated by inherent problems with the pin-on-disc test and the tribo-chemical phenomenon that dominates in these materials under the conditions used. It is clear that the tribofilm that forms acts as a protective layer between the pin and the disc. It appears that once this film has formed then the majority of the measurable wear takes place at the pin and is by further tribochemical reactions and a third body polishing action as wear debris are passed through the contact zone. Since the debris is primarily in an oxidised state this is a similar situation to polishing with colloidal silica and achieves a polished

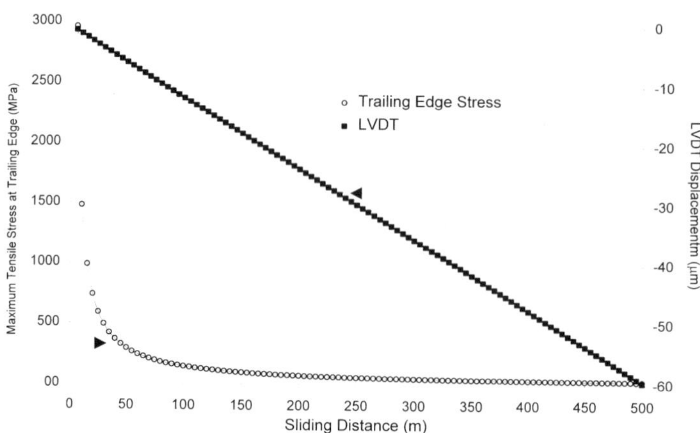

Figure 13 The variation in the tangential trailing edge stress of a circular contact with sliding distance. The linear plot shows the signal from the LVDT which monitors the rate of wear of the pin.

finish on the pin wear scar. This situation leads to the measurement of the wear coefficient (k) for the pin material only since wear on the disc is small (at 500 m sliding distance) and obscured by tribofilms. In addition the stress experienced by the disc (σ_s) is not constant throughout the test under a constant load. As the pin wears then the radius of contact increases and hence the applied stress drops (see eqn 3 and eqn 7) as demonstrated in Fig. 13, an example taken from test data. The wear coefficient, k, can only be reliably quoted under constant conditions since it assumes a constant load and wear mechanism throughout the test. A rapidly decreasing stress may result in two wear modes being active, the first, more severe wear mode, may take place under the initial high stresses but a less severe mechanism may be active at the reduced stresses. The damage initially caused by the high stress (e.g. cracking, grain pull out) may be obscured by the prolonged, less severe wear (i.e. polishing and tribofilm formation).

Materials containing TiB_2 exhibit a lower wear coefficient than the base line sialon material. These composites are harder and tougher and it is expected that this property improvement can influence wear behaviour. However, as has been demonstrated, the primary wear occurs by a tribochemical mechanism and no cracking or fracture of the disc material is observed. Hence, it is unclear how hardness and toughness can directly influence this wear mode. There are currently two possible explanations for the role of TiB_2 in the wear process. Firstly, it is possible that TiB_2 is less susceptible to the tribochemical reactions than Si_3N_4. In this case the more TiB_2 present (i.e. the less Si_3N_4) the less wear will occur. It is also possible that the presence of Ti in the amorphous tribofilms influences their role as a protective layer. If the tribofilms containing Ti adhere more strongly or are less easily deformed, cracked and removed then they may form a more wear resistant layer. Further work is required to determine the exact role of TiB_2, including TEM of wear debris and the more detailed study of the tribofilms themselves in order to determine their role in the wear process.

5. CONCLUSION

Sialon/TiB_2 composites have been produced by several methods, including hot pressing and pressureless sintering, but require a dual atmosphere (Ar/N_2) processing route. However, the presence of TiB_2 inhibits the liquid phase sintering mechanism normally used to densify sialons. In order to produce fully dense composites, with the amount of TiB_2 needed to influence material properties, hot pressing is required. *In-situ* reaction sintering has been successful in producing sialon/TiB_2 composites but is currently limited by the low vol% of TiB_2 that can be introduced in this way.

The addition of >40vol.% TiB_2 significantly increases hardness (20%) and toughness (45%) and also produces a conductive material, allowing the use of electro-discharge machining.

The materials exhibit competitive wear properties and are seen to form tribofilms by tribochemical reactions between the test atmosphere and the specimen. These tribofilms protect the disc from wear but complicate the measurement of wear itself, making determination of the wear coefficient of the disc specimen impractical at short sliding distances (<500m). Much longer wear tests (>1000m) need to be carried out in order to quantify disc wear. Wear takes place by fracture and eventual breakdown of the tribofilm and by a third body polishing action on the pin. The wear mode at the beginning of the test may be more severe but there is a rapid transition to less severe wear modes as the stress experienced at the contact zone decreases as the pin wears. The role of TiB_2 in the improvement in wear properties is as yet unclear but it is likely that it's influence on the tribochemistry is as significant as its effect on mechanical properties.

ACKNOWLEDGEMENTS

This work has been carried out partially with the financial support of the EPSRC and Cookson Syalon plc and more recently under a BRITE–EURAM project (Project BE96-3356, Contract no. BRPR-CT96-0304) in partnership with; **SKF Engineering & Research Centre**, The Netherlands, **CEIT**, Spain, **Céramétal**, Luxembourg, **Katholieke Universiteit Leuven**, Belgium and **The Nottingham Trent University**, UK.

REFERENCES

1. Y. G. Gogotsi, *J.Mat.Sci.* 1994, **29**, 2541–2556.
2. Jow-Lay Huang et. al., *Mater. Sci. Eng.*, 1994, **A174**, 157–164.
3. G. Hillinger et. al., *J. Am. Ceram. Soc.*, 1995, **78** (2), 495–496.
4. *Engineered Materials Handbook,* Vol. 4, 'Ceramics and Glasses', pp.787-803, ASM International, 1991.
5. H. Czichos et. al. *Wear*, 1990, **190**, pp. 155–61.
6. G. R. Anstis et. al., *J. Am. Ceram. Soc.*, **64**, (1981), 539.
7. K. T Faber and A. G. Evans, *Acta Metall.*, 1983, **21** (4), 565–576.
8. P. F. Beecher, *J. Am. Ceram. Soc.,* **74** (2), (1991), 255-269.
9. G. C. Wei and P. F. Beecher, *J. Am. Ceram. Soc.*, 1984, **67** (8), 571–574.
10. T. E. Fischer and H. Tomizawa, *Wear*, Proc. Int. Conf. Wear Materials, ASME, New York, 1985, 22–32.
11. T. Ekström and P. O. Olsson, *J. Eur. Ceram. Soc.*, 1994, **13**, 551–559.

NANOCOMPOSITES

Microstructural Requirements for Alumina SiC Nanocomposites

A. J. WINN and R. I. TODD

Manchester Materials Science Centre, University of Manchester & UMIST, Grosvenor Street, Manchester, M1 7Hs, UK

ABSTRACT

The relative ease with which alumina/SiC nanocomposites polish, compared with monolithic alumina, has been noted in the literature and taken to be an example of the beneficial 'nanocomposite effect' stemming from an apparent strengthening of the grain boundaries. The percentage of 'polished' surface, as opposed to areas dominated by grain pullout or fracture, following grinding or polishing therefore provides an easily obtainable figure of merit for different nanocomposite materials. A variety of alumina/SiC nanocomposites have been fabricated by co-milling commercially available powders and hot pressing in an inert atmosphere. Using variations in starting powder size and heating schedules, it has been possible to produce nanocomposites with similar matrix grain size but with different proportions of particles on the grain boundaries relative to those within the grains. The microstructures were examined using TEM and FEG-SEM, and the effect of microstructural variations on the mechanical properties of the materials was investigated. Fracture surfaces were examined, and a study was made of the relative amounts of polished surface and grain pullout after both grinding and polishing the materials using standard metallographic techniques. The performance of the nanocomposites was compared with that of a pure alumina of the same grain size.

The results indicate that the SiC has a direct effect on crack initiation and propagation in alumina, and that it is the particles on the grain boundaries which are responsible for the beneficial effect of the SiC additions. If sufficient SiC is present on the grain boundaries, crack initiation during abrasive wear can be prevented, resulting in a smoother, plastically deformed surface.

1. INTRODUCTION

Improving the mechanical properties of advanced ceramic materials by producing ceramic nanocomposites was proposed by Niihara in 1991.[1] Significant improvements in the strength and fracture toughness of alumina by the incorporation of SiC 'nano' particles, typically ≤ 300 run in size, were reported, and while other researchers in the field have failed to match the level of improvement seen by Niihara, there is general agreement that a modest improvement can be achieved by the addition of as little as 5.0 vol.% of SiC particles to alumina.[2-4]

The SiC addition causes a transition in fracture from a primarily intergranular mode for pure alumina to a transgranular mode for the nanocomposite. The nanoparticles are situated both within the alumina grains and on the grain boundaries, and disagreement exists both as to the mechanism by which the change in fracture behaviour is achieved and over which is the most effective location for the reinforcing particles. It is, however, generally accepted that an apparent strengthening of the grain boundaries occurs which reduces grain pullout during grinding and polishing, thereby increasing the wear resistance. Results showing the enhanced erosive wear resistance of nanocomposites compared with alumina were first presented by Walker *et al.*,[5] who showed that the nanocomposite wore at less than 1/3 of the rate of alumina.

The reduction in grain pullout is also responsible for the relative ease with which alumina/SiC nanocomposites can be polished, compared with monolithic alumina. The percentage of 'polished' surface as opposed to fracture following grinding or polishing therefore provides an easily obtainable figure of merit for different nanocomposite materials. In the current work, alumina/SiC nanocomposites with various volume fractions of SiC, and varying proportions of that SiC on the grain boundaries, have been made using the mixed powder route and assessed in this way.

A preliminary examination of the microstructures of some of these materials, together with the results of a simple grinding test used as an empirical measure of the grain boundary strengthening in the nanocomposites,[6] showed that the resistance of the surface to rapid material removal by grain pullout and fracture increases as more SiC is added. It remains necessary, however, to determine whether the increase of observed plastic deformation in these materials is due to an improvement in mechanical properties due to a 'nanocomposite effect' such as grain boundary strengthening, or whether the result is due simply to the difference in grain size, which diminishes with increasing SiC content. This paper addresses this issue by comparing materials with different volume fractions of SiC but with similar grain sizes. The results indicated that the improvement in the performance of the materials is due primarily to the SiC additions.

This paper also extends the previous work by refining the grinding wear test, examining fracture surfaces and extending the range of materials studied to include a predominantly intergranular nanocomposite. The results showed that it is the particles on the grain boundary which are the most important in achieving grain boundary strengthening in nanocomposites.

2. EXPERIMENTAL METHODS

A range of alumina/SiC nanocomposites was made for the current study. 1.0, 5.8 and 11.0 vol.% SiC materials were made with Sumitomo AKP50 alumina and Lonza UF 45 SiC (both approximately 200nm mean particle size). In order to produce a 1.8 vol.% 'intergranular' material, AM22 alumina (3.5μm mean particle size) supplied by Mandoval Ltd, was used with Lonza UF 45 SiC, such that the initial alumina grain size was already close to the Zener pinning limit and hence grain growth would be suppressed, resulting in a high proportion of the SiC remaining on the grain boundaries. Pure alumina specimens were made using AKP50O. The production method for all materials was as follows, pure alumina specimens being made by omitting the SiC addition.

SiC powder was ultrasonically dispersed in distilled water using a GEX 600 Ultrasonic Processor, with Dispex A40 dispersant, for 10 minutes. Alumina powder was added and the mixture was ultrasonically processed for a further 5 minutes, 0.3 wt.% of MgO (as magnesium nitrate hexahydrate) having been added to the AM22 slurry to further limit grain growth in the intergranular material. The slurry was poured into a zirconia lined bucket containing zirconia milling media and milled for 2 hours using a Szegvari 01 HD attritor. The slip was then passed through a 150μm sieve and diluted to between 4 and 6:1 liquid to solid volume ratio before being freeze dried for a minimum of 18 hours using an Edwards Shell Freezer and Edwards Micro Modulyo freeze drier. A minimum of 4:1 liquid to solid volume ratio for freeze drying is necessary to ensure a fine powder, free from hard agglomerates. All powders were calcined at 600°C for 1 hour.

Specimens of 30nm diameter and 4mm thickness were made by hot pressing the prepared powders in a graphite die in argon atmosphere, at a pressure of 25MPa. The 5.8 and 11.0 vol.% SiC materials were pressed at a temperature of 1700°C, and the pure alumina, 1.0 vol.% SiC and the intergranular materials were pressed at 1550°C. The specimens were held at maximum pressure and temperature for 30 minutes.

Specimens were prepared for FEG-SEM examination using standard metallographic techniques and thermally etched in a SiC powder bed in argon atmosphere. Grain sizes were calculated using the linear intercept method with a minimum of 300 intercepts and a correction factor of 1.57. Specimens for TEM were prepared by cutting thin slices using a Struers Accutom 50 and then dimpling and ion beam milling. TEM micrographs were used for obtaining SiC particle position data.

The grinding 'wear' tests, to assess the extent of grain pullout in the materials under reproducible conditions, were performed using standard metallographic techniques. A Buehler semi-automatic grinding and polishing machine was used with a Metlap 10 grinding wheel for the 45μm diamond grinding test. Point counting for highly reflective 'smooth' regions, as opposed to darker rough regions due to grain fracture and pullout, was performed using an optical microscope with crosshairs in the eyepiece and a step moving stage. Further grinding was performed with 6μm diamond on a Metlap 4 wheel until all the damage from the previous grinding operation had been removed. The specimens were then polished with 1μm diamond on a Texmet cloth before point counting for the second time and further preparation for electron microscopy. A constant wheel speed of 300 rpm was used throughout, with counter rotation of the specimen holder for the 45μm grinding only. Specimens were cut to give 50mm^2 surface area for each material, and 5lbf was applied per specimen.

3. RESULTS AND DISCUSSION

The materials are discussed in terms of their microstructure, performance in the grinding wear test and fracture behaviour. The influence of particle position is also considered. Grain size, proportion of particles on the grain boundary and the results of the grinding wear test are presented in Table 1. The proportion of particles on the grain boundary is expressed both in terms of percentage of total SiC and percentage of the bulk nanocomposite (e.g. for the 1.0 vol.% SiC material, 23% of the SiC particles were on the grain boundaries, and this is equivalent to 0.2% of the nanocomposite).

3. 1 Microstructure

Figure 1 shows FEG-SEM micrographs of polished and etched specimens. Noticeable in the central region of the micrograph of alumina (Fig. la) are several light coloured zirconia particles. The presence of zirconia was noted in all the materials studied and occurs as a

Table 1 Results of microstructural analysis and grinding and polishing tests. Particle position data are based on between 131 and 296 counts per material. Percentage polished areas after grinding and polishing are averages of at least 3 experiments. The vol.% of particles on the grain boundaries is an estimate based on the number of particles observed, their approximate size (i.e. ≤50 <300, >300nm), and assumes spherical particle geometry.

Material	Grain size μm	Vol.% particles on boundary		% polished area after grinding with:	
		% of SiC	% of composite	45μm diamond	1μm diamond
Al_2O_3	2.6	–	–	40 ± 5	83 ± 3
Al_2O_3 1.0% SiC	3.8	23	0.2	37 ± 2	95 ± 1
Al_2O_3 5.8% SiC	3.8	35	2.0	70 ± 1	97 ± 1
Al_2O_3 11% SiC	2.6	41	4.5	81 ± 2	99 ± 1
Al_2O_3 1.75% SiC	2.6	58	1.0	72 ± 1	96 ± 2

result of the wear of zirconia milling media and the attritor bucket during milling. These small wear particles usually stay on the grain boundaries during sintering, resulting in the formation of larger zirconia inclusions, often at triple points, by coalescence. Figure 1(b) is a micrograph of the 11.0 vol.% SiC material, showing an agglomerate of particles on a grain boundary. SiC agglomerates were often present in high volume fraction nanocomposites. Zirconia can also be seen in this agglomerate.

The 1.0 and 5.8 vol% materials had a grain size of 3.8μm, while the alumina, 11.0 vol.% and the 1.8 vol.% intergranular material all had a grain size of 2.6μm. For the conventional AKP50 alumina nanocomposites, increasing the vol.% of SiC particles increased the fraction of those particles which were on the grain boundaries (Table 1). In the case of the 11 vol.% SiC material, the increase is attributable to its smaller grain size compared to the more dilute nanocomposites. The difference between the 1.0 and 5.8 vol.% materials, however, may indicate that the grain boundaries increasingly take up positions which preferentially pass through the SiC particles as more SiC is added. This is not consistent with the assumption that the grain boundaries are randomly positioned which is made in the Zener theory of grain boundary pinning.

The significantly higher percentage of intergranular particles in the 1.8 vol.% material is apparent in Fig. 1(c). The size distribution of the as-milled AM22 alumina with SiC addition can be seen in Fig. 2. The fines in the alumina powder caused some grain growth, resulting in a proportion of the SiC particles becoming intragranular. Work is in progress to increase the proportion of intergranular particles further by removing the alumina fines by

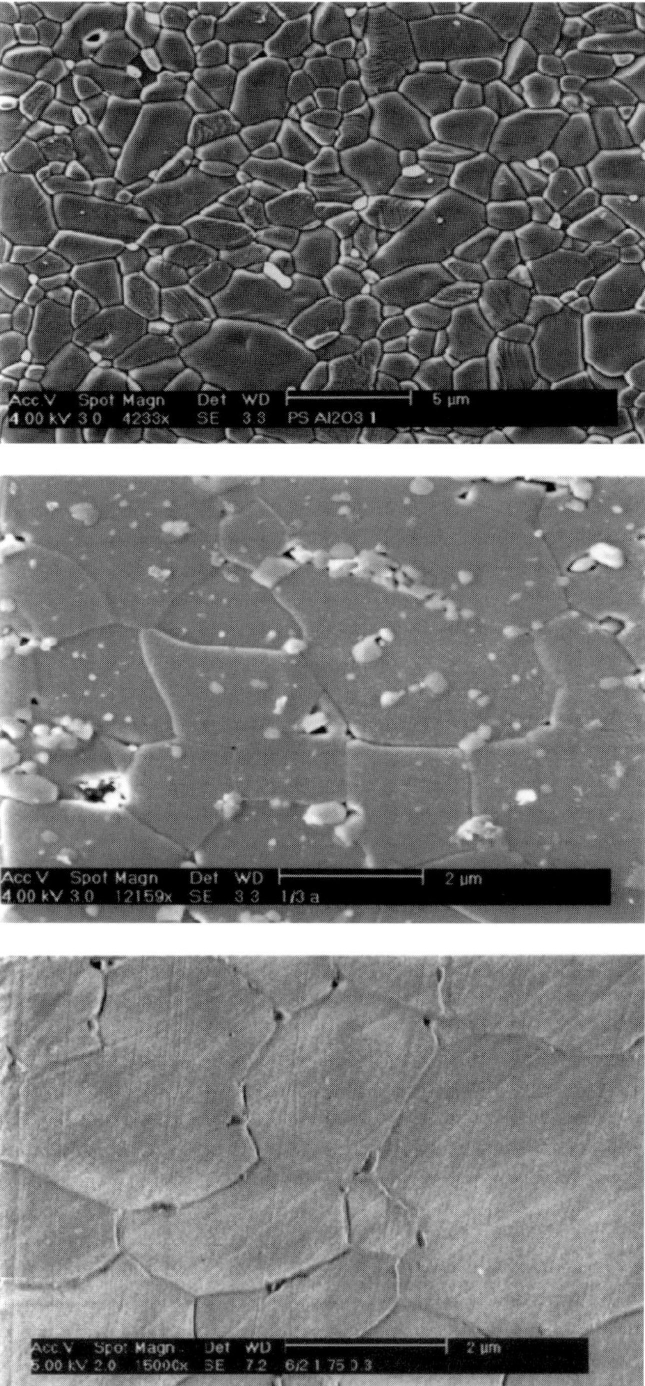

Figure 1 SEM photomicrographs of polished and etched specimens, (a) alumina, (b) 11 vol.% SiC nanocomposite and (c) 1.8 vol.% SiC intergranular nanocomposite.

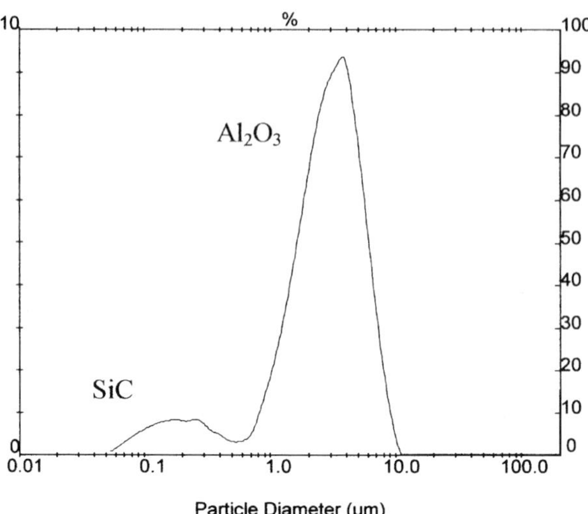

Figure 2 Size distribution of as-milled AM22 alumina with SiC addition.

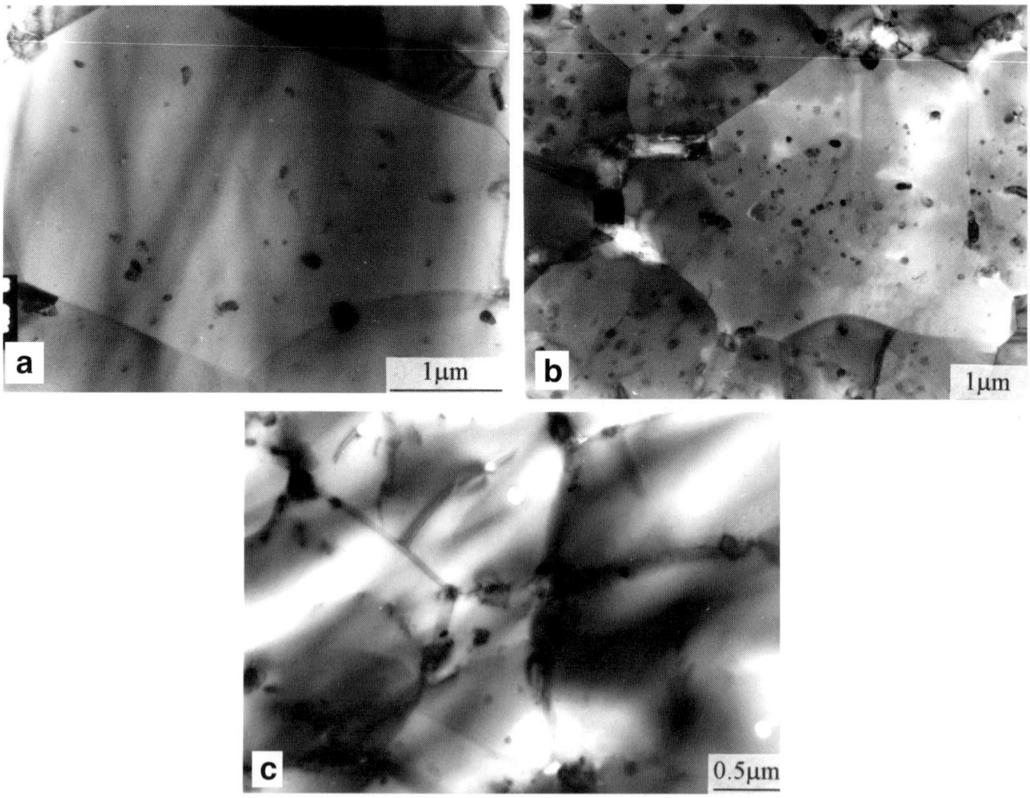

Figure 3 TEM photomicrographs of nanocomposite specimens, (a) 1.0 vol% SiC nanocomposite, (b) 11.0 vol.% SiC nanocomposite and (c) 1.8 vol.% SiC intergranular nanocomposite.

sedimentation. Figure 3 shows TEM micrographs of the 1.0, 11.0 and 1.8 vol.% SiC materials, again highlighting the greater tendency for particles to be on the grain boundary in the 1.8 vol.% SiC intergranular nanocomposite. The white circles in the micrograph of the 1.8 vol.% material may be holes where SiC particles fell out during ion beam milling. Microstructural features can be seen associated with the holes nearest the top of the image.

3.2 Grinding and Fracture Behaviour

Figure 4 shows optical micrographs of specimens after grinding with 45μm diamond, taken at the magnification used for point counting the relative proportions of grain pullout. These clearly demonstrate the dramatic improvement in the resistance to grain pullout shown by the nanocomposites, with the exception of the 1.0 vol.% SiC material, in comparison with pure alumina. All the materials had a macroscopically uniform surface texture following grinding apart from the 1.8 vol.% material which, in approximately 50% of the grinding tests performed, had a rougher region covering up to a maximum of 30% of the specimen surface. This rougher region gave similar point counting results to the pure alumina. Figure 5 shows SEM micrographs of 45 μm ground samples of all the materials. All the micrographs show representative regions of the ground surfaces. The alumina (Fig. 5(a)) has regions of grain pullout by intergranular fracture, streaks of plastically deformed material and particulate wear debris which has collected in recesses in the surface. In contrast, the nanocomposites show little or no evidence of intergranular fracture and a small amount of wear debris is visible only on the 1.0 and 5.8 vol.% specimens. Any surface fracture on these materials was transgranular.

While the point counting revealed little difference between the behaviour of the 1.0 vol.% SiC nanocomposite and the pure alumina (Table 1), a difference in response to grinding was apparent in the SEM. The addition of as little as 1.0 vol.% of SiC was sufficient to prevent the grain boundary cracking usually associated with alumina materials and to promote the transgranular fracture characteristic of nanocomposites.

Plastically deformed material accounts for more than 50% of the surface of the 5.8 vol.% material, providing evidence for grain boundary strengthening and also less of a tendency for fracture than in the 1.0 vol.% material.

Increasing the volume of SiC to 11.0% (Fig. 5 (d)) caused a dramatic improvement in the surface finish of the nanocomposite. Some deeper plastically deformed grooves can be seen which, together with the greater amount of plastic deformation observed over the entire surface, demonstrates that this material has greater resistance to grain pullout and fracture than the 1.0 and 5.8 vol.% materials. A lack of deep grooves in the 1.0 and 5.8 vol.% materials indicates that the abrasive particles responsible for the deeper grooves in the 11.0 vol.% material were probably causing much of the grain pullout and fracture in the alumina and the lower vol.% 'standard' nanocomposite materials.

The 1.8 vol.% intergranular nanocomposite has a surface finish similar to that of the 11.0 vol.% material, sharing the characteristics of a highly plastically deformed surface with deep grooves. Analysis of the positions of particles for these nanocomposites (Table 1) reveals that the intergranular material had a volume of particles on the grain boundaries 5 times that of the 1.0 vol.% material, but a similar volume within the grains. If it were the particles within the grains which were responsible for the 'nanocomposite effect', the per-

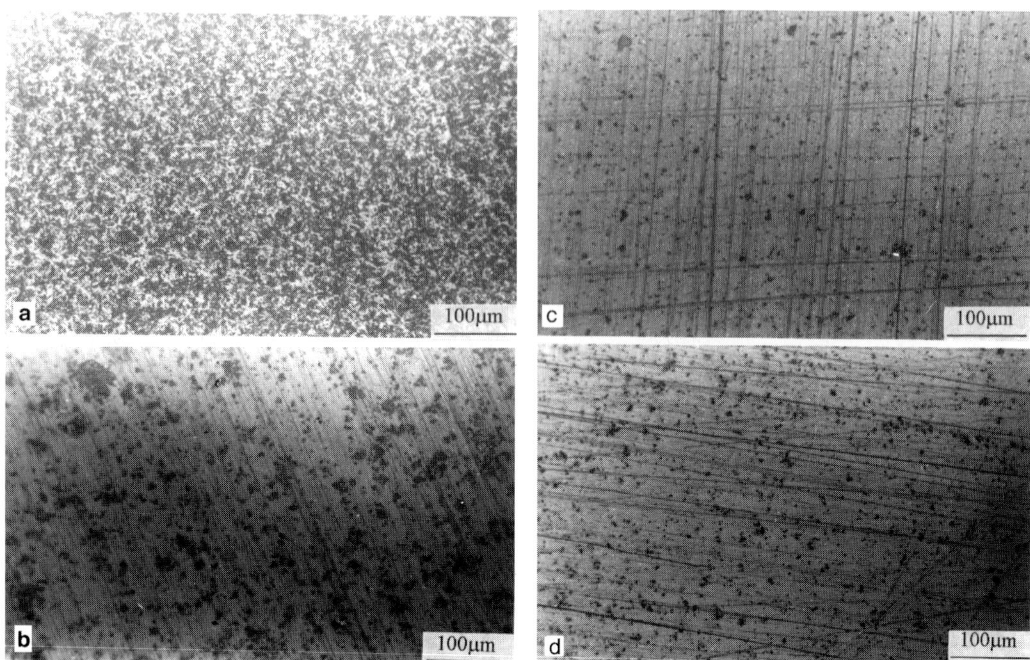

Figure 4 Optical microscope photomicrographs of specimens after grinding with 45μm diamond, (a) alumina, (b) 5.8 vol.% SiC nanocomposite, (c) 11.0 vol.% SiC nanocomposite and (d) 1.8 vol.% SiC intergranular nanocomposite.

formance of the 1.8 vol.% material in the grinding wear test should have been similar to that of the 1.0 vol.% material. That the performance of the 1.8 vol.% material is clearly improved over that of the 1.0 vol.% material must be ascribed to the greater number of particles on the grain boundaries, and the results therefore suggest that it is the particles on the grain boundaries which are primarily responsible for the beneficial effect of the SiC additions.

The fracture surface results are consistent with the surface observations reported above. Figure 6 shows SEM micrographs of the fracture surfaces of the alumina, 1.0 and 11.0 vol.% nanocomposites. Pure alumina (Fig. 6 (a)) underwent primarily intergranular fracture. The addition of only 1.0 vol.% SiC particles changed the morphology to almost completely transgranular (Fig. 6 (b)) fracture, and the higher vol.% nanocomposites also underwent transgranular fracture as shown by the example in Fig. 6(c).

Two modifications to the response of alumina to abrasion on adding SiC have been described above. The first modification was the transformation of the fracture mode from inter- to trans-granular by the addition of 1.0 vol.% of SiC. The second modification appeared when higher volume fractions of SiC were added, when the occurrence of fracture became suppressed and the surface was plastically deformed instead. The experimental observations can be rationalised by considering surface fracture to consist of two distinct processes: (i) crack initiation at alumina grain boundaries, followed by (ii) crack propagation

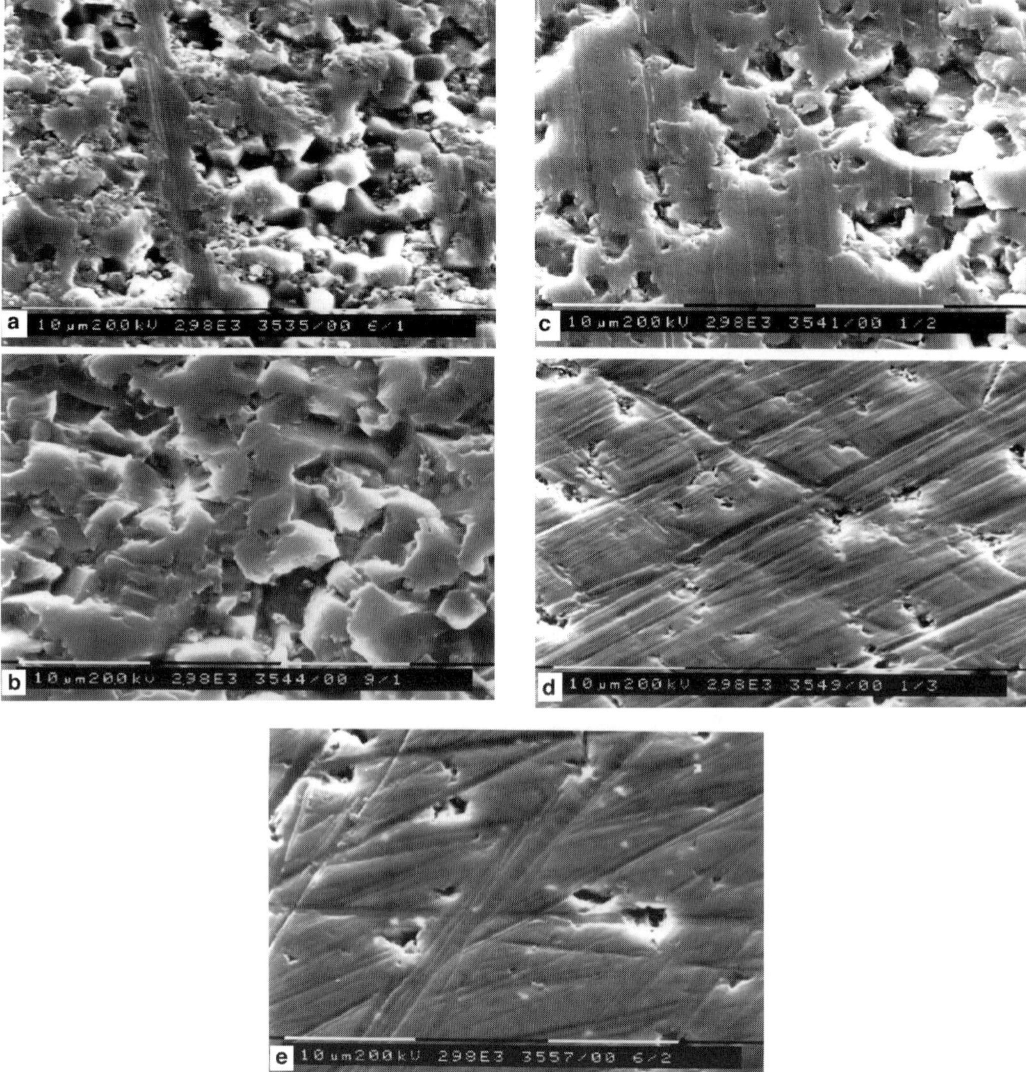

Figure 5 SEM photomicrographs of 45µm ground specimens, (a) alumina, (b) 1.0 vol.% SiC nanocomposite, (c) 5.8 vol.% SiC nanocomposite, (d) 11.0 vol.% SiC nanocomposite and (e) 1.8 vol.% SiC intergranular nanocomposite.

along the preferred fracture route. The addition of a small amount of SiC (1.0%) strengthened the grain boundaries sufficiently to deflect any cracks away from the grain boundaries and into the grains, but was not sufficient to prevent crack initiation in all boundaries. The addition of further SiC strengthened all the grain boundaries, such that crack initiation was prevented altogether, the material then having no option but to deform plastically when abraded by a harder material.

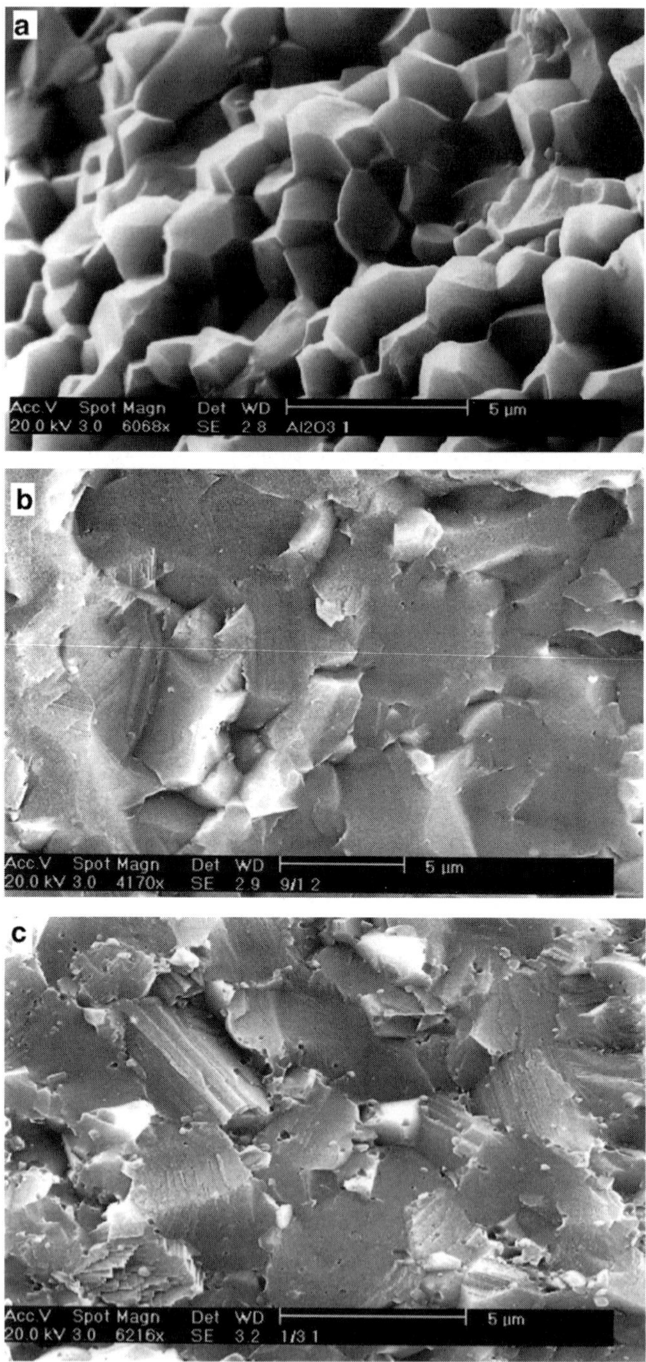

Figure 6 SEM photomicrographs of fracture surfaces, (a) alumina, (b) 1.0 vol.% SiC nanocomposite and (c) 11.0 vol.% SiC nanocomposite.

The point counting results for the 1.0 vol.% material after grinding with 6 and 1µm diamond (Table 1) showed that the strengthening of the grain boundaries was sufficient to prevent grain boundary fracture when subjected to this less abrasive condition. All the nanocomposites demonstrated a similarly improved performance over alumina in this test.

If the presence of ≥1.0 vol.% of SiC particles on the grain boundaries is indeed the cause for the improved response to grinding of materials studied in the current work, it remains to be explained why the 5.8 vol.% material, which has 2.0 vol.% SiC on the grain boundaries, undergoes a higher proportion of surface damage during grinding compared to the 1.8 vol.% intergranular nanocomposite, which has only 1% on the grain boundaries. The regions of grain pullout on the 5.8 vol.% specimen were frequently of the order of >50µm, and poor dispersion of the SiC in this material may be responsible for this behaviour, by reducing the amount of SiC on the grain boundaries to less than 1.0 vol.% in these regions. The mean grain size of the 1.0 and 5.8 vol.% materials is 3.8µm whereas that of the alumina, 11 and 1.8 vol.% materials is 2.6µm. It is also possible, therefore, that the poorer performance of both the 1.0 and 5.8 vol.% materials is due in part to their larger grain size. It is notable that the 1.8 vol.% material did not always polish consistently over the whole surface, and so did not perform as well as the standard 5.8 vol.% material in that respect. 1.0 vol.% SiC on the grain boundaries could therefore be the lower limit at which the 'nanocomposite effect' only just works for this grain size.

4. CONCLUSIONS

- Alumina, conventional alumina/SiC nanocomposites with a range of SiC contents, and a 1.8 vol.% SiC 'intergranular' nanocomposite with 58% of the SiC situated on the grain boundaries, have been fabricated. A 'grinding wear test' has been used to assess the response of the materials to abrasion. Fracture surfaces have been examined.
- The addition of only 1 vol.% SiC to alumina is sufficient to change the fracture mode of alumina from primarily intergranular to transgranular. This not enough, however, to prevent the initiation of surface damage due to abrasion by 45 µm diamond, although an improvement is seen when abraded with 6µm diamond.
- The performance of the 1.8 vol.% intergranular material in the grinding wear test is similar to that of the 11.0 vol.% material. This is ascribed to the high volume of SiC on the grain boundaries and the results suggest that it is the particles on the grain boundaries which are responsible for the beneficial 'nanocomposite effect'.
- The results can be explained in terms of the two distinct processes of (i) crack initiation and (ii) crack propagation. More SiC on the grain boundaries is necessary to suppress crack initiation than is required to change the fracture mode from intergranular to transgranular.

ACKNOWLEDGEMENT

This project was funded by the EPSRC on grant numbers GR/K 61548 and GR/K 71325.

5. REFERENCES

1. K. Niihara, 'New design concept of structural ceramics, ceramic nanocomposites', *J. Ceram. Soc. Jpn*, 1991, **99**, 974–982.
2. J. Zhao, L. C. Stearns, M. P. Harmer, H. M. Chan, G. A. Miller and R.F. Cook, 'Mechanical behaviour of alumina-silicon carbide nanocomposites', *J. Am. Ceram.Soc.*, 1993, **76**, 503–510.
3. C. E. Borsa, S. Jiao, R. I. Todd and R. J. Brook, 'Processing and properties of Al_2O_3/SiC nanocomposites', *J. Microscopy*, 1995, **177**, 305–315.
4. M. Sternitzke, 'Review: Structural ceramic nanocomposites', *J. Eur. Ceram. Soc.*, 1997, **17**, 1061–1082.
5. C. N. Walker, C. E. Borsa, R. I. Todd, R. W. Davidge and R. J. Brook, 'Fabrication, Characterisation and Properties of Alumina Matrix Nanocomposites', *Brit. Cer. Proc.*, 1994, **53**, 249–264.
6. A. J. Winn, Z. Wang, R. I. Todd and F. R. Sale, 'Alumina/SiC Nanocomposites by Mixed Powder and Chemical Process Routes – A Preliminary Examination of Microstructure and Response to Abrasion', *Silicates Industriels*, 1998, **63**, 147–152.

Microstructure Characterisation and Superplasticity of an Alumina/Carbon Nanocomposite

A. BATAILLE, J. CRAMPON and R. DUCLOS

Laboratoire de Structure et Propriétés de l'Etat Solide, U.A. CNRS 234, Bat C6, 2ème Etage, Université des Sciences et Technologies de Lille, F–59 655 Villeneuve d'Ascq Cedex, France

ABSTRACT

A nanocomposite formed of a fine-grained alumina matrix and of carbon nanoparticles was produced by hot-pressing. Compressive deformation tests were performed on this material at T = 1400 °C up to true strains of –70 %. In all cases true strain rates of up to 2×10^{-4} s^{-1} corresponded to low flow stresses (< 50 MPa). No cavitation was observed at grain boundaries and few microstructural changes were found to accompany the deformation. Grain boundary sliding accommodated by diffusional mechanism contributed to the main part of deformation. It is shown that the deformation results are consistent with a diffusivity of the grain boundaries not being affected by the presence of the carbon nanoparticles. The effect of carbon particles lies primarily in a reduction in grain growth without increase in flow stress.

1. INTRODUCTION

Single-phase alumina ceramics have usually been reported to present poor performances in terms of superplasticity.[1,2] On the one hand, in fully densified materials grain sizes are generally micrometric and on the other hand grain growth is often significant during deformation. In order to limit the grain growth several authors[3-5] have introduced ZrO_2 particles in an alumina matrix. However deformation and specially grain boundary sliding can change the distribution of those particles and therefore their effectiveness in the maintenance of a fine alumina grain size can be lowered. Moreover the addition of ZrO_2 results in an increase in the flow stress.[5]

During a superplastic process, grain boundary sliding accommodated by various diffusion mechanisms contributes to a large amount of total deformation[6] (up to 77 %). Hence any segregation phenomenom or any particle in the grain boundary may affect the grain boundary sliding one way or another. For example ZrO_2 addition results in a change in grain boundary diffusivity. As a matter of fact the related activation energy increases as does the flow stress.[5,7] In another way SiC particles pinning effect of alumina grain boundary is significant and considerably lowers the creep rate.[8]

The present study examines the effect of carbon particles distributed in an alumina matrix on the microstructure evolution during compressive deformation at T= 1400°C under stresses lower than 50 MPa at true strain rates up to 2×10^{-4} s^{-1}.

Figure 1 TEM image of the hot-pressed Al_2O_3/carbon nanocomposite prior to deformation testing.

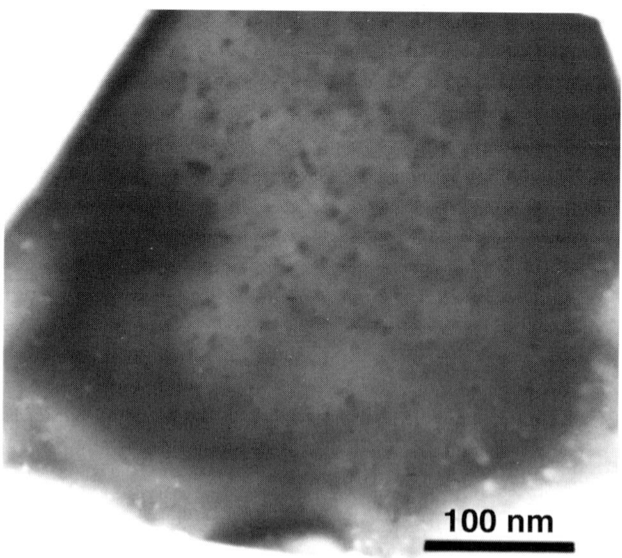

Figure 2 TEM micrograph showing the fine homogeneous distribution of carbon nanoparticles in the grain bulk.

2. EXPERIMENTAL DETAILS.

A commercial high-purity α-alumina powder (Baikowski SM8) was used. The powder was atomised and nominally doped with 500 wt ppm MgO. The carbon was determined to be present in the amount of 1.3 wt% (about 2.6 vol.%). The analyses of the as-received powder, given in wt ppm, are as follows: Fe = 10, Na 17, K = 38, Si = 38, Ca 8 and Cr = 4. The as-received powder had an average particle size of 0.3 μm.

The powder was cold-pressed at 20 MPa in the form of 30 mm diameter discs, 8 to 12 mm thick. Hot pressing was performed at T = 1450 °C in a carbon die at 45 MPa for 30 mm under vacuum, the constant heating-rate being 10 °C. min^{-1}.

Density of the resulting billets was determined by weighing in alcohol using the Archimedes' method. The final density was greater than 99.5 % of theoretical. Deformation specimens typically in the form of 4.5 × 4.5 × 10 mm^3 parallelepipeds were diamond machined from the hot-pressed discs. Specimen compression axis corresponded to the pressing one.

Compressive deformation experiments were performed under vacuum (less than 10^{-3} Pa on average) on an universal screw-driven testing machine. Typically, specimens were heated to the required temperature and experiments were set-up after 30 min to ascertain thermal homogeneity.

Specimens were deformed between parallel nuclear grade carbon rods at a constant crosshead displacement rate (30, 60 and 90 μm min^{-1}, corresponding to an initial true deformation rate $\dot{\varepsilon}_0$ of 5 × 10^{-5}, 10^{-4} and 1.5 × 10^{-4} s^{-1}). The evolutions of load versus time were monitored and true stress versus true strain curves were calculated from these data. Separate annealings of the nanocomposite allowed us to assess the static grain growth. We compared the microstructure evolution with the one of deformed specimen (the latter involving both static and dynamic grain growth).

Microstructural characterisations were carried out by transmission electron microscopy (TEM). From TEM micrographs the average linear grain size was measured as $d = 1.38\ s^{1/2}$ (where s is the average grain surface taking into account the true volume fraction present in the material).

3. RESULTS

3.1 Microstructural characterisation of hot-pressed nanocomposite

Hot-pressing yielded a fully dense black-colored composite. At low-magnification TEM micrographs (Fig. 1) exhibit a finer microstructure than those of MgO-doped aluminas. The mean alumina grain size is 0.56 μm. The aspect ratio, defined here as the mean grain size in a direction parallel to the pressing axis over that in a perpendicular direction, is about one. Figure 2 shows the homogeneous distribution of the carbon particles through the alumina grains. The mean distance between two carbon particles is in the range 30 to 60 nm.

These particles were statistically located both inside grains and at grain boundaries (Fig. 3). According to observations, the particle size ranges between 10 and 20 nm, their indi-

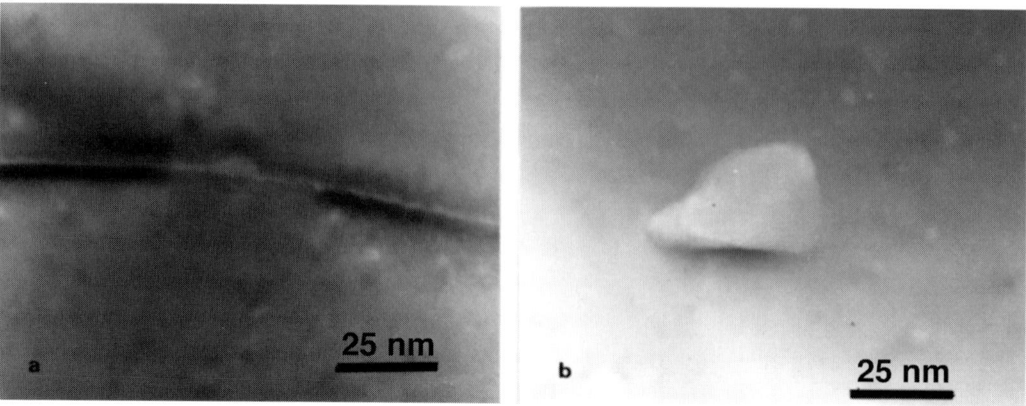

Figure 3 TEM high magnification of carbon particles at grain boundary (a) and into the grain (b).

vidual volume being estimated at about 10^3 nm^3. No dislocation were observed in alumina grains in the vicinity of carbon particles.

3.2 Grain growth and microstructural changes

The alumina grain growth was examined by determining the final grain sizes of deformed specimens. In addition, annealing of nanocomposite sample at $T = 1400$ °C was also carried out for duration up to 4.5 hours. Note that for each deformation test, the total time at the required temperature was always less than 2 hours.

A weak increase in grain size was generally observed for annealed specimens. As a matter of example, the alumina grains grew from an initial size of 0.56 μm up to a size of 0.7 μm after 4.5 hours at 1400 °C. By contrast, a nanocomposite specimen deformed at an initial true strain rate $\dot{\varepsilon}_0 = 10^{-4} s^{-1}$ up to $\varepsilon = -0.65$ had a final alumina grain size $d = 1.12$ μm at $T = 1400$ °C (total time being less than the 2 hours). Hence the examination showed that grain growth was evident during deformation tests and that the dynamic grain growth prevailed over the static one. Additional TEM investigations of deformed specimens did not reveal any drastic microstructural evolution:

(i) carbon nanoparticle distribution was not influenced by deformation,

(ii) dislocations were rarely observed and, in particular, no association of those dislocations with the presence of nanoparticles could be found,

(iii) cavity formation was not observed, neither at alumina/alumina boundaries nor at carbon/alumina interfaces,

(iv) no marked distorsion of alumina grain boundaries resulting from a possible strong particle pinning was noticed.3.3.Deformation tests

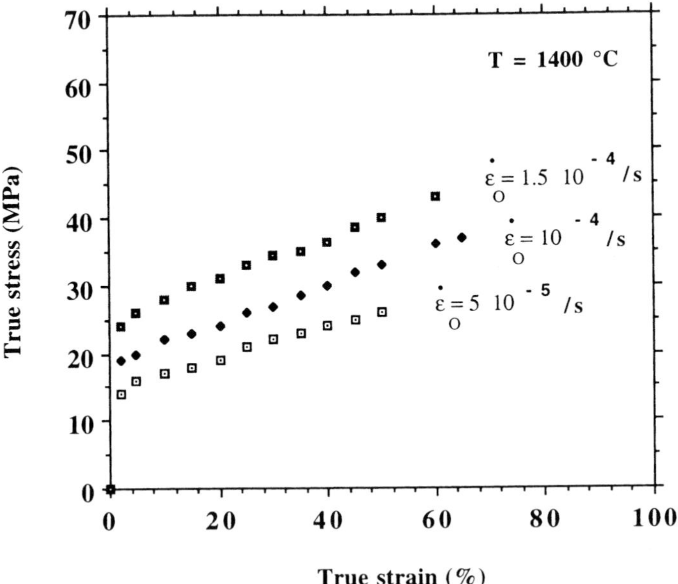

Figure 4 True stress versus true strain curves at $T = 1400$ °C at different true strain rates.

3.3 Deformation tests

The true stress–true strain curves obtained under vacuum at $T = 1400$ °C are plotted in Fig. 4. In general, deformation tests were carried out up to true strains higher than $\varepsilon = -0.50$. Note that no samples broke during the tests.

The flow stress at the test onset was rather low and then continuously increased with strain. Note these curves are not corrected of the increase in true strain rate with decrease in specimen height h. The current true strain rate $\dot{\varepsilon}$ depends on the initial value $\dot{\varepsilon}_0$, according to the relation: $\dot{\varepsilon} = \dot{\varepsilon}_0 / (1 \pm \dot{\varepsilon})$ where $\dot{\varepsilon}_0$ is the nominal strain $\dot{\varepsilon}_0 = (h - h_0) / h_0$. The true strain rate has increased two times when a true strain rate $\varepsilon_0 = -0.05$ is reached. This correction requires the knowledge of the applied true stress σ and the true strain $\dot{\varepsilon}$. To this end the steady state deformation rate was assumed to be represented by the usual equation:

$$\dot{\varepsilon} = C.\sigma^n.d^{-p}.\exp\left(-\frac{\Delta H}{RT}\right) \tag{1}$$

where d is the grain size, n and p are the stress and the grain size exponents respectively and ΔH is the activation energy of the rate-controlling diffusion process. The isostructural stress exponent was determined at constant temperature by a displacement rate change method. The stress exponent was evaluated at $n = 2.4 \pm 0.2$.[9] Considering the equation (1) the cor-

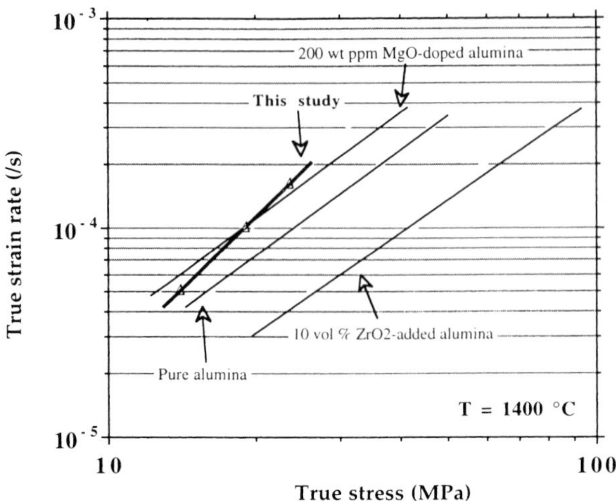

Figure 5 True strain rate versus true stress at –2% true strain. Present results and additional data.

rected $\sigma(\varepsilon)$ curves still show a strain hardening that can be related to the microstructural changes reported above.

Finally an estimate of the activation energy[9] gave: $\Delta H = 515 \pm 50$ kJ.

4. DISCUSSION

The role of carbon particles on the deformation of the nanocomposite seems to be beneficial and is now discussed by first comparing the behaviour of this material to those of three aluminas[10] of similar initial grain sizes (0.5 µm): an undoped alumina, a 200 wt ppm MgO-doped one and a 10 vol.% ZrO_2-added alumina. Figure 5 presents the relative variation of true strain rate versus true stress of the different materials. The true stress values correspond to the flow stresses at –2 % true strain. In Fig. 6 the grain growth of deformed specimens is presented as a function of true strain for the nanocomposite, the ZrO-added alumina plus two other aluminas: our alumina matrix and a material doped with 500 wt ppm MgO.[11] From these two sets of curves different informations can be drawn:

(i) at a given strain rate ZrO_2 addition increases the flow stress. In contrast the present nanocomposite has a behaviour similar to that of a 200 wt ppm MgO-doped alumina (Fig. 5). This can be related to the increase in the activation energy for deformation associated with the introduction of ZrO_2[5,7] (from about 540 kJ mol^{-1} up to about 720 kJ mol^{-1}). Here no increase in the activation energy has been observed. The diffusion shall not be affected by the carbon particles.

(ii) it is evident the carbon nanoparticles decrease the grain growth in a factor higher than two in comparison to the alumina matrix one (Fig. 6). This is definitely an advantage for

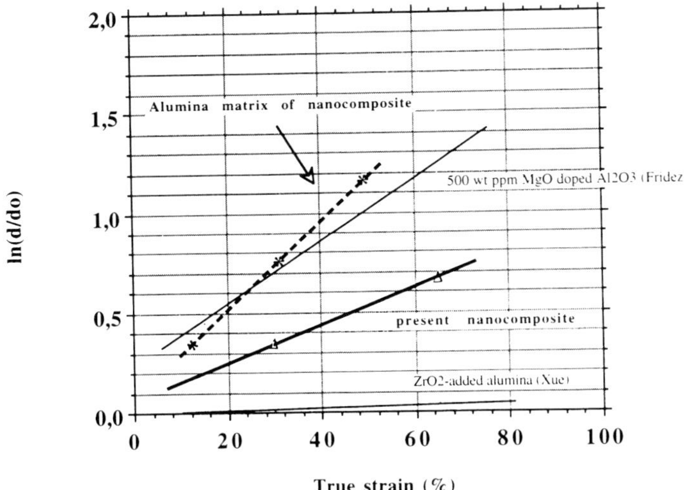

Figure 6 Normalised grain size of deformed specimens versus true strain at T = 1400 °C (except for Fridez's line for which T = 1450 °C).

the prospect of superplastic forming. As for the 10 vol% ZrO_2-added alumina the dynamic grain growth is reduced to a considerable degree and consequently the second-phase pinning by ZrO_2 operates drastically. Duclos and Crampon[12] reported about the interactions between ZrO_2 inclusions and alumina/alumina boundaries in zirconia/alumina composites. The interaction with the grain boundary is efficient and grain-growth reducing when inclusion size is more than a third of alumina grain size. Here the particle size over alumina grain size ratio is evidently lower than this limit. However the interactions with carbon particles are numerous due to the high density of the particles. Consequently the carbon particle density may compensate partly the smallness of the particles and the grain growth reduction is limited.

(iii) the carbon particles do not seem to modify neither the grain boundary sliding capacity, nor the diffusivity of the grain boundaries. At similar initial grain size, the strain rates experienced by the nanocomposite are equivalent to those of the 200 wt. ppm MgO-doped alumina (Fig. 5). In addition, a calculation of the creep rates of the MgO-doped alumina matrix in the conditions of nanocomposite tests (grain size, flow stress, temperature) results in values close to the ones experienced by the nanocomposite.[9] This would imply the carbon/alumina bonding is not too rigid and the diffusion on these interfaces shall be equivalent to alumina/alumina interface. Under these conditions, deformation is mainly intergranular, explaining the rare observation of dislocations. Moreover cavity formation is not necessary to release local stress concentration.

An opposite situation is found in silicon carbide/alumina composites.[13–15] In these composites, the bonding between SiC particle and alumina is assumed to be rigid, involving a decrease in the strain rate of several orders of magnitude. This was associated with the strong pinning effect of the silicon carbide particles. The silicon carbide reduced the grain growth and also the grain boundary sliding capacity of alumina. For these composites cavitation operates during deformation and obvious distorsion of grain boundary is associated with silicon carbide particles. These are found to exert a resistive force on moving boundaries.

Consequently, the effect of carbon nanoparticles lies primarily in a lowering of grain growth during hot-pressing and deformation. As said above, the difference of size between the carbon nanoparticles particles (ranging between 20 and 40 nm) and the SiC particles may be of some importance in exerting a resistive force on a boundary.

Nakano[16] compared recently the tensile deformation behaviour of one 2000 wt ppm MgO-doped alumina (initial grain size $d = 1.1$ μm) and of two sets of ZrO_2/ Al_2O_3 composite (10 vol.% ZrO_2) with two alumina initial mean grain sizes: $d = 0.50$ μm and $d = 1$ μm. The final elongations are representative of the difference of tensile superplastic behaviours. The deformation capability improves with the addition of ZrO_2 (for an initial grain size of about 1 μm) owing to the reduction of dynamic grain growth.

At $T = 1450$ °C and at $\dot{\varepsilon}_0 = 1.7 \times 10^{-4}$ s^{-1} the final elongation reaches about 100 % for MgO-doped alumina and about 140 % for ZrO_2-added alumina. Further improvement of the superplastic behaviour is observed when the initial grain size is maintained at 0.5 μm (final elongation of about 340%). Taking into account the observations about the present nanocomposite, the reduced dynamic grain growth associated with the fine initial alumina grain size suggests the introduction of carbon nanoparticles would probably result in similar improvement of the tensile superplastic behaviour.

4. CONCLUSION

The introduction of a small amount of carbon in alumina microstructure resulted in a significant reduction of microstructural evolution during deformation. The fineness of initial nanocomposite microstructure associated with the reduction of the grain growth account for the observed good superplastic performances.

The carbon particles are tiny and numerous. The decrease of grain growth during deformation can be attributed to a weak but multiple pinning effect of the particles. Their individual effect on grain growth is slight but their high density counteracts the weakness of the pinning. Moreover no hardening effect of carbon particles was observed suggesting no alteration of the diffusivity of the alumina grain boundaries.

REFERENCES

1. L. A. Xue and I. W. Chen, J. Amer. Ceram. Soc., 1990, **73** (11) 3518–3521.
2. K. R. Venkatachari and R. Raj, *J. Amer. Ceram. Soc.,* 1986, **69** (2), 135–138.
3. F. Wakai and H. Kato, *Adv. Ceram. Mater.,* 1988, **3,** (1), 71–76.
4. K. Okada, Y. I. Yoshizawa and T. Sakuma, in *'Superplasticity in advanced materials',* S. Hori *et al. eds*, The Jap. Soc. Res. on Superplasticity, 1991, 227–232.
5. L. Clarisse, 'A study of alumina / zirconia composite superplasticity', PhD thesis (in french), University of Lille, 1997.
6. L. Clarisse, A. Bataille, J. Crampon and R. Duclos, in Proc. Int. Conf. *Euro Ceramics V*, Versailles,

France, June 1997, European Ceramic Society, 651–654.
7. L. A. Xue and I. W. Chen, *J. Amer. Ceram. Soc.,* 1996, **79** (1), 233-238.
8. T. Ohji, A. Nakahira, T. Hirano and K. Niihara, *J. Amer. Ceram. Soc.,* 1994, **77** (12), 3259–3262.
9. A. Bataille, J. Crampon and R. Duclos, to be published in *Ceram. Int.*
10. L. A. Xue and I. W. Chen, *J Amer. Ceram. Soc,* 1991, **74** (4), 842–845.
11. J.D. Fridez, C. Carry and A. Mocellin, in *'Advances in Ceramics',* Vol. **10,** Amer. Ceram. Soc., 1984, 720–740.
12. R. Duclos and **J.** Crampon, *Scripta Met. Mater.,* 1990, **24,** 1825–1830.
13. A. Nakahira, T. Sekino, Y. Suzuki and K. Niihara, *Ann. Chim. Fr.,* 1993, **18**, 403–408.
14. T. Ohji, T. Hirano, A. Nakahira and K. Niihara, *J. Amer. Ceram. Soc.,* 1996, **79** (1), 33–45.
15. L. C. Stearns and M. P. Harmer: *J. Amer. Ceram. Soc.,* 1996, **79** (12), 3013–3019.
16. K. Nakano, T. S. Suzuki, K. Hiraga and Y. Sakka, *Scripta Mater.,* 1998, **38** (1), 33–38.

PROCESSING

The Evaluation of Novel Interphases for Oxide–Oxide Ceramic Matrix Composites

A. TYE and M. H. LEWIS

Centre for Advanced Materials, University of Warwick, Coventry CV4 7AL

ABSTRACT

Stoichiometric MXO_4 type compounds, where M represents a rare earth or yttrium ion and a pentavalent cation, have been prepared using mixed oxide and liquid precursor methods. Their stability in relation to Al_2O_3, Mullite and yttrium–aluminium gamet (YAG) has been determined by examining interfaces exposed in reaction couples after heating to 1400°C.

Complex oxides of the phosphate and vanadate type are shown to possess the desired chemical stability with some of the candidate oxides and can be considered as suitable interphase materials. Close control over composition and homogeneity is shown to be important in determining their performance as potential interphases due to the possible formation of a liquid phase which can react readily with the oxide matrix or fibre.

Selected MXO_4 compounds have also been successfully deposited onto oxide substrates and woven oxide fibres using liquid precursors and RF-magnetron sputtering techniques, yielding controlled and uniform fibre coatings.

1. INTRODUCTION

The development of oxide/oxide CMCs is driven by the need for alternative materials which show high temperature stability and oxidation resistance suitable for gas turbine applications. A primary limit to application temperature of most interfaces so far developed in experimental non-oxide CMCs is their oxidative instability. Carbon-rich interfaces, produced either by natural fibre/matrix reaction (for Nicalon or Tyranno within silicate matrices) or by vapour deposition on fibre precursors (as in CVI SiC/SiC composites such as Serasep from SEP) present a low-temperature (~400°C) limit in oxidising conditions. In these CMCs the matrix, or a passive oxidation pretreatment (SiO_2 on fibre ends to prevent channelled carbon oxidation) provide protection only in the absence of matrix microcracking. An alternative of vapour-deposited BN (or mixed BN/C multilayers) provides an effective increment (to ~1000°C) in short-term survival which has been demonstrated for silicate and SIC matrix CMCs.

The trend towards the development of oxide CMCs is therefore a logical one. Earlier oxide fibres were based largely on Al_2O_3 and are fabricated relatively economically by sol-spinning and pyrolysis. The commercial 'Nextel' series of fibres from 3M are based on Al_2O_3 with variable amounts of SiO_2 for grain size control and the promotion of mullite crystallisation in order to improve upon the poor creep resistance and stress-rupture properties of the single oxide fibres. This trend has led to the search for suitable 'interphases' which are compatible with these oxide fibres and matrices, a critical factor determining mechanical performance of CMCs.

An optimised interface microstructure is based on the micromechanical parameters *t* and G_i which represent the shear stress in the interface and its de-bond energy, respectively. It must also include the constraints imposed by the thermal and environmental conditions during composite fabrication and service. The interphase zone should act as a reaction barrier between fibre and matrix, inhibit diffusion of atmospheric species to the fibres, and remain intrinsically stable at high temperatures in differing atmospheres e.g. to avoid 'channelled oxidation from exposed fibre ends or via matrix microcracks.

Strategies which are being used for oxide/oxide systems have attempted to replicate some of the microstructural features of the non-oxides using phases with a low-cohesion 'layer-plane' by 'in-situ' reaction between a simple fibre coating and surrounding matrix. A second strategy is an empirical search for complex oxides which have a high probability of low cohesion with matrix or fibre due to polarisation of oxygen bonds by high valence cations, exemplified by vanadates, phosphates or niobates of type MXO_4.

An example of the first strategy is that of a cerium-doped zirconia layer applied to the fibre surface reacting with an alumina matrix to produce oriented crystals of Ce-doped, β–Al_2O_3 (Ce_2O_3.11 Al_2O_3). The reaction is initiated by reducing conditions, to effect a $Ce^{4+} \rightarrow Ce^{3+}$ valence state change, and temperatures above ~1200°C to promote diffusion into (and reaction with) the pure Al_2O_3 layer.[1] The preferred orientation, with β–alumina layer plane tangential to the filament surface, favours interfacial de-bond and is an improvement on similar (La–β alumina) interphases formed by liquid-precursor deposition. However, debond shear stresses in filament 'push-through' tests are relatively high even though subsequent sliding stresses (τ ~40–50 MPa) are within reasonable limits for damage-tolerance.

Recent research[2–5] has used the second strategy, and examples of potential 'interphase' oxides which have been studied in current UK and EC projects are covered in this paper, including their thermal and chemical stability with selected oxides and the evaluation of various methods of deposition of the interphases onto commercially available woven oxide fibres.

2. MXO_4 TYPE INTERPHASES

The generic formula MXO_4 is used to describe those compounds in which X is a pentavalent cation, tetrahedrally or octahedrally co-ordinated by oxygen, and M is a rare earth or ytrrium ion also in 8- or 9-fold co-ordination with oxygen. In general, these compounds have related structures (xenotime, monazite, fergusonite) and exhibit high temperature congruent melting. Those selected for study here are summarised in Table 1. With the exception of $NdVO_4$ and $NdNbO_4$, the compounds have been compared in relation to ease of stoichiometric synthesis, deposition method, reactivity and debond-energy with candidates for CMC fibre and matrix phases (Al_2O_3, mullite, YAG).

The comparisons between compounds with varying M and X within the same structure-type and between compounds with the same X but different M (and structure-type) might enable the critical parameters for variable debond condition to be identified. For example, within the vanadate and phosphate series, the crystal structure changes from monazite (monoclinic) to xenotime (tetragonal-zircon iso-structure) with changing M in the rare earth series. An additional selection factor is based on thermal stability and most compounds melt congruently at temperatures above 1600°C.

Table 1 Summary of crystal structures and known melting points (T_m) of selected MXO_4 compounds

	Phosphate	Vanadate	Niobate
La	Momoclinic (Monazite) T_m=2072° C	Momoclinic (Monazite) T_m=1810° C	Momoclinic (Fergusonite) T_m=1620° C
Nd	Momoclinic (Monazite) T_m=1975° C	Tetragonal (Zircon) T_m>1500° C	Momoclinic (Fergusonite) T_m>1000° C
Y	Momoclinic/hexagonal T_m=2000° C	Tetragonal (Zircon) T_m>1810° C	Momoclinic (Fergusonite) T_m=2000° C

Table 2 Interphases considered compatible with selected oxides at 1400°C

Al_2O_3		YAG	Mullite
$LaPO_4$	$LaNbO_4$	YVO_4	No reactive couples detected
$NdPO_4$	$YNbO_4$	YPO_4	
$LaVO_4$			

It is important to consider the ease of synthesis and tolerance of compositional deviations from the exact 1:1 ratio of M_2O_3/X_2O_5 in each binary system. For example, in earlier work, an apparent reactivity between Al_2O_3 and $LaPO_4$ was later shown to result from liquid formation above 1050°C in P_2O_5 rich compositions, consistent with a peritectic isotherm in this system.[2] In the Nd_2O_3–P_2O_5 system, however, the isotherm in P_2O_5–rich $NdPO_4$ compositions is at 1270°C and hence $NdPO_4$ is a favoured interphase material. In addition, Nd_2O_3 is less hygroscopic than La_2O_3 and, as a more readily handled precursor, facilitates better control over the composition of the resultant interphase.

Reaction couples were prepared by mixing single crystal filaments (Saphikon) or substrate fragments (YAG) with the MXO_4 powders. The mullite source is a commercial sol. The mixtures were then isostatically pressed and co-sintered at 1400°C for up to 8 hours. The MXO_4 powders were synthesised from liquid precursors or by hydrolysis of colloidal mixtures of the consituent oxides.[6, 7]

Compatibility is characterised by interfacial cross diffusion or the appearance of additional phases, detected by X-ray diffraction, electron microscopy and interfacial microprobe analysis of polished and fracture surfaces. Of the interphases shown in Table 1 ($NdVO_4$ and $NdNbO_4$ not studied), those considered compatible with each of the candidate fibre/matrix oxides, after co-sintering at 1400°C, are shown Table 2.

An example of a non-reactive interface is that of $Al_2O_3/LaVO_4$ and Fig. 1 shows a typical interface exposed in a fracture surface. It can be seen that the only interactive features on the Saphikon debond surface are small ridges formed by grain boundary/interface energy equilibration during sintering.

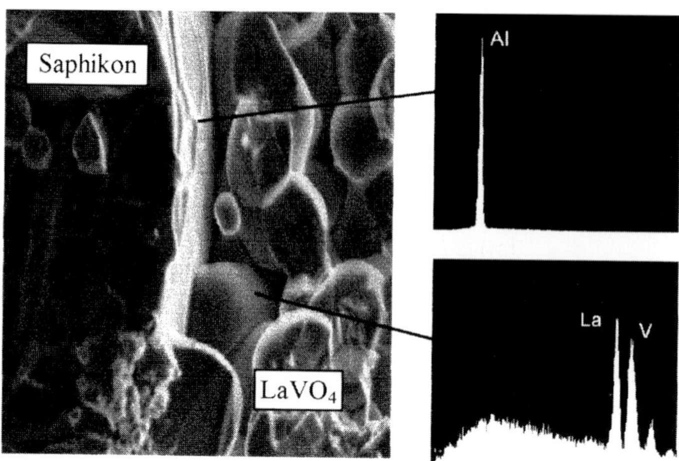

Figure 1 $LaVO_4$ debond interface with Saphikon and EDAX analysis either side of interface.

In the other cases, one or more reaction products can be detected, for example, YVO_4 and YPO_4 in conjunction with Al_2O_3 both result in the appearance of a reaction layer at the interface, thought to be YAG and possibly with some substitution by the V^{5+} or P^{5+} ions. The niobates, whilst shown to be stable in contact with alumina and mullite, exhibited reactions with YAG resulting in the appearance of Al_2O_3 as well as at least one other phase. Another example is the compatibility of $LaVO_4$ with Al_2O_3 but a reactivity with YAG such that the YVO_4 interphase is favoured. Consequently, candidate interphases for some diphasic fibres or matrices (e.g. Al_2O_3 /YAG) are not always identifiable.

Some caution is, however, required in using these model reaction couples since in real fabrication of CMCs, fibres or matrices may not be pure single or diphasic ceramics but contain residual grain boundary liquids or segregates. For example, some reaction has been identified for vanadates and niobates with Nextel 720 fibres (which have mullite/Al_2O_3 microcrystalline structures) whereas no reaction is detected with the separate phases (Table 2). This in part might be attributable to a known time-dependent microstructural instability exhibited by the Nextel 720 fibres, associated with a change from metastable pseudotetragonal $2Al_2O_3.SiO_2$ mullite to the orthorhombic 3:2 form via Al_2O_3 precipitation and shown by the XRD data in Fig. 2.[8] This instability becomes apparent at a temperature of 1200°C and therefore is expected to influence the performance of the interphases studied at a higher temperature in this work.

3. INFILTRATION OF WOVEN OXIDE FIBRES WITH LIQUID PRECURSOR INTERPHASE

Although there are many different ways of depositing an interphase coating (CVD, PVD etc.) onto a ceramic fibre, the preferred method in this work is by deposition from the liquid phase (sol). In addition, attention has been focused on developing methods appropriate for coating fabric layers rather than individual fibres or fibre tows. This approach is based on

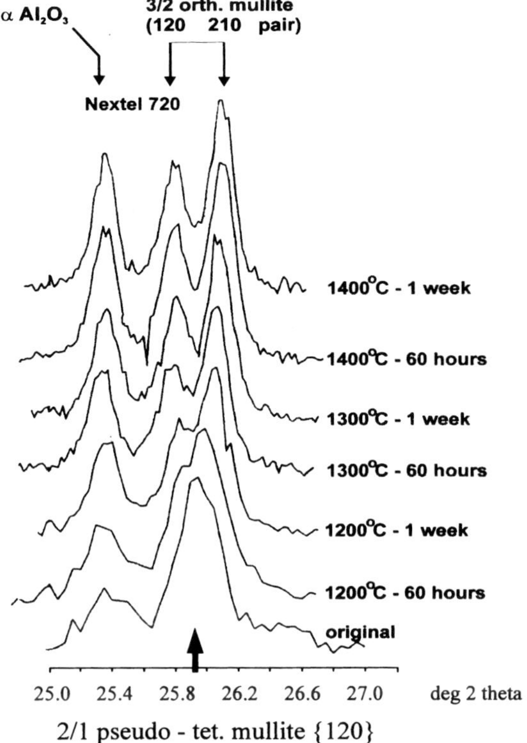

Figure 2 XRD data depicting the microstructural instability of Nextel 720 fibres at 1200°C.

the need to enable scale up of the infiltration process in order to meet the requirements of real component fabrication using CMCs.

Extensive use has been made of the immiscible liquid bi-layer approach.[9] In this technique, the individual layers of fabric are immersed in the sol precursor and a layer of low density immiscible liquid carefully poured on top of the sol. The sol is water based and the immiscible liquid can be octanol or hexane. The layer of fabric is drawn steadily through the liquid bi-layer into air. The surface energies of the system encourage the water based sol to wet the fibres at the sol/octanol interface but at the octanol air interface excess sol is removed from around the fibres minimising bridging by the interphase sol. The fabric is then air dried prior to matrix infiltration.

The method has been succesfully adapted to allow infiltration of Nextel weave with a $NdPO_4$ interphase and has been optimised following comparison of a range of dipping sequences based on the many permutations available when using two precursor solutions and an immiscible liquid bi-layer inbetween infiltrations. Earlier experiments had shown the necessity to develop an 'in-situ' precipitation technique following the poor levels of infiltration shown when a pre-reacted $NdPO_4$ precipitate was used. This is attributed to the needle-like morphology and particle size of $NdPO_4$ inhibiting the infiltration process.

The procedure yielding the most promising results so far is described in the flow chart

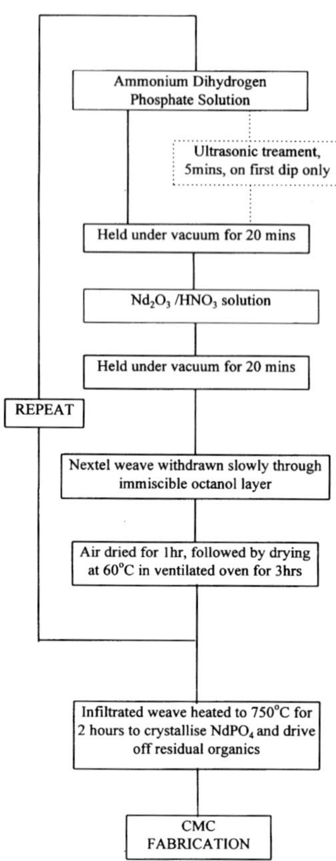

Figure 3 Flow diagram of optimum procedure developed for infiltration of Nextel 720 satin weave with $NdPO_4$ interphase.

shown in Fig. 3 and involves a multiple dipping operation using alternate aqueous solutions of the P and Nd precursors. The ultrasonic treatment during the first dipping step helps to open out the weave in order to facilitate sol penetration between individual fibres, but is not repeated in subsequent stages as it would serve only to remove the interphase already deposited.

XRD analysis of the interphases deposited in this manner, performed on samples obtained by ultrasonically removing the dried interphase from the woven fibres after heating to 750°C for 2 hours, shows broad diffraction peaks all of which can be attributed to $NdPO_4$ having the monoclinic monazite structure. Figure 4 shows a comparison of the XRD trace obtained with that given for $NdPO_4$ (monazite) in JCPDS PDF #25–1065.

An example of the degree of infiltration which can be achieved is seen by comparing the SEM micrographs shown in Fig. 5. Figure 5(a) shows an infiltrated weave prepared by using the octanol layer at the end of the repeated dipping process only. The $NdPO_4$ can be seen to have penetrated the weave relatively thoroughly, but to also remain in excess, filling some of the intra-tow voids. This would lead to problems with infiltration by the matrix in

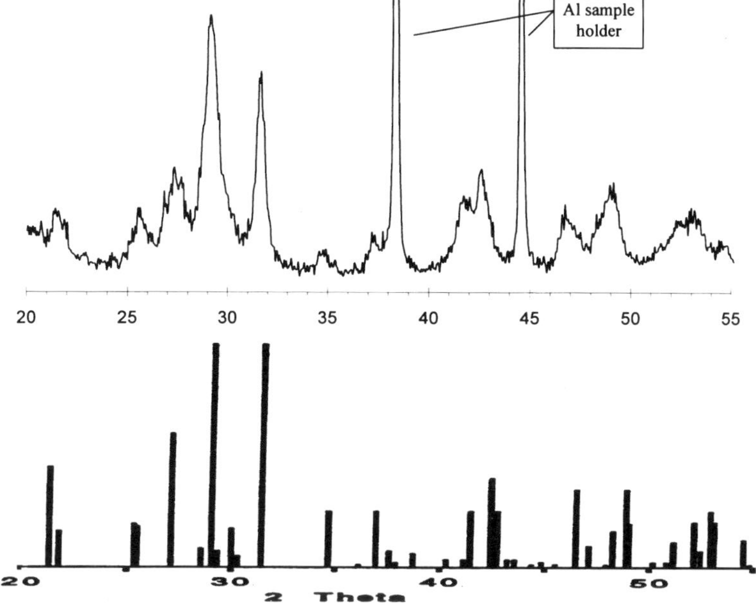

Figure 4 XRD data showing formation of monoclinic NdPO4 phase in interphase deposited on Nextel woven fibres, by comparison with JCPDS PDF #25–1065

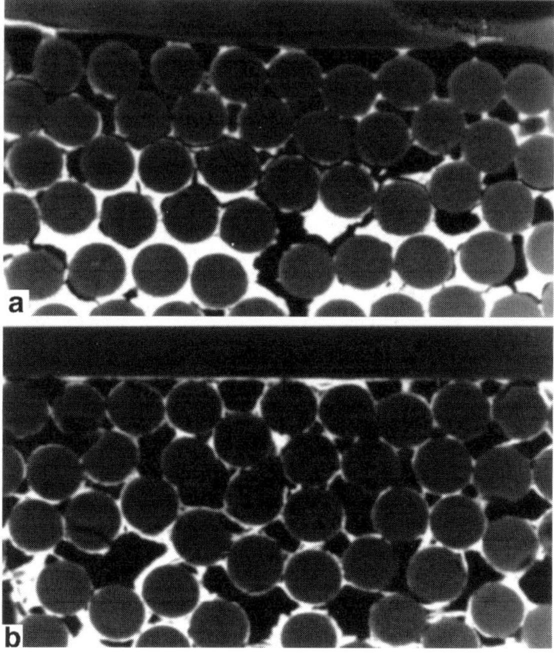

Figure 5 SEM backscattered images of Nextel 720 weave infiltrated with a NdPO4 interphase using the octanol bi-layer method. (a) Octanol used at end of repeated infiltrations only, (b) octanol used in between successive infiltration steps.

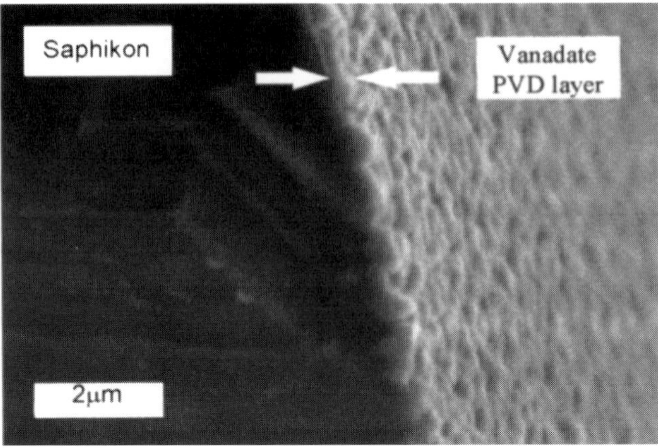

Figure 6. Appearance of fine crystalline PVD vanadate layer on Saphikon filament after annealing.

the subsequent stages of CMC fabrication. By contrast, the distribution of the interphase is shown in Fig. 5(b) to be reasonably uniform around individual fibres and along their length yet with relatively few of the intra-tow voids being filled. An adequate level of deposition is achieved after 3–4 infiltrations.

4. PHYSICAL VAPOUR DEPOSITION (PVD) OF INTERPHASES

Whilst the liquid precursor dipping technique, described in section 3, is shown to be a suitable method for yielding a relatively good degree of interphase coating on woven oxide fibres, PVD offers the advantage of being able to achieve a more uniform coating with greater control of the layer thickness and composition. RF magnetron sputtering of vanadates and phosphates has led to the preparation of fine grained, uniform coatings on oxide fibres and flat substrates, an example of a vanadate coating on Saphikon fibre being shown in Fig. 6.

Compositional and phase analysis of thin films sputtered using targets made from the liquid precursor derived powders used for the reactivity studies show that stoichiometric transfer of the constituent ions is possible, resulting in the formation of an amorphous layer. This can be crystallised upon heating to around 800°C or above in air to yield the MXO_4 type compounds as characterised, for the example of $LaVO_4$ on Al_2O_3, by the XRD traces shown in Fig. 7.

5. CONCLUSIONS

Lanthanum vanadate, yttrium vanadate and neodymium phosphate have been shown to be stable interphases with at least one of the candidate oxides for CMC fabrication considered in this work, i.e. alumina, YAG or mullite. Their phase compatibility at temperatures up to

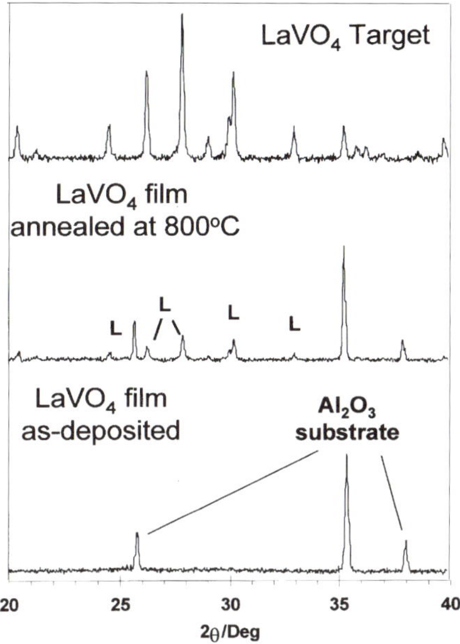

Figure 7. XRD data showing development of monoclinic $LaVO_4$ phase (peaks labelled L) after annealing of amorphous sputtered films.

1400 °C for several hours is an indication that these compounds might offer an alternative to lanthanum phosphate previously reported as a suitable interphase for oxide–oxide CMCs.

These novel interphases have been successfully deposited by PVD via RF-magnetron sputtering onto fibres and flat substrates to yield uniform coatings. A liquid phase deposition route has also been developed which has been used to infiltrate Nextel 720 satin weave with the neodymium phosphate interphase.

It remains to be seen whether these compounds will exhibit the desired debonding behaviour required of a successful CMC interphase. Further work is under way to investigate this and to elicit further information about the exact nature of the interfaces, e.g. preferred orientation, between these vanadates and phosphates and the oxides likely to be used as CMC fibres/matrices.

ACKNOWLEDGEMENTS

The main financial support for this project has been gratefully received from an EC BRITE-EURAM contract BRPR–CT95–0110 and involves some work carried out in conjunction with partners at the I.R.C., Birmingham University.

REFERENCES

1. M. G. Cain, R. L. Cain, M. H. Lewis, and J. Gent, *J. Am. Ceram.Soc.*, 1997, **80**, 1873
2. M. G. Cain, R. L. Cain, A. Tye, P. Rian, M. H. Lewis and J. Gent, *Proc. 1st Int. Conf. Ceramic and Metal Matrix Composites '96*, M. Fuentes, J.M. Martinez-Esnaola and A.M Daniel eds, 37.

3. M. H. Lewis, A. Tye, E. Butler, B. Lerazer and A. G. Razzell, *Proc. 6th Int. Symp.Ceramic Materials and Components for Engines*, October 1997, Arita, Japan.
4. P. E. D. Morgan and D. B. Marshall, *J. Am. Ceram. Soc.* 1995, **78**, 1553–1563.
5. D.-H. Kuo and W. M. Kriven, *Mater. Sci & Eng.*, 1996, **A210**, 123–134.
6. JCPDS, PDF # 32–199.
7. S. Erdei, *J. Mater. Sci.,* 1995, **30**, 4950–4959.
8. C. P. Wong, and M. H. Lewis, *Univ. of Warwick research report*, 1997.
9. R. S. Hay, *Ceram. Eng. Sci. Proc.*, 1991, **12**, 7.

Multi-Phase Piezoelectric Struct Co-Extrusion

188

D. H. PEARCE and T. W. BUT

Functional Materials Group, IRC in Materials and School
The University of Birmingham, Edgbaston, Birmingham

Email:d.h.pearce@bham.ac.uk WWW: http://www.fmg.bham.ac.uk/

ABSTRACT

A novel processing route has been developed which will allow electrode structures to be applied around and within structures of Lead Zirconate Titanate (PZT) before sintering, without the use of expensive metal electrodes such as palladium or platinum. The processing methods which have been applied to obtain these results have involved Viscous Processing and co-extrusion. De-agglomerated pastes of PZT and different levels of Silver (I) Oxide have been formed and co-extruded into cylindrical sections to demonstrate the feasibility of this technology. The microstructural and functional properties of these structures are shown, and the potential for this technology to generate complex novel piezoelectric devices illustrated.

1. INTRODUCTION

All piezoelectric components must possess two distinct phases in order to be functional. One is the functional material itself. This is most commonly based on the lead zirconate titanate (PZT) system which, with suitable choices of composition and dopants, can exhibit a wide degree of tailoring to obtain the desired functionality. The second phase always present is some kind of electrode element. In its simplest form this consists of a parallel plate capacitor arrangement with the piezoelectric phase between. This electrode arrangement is most easily applied in an ink form after the initial firing stage. For other devices such as multi-layer actuators, the conductive phase must be applied in the green ceramic state, since many electrodes must be placed within the final compact. This is achieved through layering of tape cast sheets combined with printing techniques for the electrodes. These two basic configurations are shown in Fig. 1.

The clear difference between the two examples is the requirement for different electrode materials. In the first case, this need only be a low temperature fired system, and can therefore be a glass-bonded metallic thick film, most commonly silver based. For the other example, the electrodes must withstand the same firing temperature as the functional ceramic and cannot therefore usually be silver. Instead, the material used for multi-layer actuator electrodes is a 70:30 palladium/silver alloy, which melts at a higher temperature than pure silver and can be fired with PZT at up to 1190 °C. At this elevated temperature, however, silver ions are very mobile, and inter-diffusion does occur.[2,3] This has been seen to be a problem, but one which must be withstood, since the only other option would be to use platinum based electrodes, which would be even more expensive. In any case, palladium is

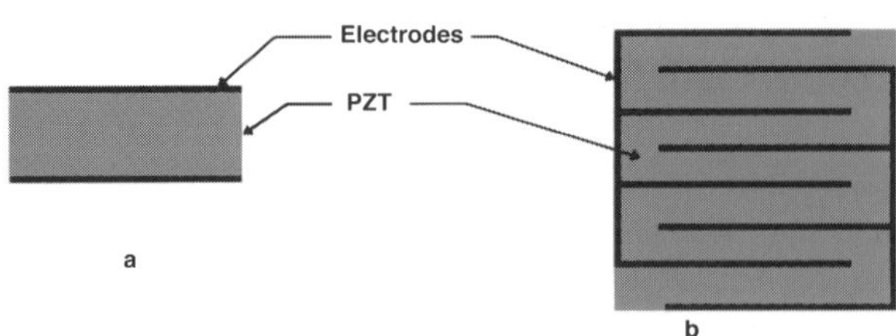

Figure 1 Schematic representation of electrode configurations for (a) post-sintered and (b) co-fired electrodes.

vastly more expensive than silver, and therefore a method which would allow functionality of a piezoelectric component without using the more expensive metallic components should show significant cost benefits.

Recent work by Hwang et al. has shown that, contrary to conventional views, silver itself can be used to promote beneficial effects in PZT.[4] By using silver oxide as a precursor, considerable increases in sintering rates and improvements in mechanical strength have been shown. The starting point for the current work therefore was to attempt to answer the following questions:

- Can silver be used in some way to aid the formation of conductive and piezoelectric phases in a component, through its ability to reduce the sintering temperature of PZT?
- Can the known effects of silver in depressing the piezoelectric properties of PZT be avoided in some way?
- Could the technique allow entirely novel piezoelectric structures to be made which would enable hitherto unrealisable devices to be fabricated?

This paper therefore describes the progress towards answering these questions, and will relate the results achieved in solving the first. Co-extrusion was judged to be a suitable method for the incorporation of the various phases which would be assembled in the green state for the proof of the first question. This is a technique which is not widely used in the area of ceramics processing, and has been most commonly associated with food and plastics technology, related examples being Blackpool rock and pultruded glass fibre reinforced plastics. More recently, work by Halloran *et al.* has shown that the technique can be used to create very fine detailed two-dimensional structures through multiple co-extrusions of thermoplastic based ceramic systems.[5]

One important consideration when combining two or more paste phases is how closely the rheology of each phase is matched. Normally this would be a significant problem for an extrudable system, since the rheology will be largely determined by the properties of the binders, and therefore the binder systems must be tailored to each other. To some extent in this case, however, this can be overlooked, for two reasons. Firstly, the compositions chosen for co-extrusion were very similar in terms of the functional components. The rheology for all the formulations, using the same binder levels, was found to be very similar, through normal techniques based on measurements of applied load during extrusion. This leads to the second reason, which is closely linked to the fundamentals of Viscous Processing (VP). If a suitable paste can be formed through VP it can almost always be assumed that the rheology conforms to a certain set of criteria laid down in the original work.[6] Since suitable pastes were formed in all cases for this work, and pure PZT pastes have been shown previously to be suitable for VP, the rheological behaviours were taken to be largely similar. In addition, since VP pastes are by definition stiff, effects which are not directly related to rheology become more important, such as die wall friction. This will be made clearer in later sections.

2. EXPERIMENTAL METHOD

For this work, pastes based on the previously characterised PVA–water binder system were used, formed via Viscous Processing, which is described in more detail elsewhere.[7,8] The pastes were formed using a 'soft' doped PZT powder (PZT4050, Advanced Ceramics Ltd, Stafford) and a commercial silver (I) oxide powder (Johnson Matthey). For single phase measurements, compositions of 0%, 2%, 5%, 10% and 25% Ag_2O by weight were processed. Two-phase experiments were conducted by combining pastes containing 5% and 25% Ag_2O, for the piezoelectric and conductive phases, respectively. High shear mixing was performed on a twin roll mill to obtain plastic formable sheets. These sheets of green ceramic material were then used as the starting material for further experiments, both in the single phase state and for multi-phase processing. Plates formed from pressed and cut sheets were created for electrical property measurements and extruded rods created by combining differing pastes for co-extrusion experiments. Samples were dried to remove water at 40 °C. Binder burnout was performed in flowing air through controlled heating to 600 °C at 1°C per minute. Sintering was carried out in enclosed crucibles at a temperature of 1000 °C for the silver oxide loaded compositions and at 1200 °C for the pure PZT samples, for 1 hour in both cases. More detailed descriptions of the experimental techniques can be found elsewhere.[9]

Dilatometry analysis was carried out on extruded rods of compositions containing 0%, 5% and 30% of Ag_2O, in order to assess the feasibility of combining the two phases, and in particular to measure the mismatch occurring during co-sintering of the phases. This was performed on previously burnt-out samples, at a heating rate of 10°C per minute up to the desired temperature, again at 1000 °C and 1200 °C for the silver oxide loaded and pure PZT phases respectively.

Co-extrudable configurations were designed to show the feasibility of this technique in forming the types of structures shown in Fig. 2. The first type (a) illustrates a simple coaxial radially symmetric extrudate, which would illustrate the closeness of rheological matching and the possibility of co-sintering. The second (b) shows another coaxial structure, though in this case not radially symmetric, in order to form a bimorph structure. The third (c) illustrates one simplified embodiment of a final goal which allows for torsional actuation. If this section is subjected to a continual 45° twist as it exits from the extrusion die, and after sintering is poled in the directions indicated, torsion is generated in a straight section. This

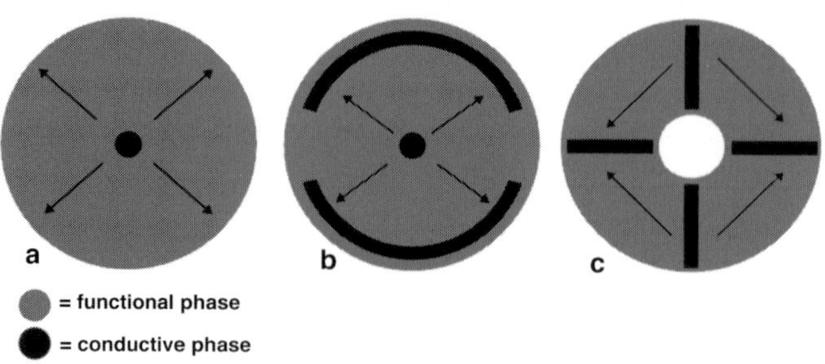

Figure 2 Schematic diagrams of various cross-sections for co-extrusion (arrows indicate dirtection of polling).

results in an axial force when this section is coiled into a spiral or helix, and is the subject of past and recent patents.[10,11] Experiments were carried out in generating the first two forms shown, and the results will concentrate on the second type. Plugs with the required electrode configurations taken from Fig. 2 were fabricated by hand to fit a 20 mm diameter extrusion barrel by combining sections of VP sheets of different compositions. Extrusion was performed through a 90° straight section die of diameter 2.5 mm.

3. RESULTS

3.1 *Single Phase Results*

Microstructural observations were most readily carried out through optical microscopy of polished sections. This enabled clearer observations of the free metallic silver phase to be made. Micrographs of silver oxide loaded compositions are shown in Fig. 3. These show the increasing content of free silver in the material (shown as the lighter dispersed phase), and that even at 2% initial loading some free silver is formed. The solubility limit of silver in this type of PZT is therefore below the equivalent of 2 wt% Ag_2O, equating to 1 .9 wt% of silver after reduction. This is consistent with electrical property measurements which show the major effect of silver addition occurring below this level, while above this the metallic dispersion effects of silver become clearer. This is clear by considering the dielectric constants of the compositions before and after poling, shown in Table 1. While the unpoled permittivity shows a slight drop followed by a gradual increase, the poled values show a large initial drop, followed by a similar steady increase. The large reduction in the increase in permittivity on poling indicates further the effect of the silver on affecting the soft properties of the PZT.

The results from dilatometry of silver oxide loaded PZT are shown in Fig. 4. The main point from these results is the large difference in sintering behaviour between the unloaded PZT and the materials containing silver oxide. The sintering temperature is effectively reduced by 200 °C through the addition of 5 wt% Ag_2O. This result shows that it is possible to sinter silver loaded PZT at a temperature low enough for the liquid phase Ag metal not to be

Table 1 Dielectric constant measurements, taken at 1kHz on sintered plate samples of soft PZT with varying levels of silver oxide.

	Dielectric constants ε_r		
Wt% Ag_2O	Before poling	After poling	Increase
0	1128	1627	44%
2	1052	1284	22%
5	1144	1366	19%
10	1188	1417	19%

a large problem in terms of mobility and volatility. The other important point to note is that the addition of up to 30 wt% does not significantly affect this increase in sintering rate. The further increase in this case can be largely attributed to the difference in final density, since silver metal is more dense than PZT (ρ_{Ag} = 10.5 gcm^{-3}, ρPZT[7,8] = 7.1 gcm^{-3}), whereas the density of the silver oxide precursor is less than that of PZT (ρ = 7.1 gcm^{-3}). This results in a 2% linear mismatch when sintering compacts from a green density of 55% to full sintered density. This variation in sintering shrinkage is close to that measured through dilatometry. This result shows that there exists a possibility of co-sintering the two compositions containing silver oxide, provided that this mismatch can be accommodated.

3.2 *Multi-Phase Results*

Initial results on the first form of coaxial extrudates formed in straight sections and shown previously in Fig. 2a indicated two main features. Firstly, the central core, containing 25% Ag_2O was conductive after sintering, and secondly that the main phase could be made piezoelectric through radial poling, using the co-sintered central core as the high voltage electrode and an outer applied electrode as the ground. Clearly, therefore, to some extent at least the previously mentioned sintering shrinkage mismatches were accommodated in this case. Measurements of piezoelectric coefficients have not as yet been possible, due to the geometry of the samples, though clear resonances were observed on impedance-frequency traces corresponding to both the longitudinal and radial resonances.

Forming the more complicated second form shown in Fig. 2b proved more problematic, mainly in the difficulties associated with simultaneously maintaining the conductivity of the electrode phase and functionality of the piezoelectric phase. While conductivity of the outer electrodes could be maintained, that of the inner electrode would alter, normally through poling. Short sections of such structures were poled, but it was found that the central core electrode would exhibit a significant increase in resistivity, and often change to being open circuit, together with an increase in the poling current to such a level that poling became impossible. This effect could be attributed to field concentrations around the central core both creating conductive paths through the piezoelectric phase and eroding the metallic

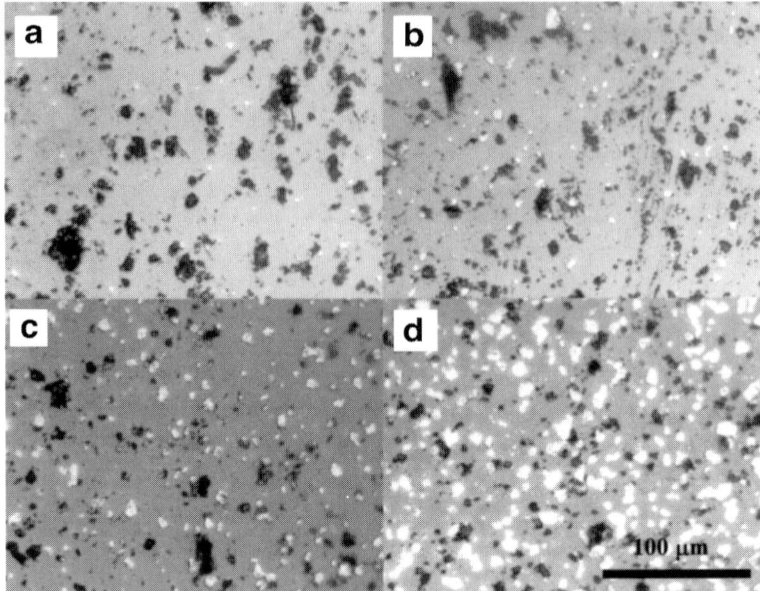

Figure 3 Optical micrographs of polished sections of sintered PZT with varying additions of silver oxide: (a) 2%; (b) 5%; (c) 10%; (d) 25%.

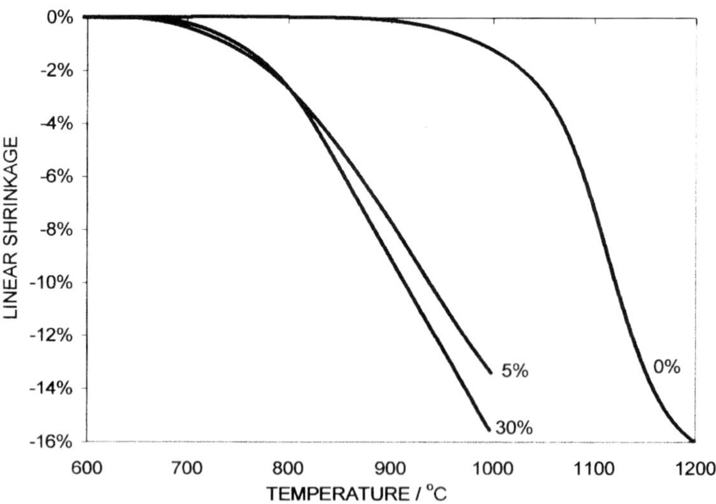

Figure 4 Dilatometry analysis on PZT4050 with varying levels of silver oxide additions.

silver particle connections. Further work is necessary to understand the mechanisms involved in processing these multi-phase structures before repeatable functional devices can be fabricated. The results obtained thus far are given in the following section in terms of the green state of these complex multi-phase extrudates.

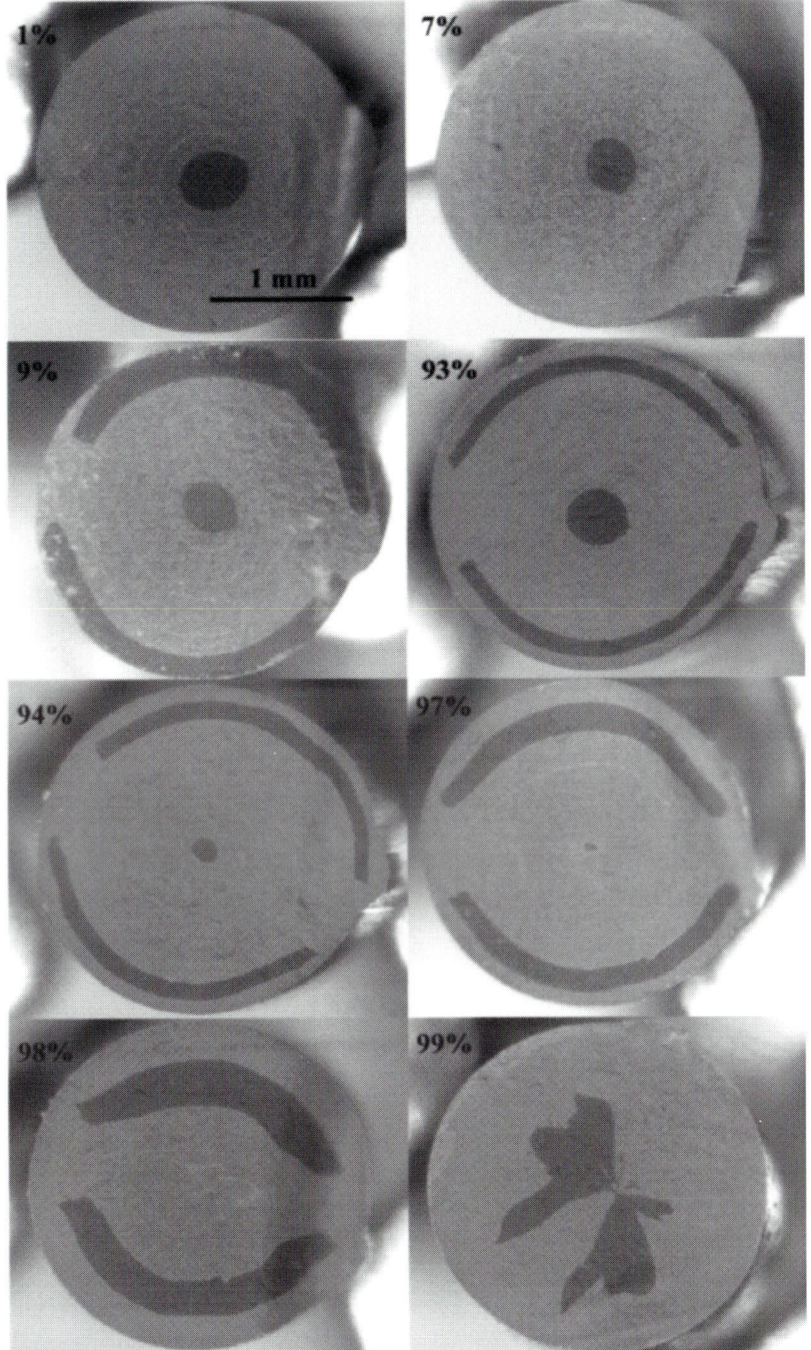

Figure 5 Optical micrographs of green sections taken throughout a co-extrusion experiment, showing the variation in section at the start and finish of extrusion. Percentages indicate the volume of material extruded.

3.3 Section Variability During Extrusion

A clear indication of the success of the green state processing in producing coextruded structures is indicated in the series of micrographs shown in Fig. 5. This illustrates the main factors involved in producing co-extruded sections by the method used. In this case the whole of the paste assembly in the barrel is able to flow, in contrast to the thermoplastic pastes used in other work where only the paste entering the heated die could flow.[5] This is the cause of the change in appearance of the extrudate observed. At the start of the extrusion the central phase can be seen to exit preferentially to the outer phase, though at 9% of the extrusion the full structure is observed. This is maintained for the majority of the extrusion, up to around 94%. At this point the result of the initial state of the extrusion starts to affect the final state, as the outer phase extends inwards and the central core disappears. By the end of the extrusion the central core has gone, and the distorted outer phase extends towards the centre of the extrudate.

The change in extruded cross-section illustrates that the whole extrusion process is, as should be expected, wholly conformal in the flow behaviour and, importantly, the whole cross-section of the extrudable plug comes through in a symmetrical form, i.e. there is no distortion across the section. What affects the uniformity of the extrusion is not the difference in rheological behaviour of the two phases, but the die and barrel wall friction. In addition, in this case, the die geometry will have had an effect on the results, being of a 90% design rather than conical. This will have resulted in the formation of a 'dead zone' at the internal corner of the bottom of the barrel, and could to some extent explain the observed behaviour. The combined effect will be to hold back the outer portion of the extrudate, and therefore result in the observed differences at the beginning and end of the extrusion. What may be surprising, however, is that the effect is only evident over a relatively small proportion of the whole extrudate, in the first and last 5 to 10%. This illustrates the uniformity of the viscous processed starting materials, allowing co-extrusion to occur without any binder segregation or phase instabilities.

4. CONCLUSIONS

Initial steps have been taken in proving the concept of processing multi-phase structures consisting of PZT and silver compositions. Both piezoelectricity and conductivity have been proved to be possible with co-extruded structures. Simple structures have been created through this technique. More complex structures are currently being attempted, though it is clear that much further work is necessary to optimise the processing and microstructural issues involved.

ACKNOWLEDGEMENTS

Thanks are due to Roy Wyatt of Basys Marine Ltd, Barnstaple, and to Dr Tony Hooley of 1... Limited, Cambridge, for ideas and impetus for this work.

REFERENCES

1. D. Berlincourt, *J. Acoust. Soc. Am.,* 1992, **91** (5), 3034–3040.
2. C. Schuh, M. Kulig and K. Lubitz, *Proc. Conf. Electroceramics V,* 1996, **1,** 201–205.
3. M. V. Siankina, G. I. Dontsov and V. M. Zhukovsky, *J. Mat. Sci.,* 1993, 28, 5189–5192.
4. H. J. Hwang, K. Watari, M. Sando, M. Toriyama and K. Niihara, *J. Am. Ceram. Soc.,* 1997, **80**, 791–795.
5. D. Popovic and J. W. Halloran, US Patent No *5,645,781,* 1997.
6. N. Alford, US Patent No. 4677082, 1985.
7. N. Alford, J. D. Birchall and K. Kendal, *Nature,* 1987, **330** (6143), 51–52.
8. D. H. Pearce, G. Dolman, P. A. Smith and T. W. Button, *Proc. Conf. Electroceramics V,* 1996, **2**, 385–390.
9. D. H. Pearce and T. W. Button, *Proc. Dielectrics Soc. 29th Annual Conference,* 6–9 April 1998, Canterbury, to be published in *Ferroelectrics.*
10. D. H. Pearce, British Patent Application No. 9618052.6.
11. R. Adler, US Patent No. 3900748, 1975.

Optimisation of Ball Valve Processing Parameters

R. J. HUZZARD* and S. BLACKBURN†*

**IRC in Materials for High Performance Applications, The University of Birmingham, Edgbaston, Birmingham B2 5TT, UK*

†School of Chemical Engineering, The University of Birmingham, Edgbaston, Birmingham B2 5TT, UK

ABSTRACT

The industrial application of ceramic injection moulding has been limited to small components because of the associated binder burnout problems. Injection moulding may improve the manufacture of alumina ball valves for the process industries by reducing costly and potentially damaging machining operations. However, these components are too large to be produced using conventional thermoplastic binders. Aqueous injection moulding allows most of the binder system to be driven off as water vapour, cutting process time, reducing burnout difficulties and causing less environmental pollution. Optimum processing parameters for the manufacture of ball valve blanks of 25.4 mm outside diameter were efficiently identified using orthogonal arrays as described by Taguchi. Factors such as barrel and die temperature (T die), injection pressure (P inj), velocity and time were evaluated for their influence on defects. The defects observed included cracking, poor make up and distortion. Ball valve blanks of significantly improved quality were successfully manufactured under the optimised conditions.

INTRODUCTION

The approach of designers and engineers towards material requirements has changed over recent years. The requirements for better properties are being fulfilled by the development of new materials and a more comprehensive understanding of the influence of microstructure. The need to control the shape and size of components is achieved by the investigation of new and existing forming technologies. Ceramic injection moulding (CIM) is providing solutions to many of the requirements for small components in structurally complex applications. However, the difficulty in burning out the binder system after fabrication puts a practical upper limit of about 15 mm on the possible cross section. Aqueous injection moulding (AIM) provides the opportunity to make larger components of complex shape. Aqueous binder systems have been introduced from which most of the binder can be removed as water vapour, reducing debinding and total process time. A much lower quantity of binder is removed giving environmental benefits over traditional ceramic injection moulding. The literature on AIM has focused on the preparation and properties of the binder solutions.[1–3] The first example of cellulose solutions being used in AIM was by Rivers[4] which considered metallic components. The literature also gives examples of agar systems for injection moulding[5–7] but again processing variables were not considered. These aqueous binder systems do not always confer the high strength required to remove moulded components without damage. To counteract this, a binder system was developed to optimise the binder gel strength using commercially available cellulose derivatives.[8] The pastes produced using alumina as a model ceramic material were characterised rheologically and were essentially power law fluids.[9] The rheological model was further improved by allowing for slip flow above a threshold shear stress.[10]

Ball valves for the processing industry provide an example of a large component with a high degree of complexity. Alumina is often used to manufacture ball valves for processes with abrasive or corrosive liquids and slurries. The spherical shape lends itself to fabrication by isostatic pressing. In the case of isostatic pressing, the central channel is usually precision drilled and a tap hole milled into the component to allow rotation. Failure of the ceramic is often costly due to high down-time costs for the process plant. These failures are often caused by a build up of stresses around uneven facets on the milled tap holes. If all of the features of the ball valve could be formed in a mould using injection moulding, the stresses could be distributed more evenly, and failure avoided. The method also promises better material utilisation as milling operations may be eliminated. Hence, ball valves for the process industries provide a useful component to consider the feasibility of AIM of ceramics.

There are few details available in the literature on the processing parameters required for aqueous injection moulding systems. The Rivers patent, for example, reported the manufacture of test bars of 19 mm cross section, but no details were given on processing parameters or on any difficulties encountered during manufacture. As a result, there are no ground rules available for manufacturing components by AIM. The processing conditions required differ sufficiently from traditional CIM to warrant a fresh approach.

The most common approach to optimising well-documented processes is to evaluate the influence of one parameter on product performance. This may prove a weak strategy where parameters are interdependent, and becomes impractical for poorly understood processes such as AIM. In a complex case such as AIM, properties may be influenced by a large number of variables. Consequently, a high level of experimental design is required to deliver reliable conclusions. This complex problem is addressed here by using the experimental design technique proposed by Taguchi.[11, 12]

The account of preliminary experiments on the processing of the ball valve blanks was presented previously.[13] The present paper details the final stage of that analysis and highlights the role of experimental design in the optimisation of the process.

EXPERIMENTAL

The alumina ball valve blanks were manufactured using a reactive calcined alumina BAX541 from BA Chemicals Ltd. The binder used was hydroxypropylmethylcellulose, grade HPM5000DS from Courtaulds Chemicals. The solids content in each paste was 51 per cent by volume. Binder solutions containing 8 weight per cent and 12 weight per cent were used in the formulations (Table 1). Both pastes were made by premixing in a Hobart planetary mixer and then in a Baker Perkins Extrumix for 20 minutes to give 4 kgs of each paste. The Extrumix is a double lobe mixer with a screw attachment which improves mixing and discharges the paste by extrusion at the end of the mixing period.

The ball valve blanks were moulded with the central channel in place as shown in Fig. 1, but the tap holes have been omitted for this exercise in order to simplify the die design. A cross section of the die is shown in Fig. 2. An Arburg 250–75–270D was used to injection mould the ball valve components. The Arburg moulder modified with hardened piston and barrel to allow the manufacture of ceramic components is typical of the equipment used

Table 1 Paste formulations

51% volume solids	8%*	12%*
BAX541	79.81	79.59
HPM5000DS	1.62	2.45
De-ionised water	18.57	17.96

*% binder soln. by weight

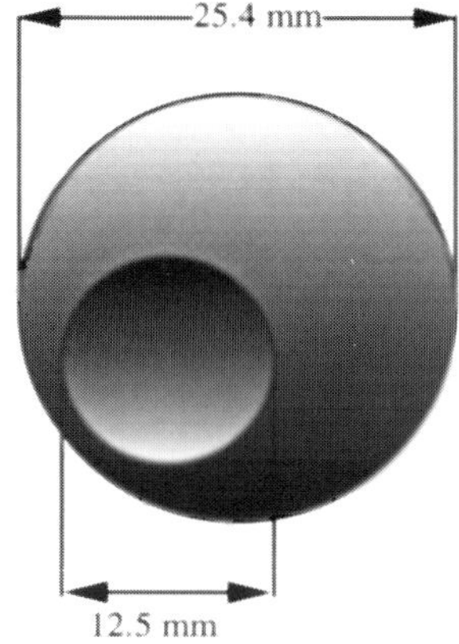

Figure 1 Ball valve blank.

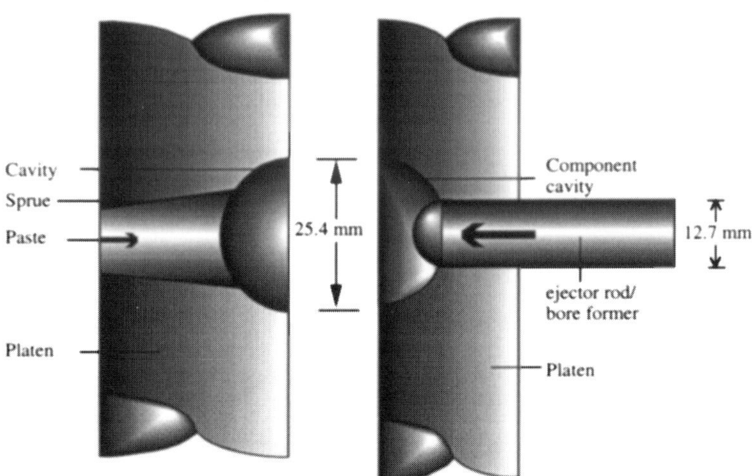

Figure 2 Ball valve die.

industrially for ceramic injection moulding. The three barrel sections and the nozzle were heated and controlled electrically. The die temperature (T die) was controlled by water circulation. Other parameters that could be adjusted include the injection pressure (P inj), injection time (t inj), injection speed (v inj), hold pressure (P hold) and cooling time (t cooling). One of the major issues for AIM when using cellulose derivatives is that the component is removed at a temperature between the gel point of the binder and the boiling point of water. In view of this, the shaped components were immediately removed from the die and placed in enclosed tubes to prevent excessive evaporation of water. The samples were dried overnight whilst enclosed in order to avoid the introduction of more variables due to drying faults. The partially dried components were placed in the open at room temperature for a further 24 hours to dry completely. Preliminary studies on the drying of cylinders made by AIM suggested that the drying time of components of 18 mm in cross section could be reduced to three hours by microwave heating. It is worth noting that even the precautionary two day conventional drying cycle used in this study, showed a considerable reduction on the binder removal times needed for such large components made with thermoplastic binders. The ball valve component would be on the upper limit of size considered with such binder systems.

The use of an orthogonal array allows several factors to be evaluated in the minimum number experiments. The selection of these arrays is complex.[11, 12] An L18 array was selected here since the preliminary work[13] cited 8 factors which strongly influenced the quality of the components. The non-linear relationship of many of these factors inferred that they should be investigated at more than one level. The fractional factorial design used for this study is shown in Table 2. The rows of the array represent individual experiments. The columns represent the factors under investigation. A number of features common to all L 18 arrays should be noted. They require 18 experiments to be carried out as the name L18 implies. The array is saturated since all 8 factors available to the array are used. The first factor, binder concentration, is only considered at two levels. Each of the two levels is used in 9 of the 18 experiments. All of the other 7 factors require 3 levels, each of the three levels being used in six of the 18 experiments.

A low die temperature of 75°C led to undesirable distortion in the preliminary exercise. Hence, temperatures of 80, 83 and 86°C were considered in the L18 array. Injection velocities (v ini) of 50, 65 and 80% of the machine's capacity were used. The injection pressures were set at 45, 55 and 65% and the hold pressure 35, 45 and 55% of the machine capacity respectively. The other three parameters considered were the injection time set at 3, 9 and 15 seconds, the barrel temperature (T barrel) set at 45, 50 and 55°C and the remaining cooling time set at 60, 120 and 180 seconds.

RESULTS

The dried components were examined for faults using two fold magnification. Material loss from between the platens of the mould occurred in some cases. Only one incident occurred where this material loss was sufficient to influence the quality of the component, but the problem is undesirable and was treated as a fault. Some of the dried components from the L18 orthogonal array in Table 2 were stretched along the direction of the bore (axial stretch). This was attributed to deformation whilst removing the ball valve from the ejector pin and is associated with insufficient gel strength in the paste. The bore was not round in other cases. The extent of both of these types of faults was measured using callipers. Axial stretch was measured simply by measuring the length of the bore. This measurement was compared with the equivalent length on the mould. Variation in shrinkage between components was

Table 2 L18 array for ball valve blanks

No.	% binder	T die	v inj	t inj	T barrel	P inj	P hold	t cooling
		(C)	%	%	(C)	%	%	(s)
1	8	80	50	3	45	45	35	60
2	8	83	65	3	50	55	45	120
3	8	86	80	3	55	65	55	180
4	8	80	50	9	50	55	55	180
5	8	83	65	9	55	65	35	60
6	8	86	80	9	45	45	45	120
7	8	83	50	15	45	65	45	180
8	8	86	65	15	50	45	55	60
9	8	80	80	15	55	55	35	120
10	12	86	50	3	55	55	45	60
11	12	80	65	3	45	65	55	120
12	12	83	80	3	50	45	35	180
13	12	83	50	9	55	45	55	120
14	12	86	65	9	45	55	35	180
15	12	80	80	9	50	65	45	60
16	12	86	50	15	50	65	35	120
17	12	80	65	15	55	45	45	180
18	12	83	80	15	45	55	55	60

small, so the difference between bore length on the component and the die was taken as a measure of axial stretch. Four measurements of the bore diameter were made at intervals of 45° to each other and the distortion was taken as

$$d = Ø_{max} - Ø_{min} \quad (1)$$

where d is the distortion to the bore, $Ø_{max}$ is the maximum diameter and $Ø_{min}$ is the minimum diameter. Many of the ball valves were not fully made up, e.g. insufficient material had been forced into the die. Make up faults were classified as inside the bore (bore) and/or outside the component (outer). Some of the components showed cracking in the bore, but three mouldings suffered a much more severe split. The cracking and splitting faults did not appear to be related and were classified separately.

The first assessment of faults in moulded compacts was made for each type of defect based on a subjective ranking system of 5 for a perfect ball valve component and 1 for a seriously defective component. The second assessment was made by placing each series of components in order, e.g. one for the worst and 18 for the best in the array. Both methods of assessment showed similar trends in practise and only the second method was reported giving a maximum value of 18. This comparative exercise was not practical on the die leakage problem, which was assigned a five where the problem did not arise. The data was

Table 3 Results of L18 array for ball valves

Experiment number	Leakage from die	Bore roundness	Axial stretch	Make up (outer)	Make up (bore)	Cracking bore	Split
1	2	1.049	19.28	16	3	10	10
2	4	1.039	19.68	14	13	7	10
3	3	1.035	19.04	3	5	2	1
4	3	1.012	20.26	9	15	12	10
5	5	1.122	19.96	2	4	6	10
6	2	1.060	19.28	6	7	4	10
7	4	1.017	20.02	5	6	3	2
8	3	1.027	19.24	11	8	9	10
9	4	1.006	19.22	15	16	16	10
10	5	1.039	19.96	1	1	1	10
11	5	1.017	19.48	8	14	14	10
12	5	1.016	19.76	13	9	5	10
13	5	1.010	20.10	4	10	8	10
14	5	1.018	19.14	10	2	15	10
15	5	1.010	19.52	12	12	13	10
16	5	1.015	19.74	7	11	11	10
17	5	1.040	19.07	18	17	18	10
18	5	1.008	18.80	17	18	17	10

rearranged to assess the influence of the processing parameters on defect formation (Table 3). These results demonstrate that the optimisation of processing parameters is a complex problem. Action taken to improve one defect had a deleterious effect on others.

The influence of percentage binder on material leakage from the die and post moulding splitting can be seen from Fig. 3(a). Each result in Fig. 3(a) represents the average from 9 components. Analysis showed that these were the strongest relationships from the whole experiment. Scrutiny of the individual results in Table 3 showed that these faults only occurred when the 8% binder solution was used. The higher concentration of binder also gave improvements in all of the other faults as shown in Fig. 3(a). Higher binder concentration improves gel strength and was therefore expected to reduce distortion. The cracking and splitting faults may also have been alleviated by the higher resultant gel strength. Axial stretch and bore roundness were improved as a result. It is not so clear why make up of the ball valve blanks was also better.

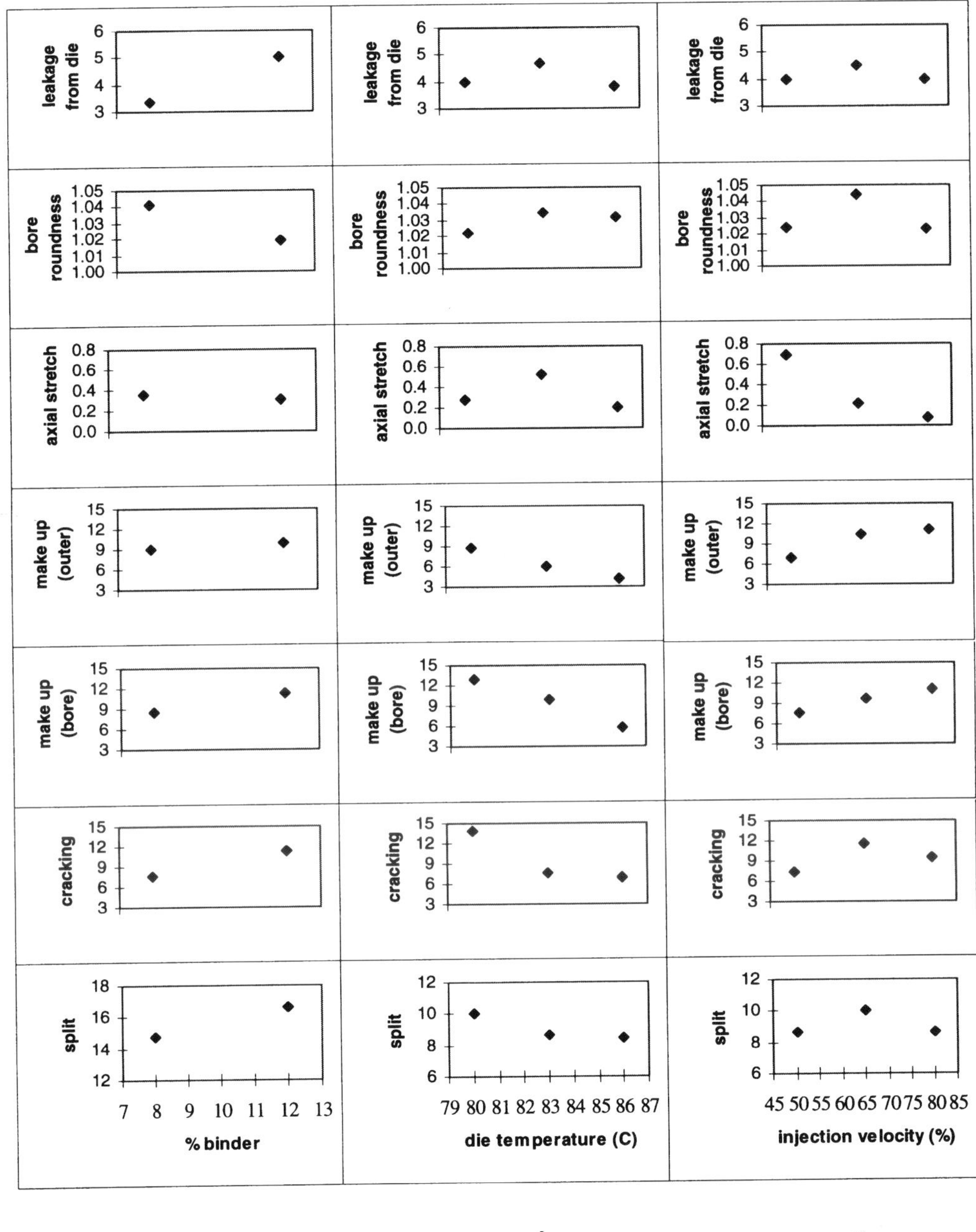

Figure 3 The influence of (a) % binder, (b) die temperature and (c) injection velocity on ball valve quality.

The lowest die temperature used in this experiment of 80°C also reduced all of the defects (Fig. 3(b)) due to the stiffer binder. A die temperature of 75°C used in the preliminary experiment[13] had a deleterious effect on faults of all types. The differential between the gel

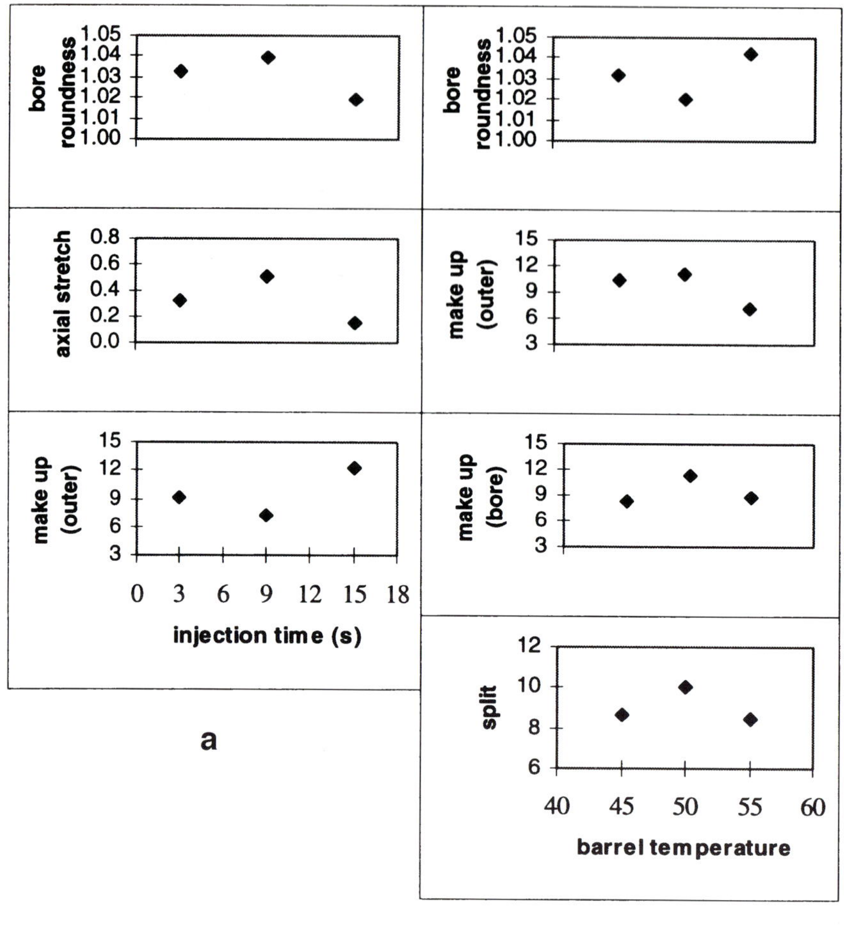

Figure 4 The influence of (a) injection time and (b) barrel temperature on ball valve quality.

point and set temperatures at 80°C was sufficiently large that high strengths were achieved quickly. The temperature was not so hot as to degrade the strength. Figure 3(c) shows that a higher injection velocity improved the make-up of the components, suggesting the importance of filling the mould quickly, before the onset of the gelling process. The high velocity was adopted despite giving worse cracking. Crack formation was avoided by using the optimum binder concentration and die temperature.

A similar methodology was used to find optimum values for the remaining parameters. The following summary gives the important factors. The longest injection time shown in Fig. 4(a), 15 seconds, gave low distortion and good make up. There was a risk that insufficient material would be transferred to the mould if the injection time was too short. Injection time and velocity were interdependent in this respect. The intermediate barrel temperature of 55°C, Fig. 4(b), reduced the magnitude of each defect. The paste was too close to the

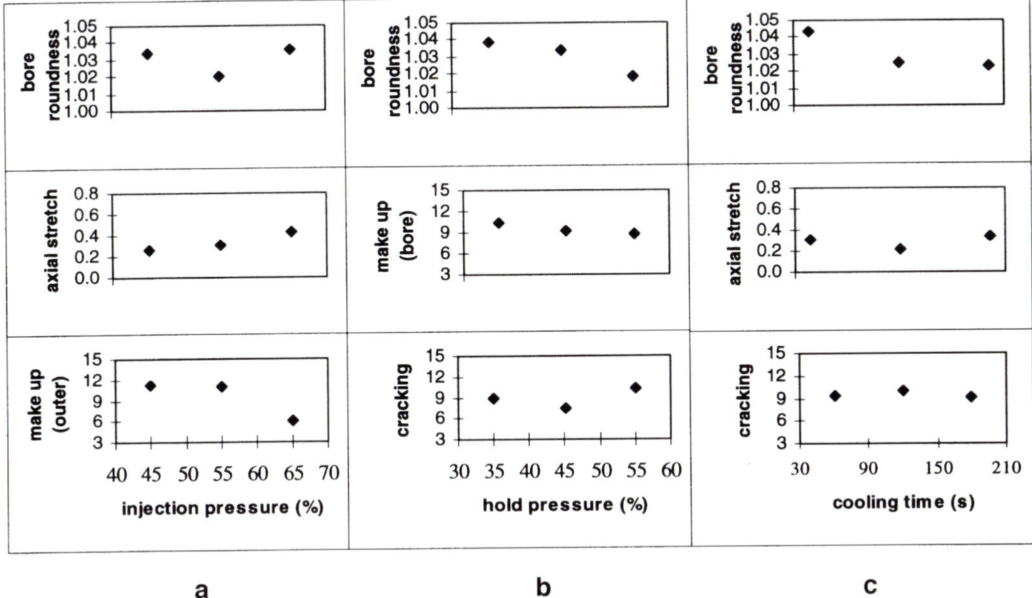

Figure 5 The influence of (a) injection pressure, (b) hold pressure and (c) cooling time on ball valve quality.

gel temperature at higher temperatures, leading to higher viscosity and poor filling of the mould. This poor filling may have contributed to other faults such as cracking and distortion. The viscosity of the AIM paste increases below 55°C. This proved to be a higher value than the optimum temperature predicted in previous work.[13] This may be because the paste did not quite reach equilibrium temperature during the relatively short time period that the paste spends in the barrel prior to moulding.

The injection pressure had less influence. The clearest indication from Fig. 5(a) was that 65% injection pressure gave the worst axial stretching and must be avoided. A hold pressure of 55% reduced the level of cracking as shown in Fig. 5(b) and the distortion in the bore. This pressure also gave marginally better make up. The cooling time of 60 seconds in Fig. 5(c) was insufficient to confer the required green strength and some distortion was observed. A cooling time of 120 seconds gave a better compromise.

Samples were made up based on these optimum parameters suggested by the L18 array. The paste with 12% binder solution was used, with a die temperature of 80°C and barrel temperature of 50°C. The injection velocity, injection pressure and hold pressure were 80, 55 and 55% of the machine capability respectively. The injection time and remaining cooling time were 15 and 120 seconds respectively. A small weld line remained close to the equator of the ball valve, but most of the samples were free from cracks and distortion, showing a marked improvement over the samples for the L18 array. It is important to emphasise that the quality of the components made using these optimum conditions was far superior to any of the ball valves made up from the L18 array. This demonstrates the power and efficiency of the designed experiment used.

CONCLUSIONS

A number of problems were encountered during the AIM of ball valve blanks; material leakage from the die seam, axial stretch, bore distortion, poor make-up, cracking and splitting. The number of combinations of parameters and levels available, and the complexity of the system required the use of designed experiments to find the optimum moulding conditions. The source of each problem was identified and the quality of the components was significantly improved using only 27 experiments in total.

Many of the individual problems proved to have simple solutions, but one of the most important factors proved to be the concentration of the binder solution. Material leakage from the die only occurred using an 8 % binder solution. The higher gel strength produced by using 12% binder level eradicated the fault. The improved gel strength of the higher concentration binder also reduced distortion both axially and around the bore. The binder concentration was expected to be limited by die filling considerations as well as binder removal constraints. In practise the die filling did not prove a problem with the higher concentration. A die temperature of 80°C proved to be optimum, above this temperature the gel strength was degraded, but below 80°C the material would not gel in the mould sufficiently. There was an optimum barrel temperature of 50°C suggesting that this gave the lowest viscosity. The viscosity of a paste falls with increasing temperature but here a limit is reached due to hydrophobic interactions which ultimately lead to gellation.

The optimum processing conditions for the ball valve blanks were; 51 per cent volume solids, 12 weight per cent binder, 55 kPa hold Pressure, 55 kPa injection pressure, 50°C barrel temperature, 80°C die temperature, 80 per cent injection velocity, 120 s residence time.

The experimental design based on Taguchi's methods was used here to improve the quality of ball valves. The method applies to all engineering and scientific problems where an optimum solution is required and a complex set of parameters can be varied.

ACKNOWLEDGEMENTS

The authors would like thank the Engineering and Physical Sciences Research Council for providing financial support for this work.

REFERENCES

1. B. C. Mutsuddy and R. G. Ford, *Ceramic Injection Moulding*, Chapman and Hall, 1995.
2. N. Sarkar and G. K. Greminger, *Ceram. Bull.*, 1983, **62**(11), 1280–1288.
3. N. Sarkar, *J. Appl. Polym. Sci.*, 1979, **24**(4), 1073–1087.
4 R. D. Rivers, 'Method of Injection Moulding Powder Metal Parts', *U.S. Pat. No. 4 113 480*, 1978.
5. Z. S. Rak and P. J. Van Tilborg, *Proc. of the 11th riso conference*, 1990, 489–492.
6. T. Zhang, S. Blackburn, and J. Bridgwater, *Brit. Ceram. Trans.*, 1994, **93**(6), 229–233.
7. A. J. Fanelli, R. D. Silvers, S. W. Frei, J. V. Burlew, and G. B. Marsh, *J. Am. Ceram. Soc*, 1989, **72**(10), 1833–1836.
8. R. J. Huzzard and S. Blackburn, *J. Eur. Ceram. Soc.*, 1997, **17**, 211–216.
9. R. J. Huzzard and S. Blackburn, *21st Century Ceramics*, D. P. Thompson, and H. Mandal eds, British Ceramics Proceedings 55, The Institute of Materials, 1996, 79–86.

10. R. J. Huzzard, and S. Blackburn, to be published in *Powder Technology*
11. P. J. Ross, *Taguchi Techniques for Quality Engineering*, McGraw-Hill, 1988.
12. G. Taguchi, *System of Experimental Design, vol. 1 and 2*, Kraus International, 1987.
13. R. J. Huzzard, and S. Blackburn, *Proc of the 1st European Conference on Powder Injection Moulding*, Munich, October 1997, 243–250.

Influence of Additives to Ag Based Electrode Pastes on Bonding to PZT

L. J. ECCLESTONE, I. M. REANEY and W. E. LEE

University of Sheffield, Dept. of Engineering Materials, Sheffield S1 3JD, UK

ABSTRACT

Metallised commercial samples of lead zirconate titanate (PZT) were examined by electron microscopy to determine the physical and chemical bonding mechanisms present. A dense interlayer was observed between the PZT and the Ag based electrode, arising from the bonding additives in the electrode composition. The sessile drop technique was used to investigate the wetting characteristics of pure Ag on PZT substrates. In general, the wetting of the PZT substrate by pure Ag increased with temperature. SEM analysis revealed localised migration of both the ceramic species (Pb, Zr, Ti) into the sessile drop, and Ag into the PZT. Ag based electrode pastes containing Ag alone, Bi_2O_3 + CuO, Bi_2O_3, CuO, glass frit, and glass frit + Bi_2O_3 + CuO, were fired on to PZT substrates using firing schedules comparable to those used commercially. The bonding additives were found to be essential for good adhesion, with poor bonding of both the pure Ag paste, and the pure Ag sessile drops on PZT.

1. INTRODUCTION

Piezoelectric lead zirconate titanate (PZT) is used extensively in sensing applications and is connected to electrical systems via metal electrodes. A major consideration in electrode systems for piezoelectric applications is that the electrode is 'compliant', able to accommodate the cyclic volume changes expected in the lifetime of such a device. Ag based electrodes are commonly used because of their high electrical conductivity and ability to be fired in air without oxidation.

Many of the properties, such as hardness and refractoriness, that make ceramics of technological importance, are due to the ionic/covalent nature of their bonding. The lack of delocalised electrons (abundant in metals) means that metal–ceramic interfaces can be energetically unfavourable.[1] A number of methods can be utilised to overcome this problem. In structural ceramic interfaces an active metal is often used as an interlayer (active metal brazing).[2] In the application of metal electrodes to electroceramics, the metal powder is combined with various additives in a viscous paste. These additions include bonding agents (glass frits, reactive oxides and fluxing agents), and organic based vehicles and dispersants to control paste rheology for deposition by screen printing or spraying.

In *frit bonded* systems the role of the glass is to melt, be drawn to the electrode–ceramic interface, wet the ceramic, and form a physical bond with the network of metal particles forming the conductive volume of the electrode. In *flux bonded* systems, the role of the flux is to melt, be drawn towards the electrode–ceramic interface, to wet and/or penetrate into the grain boundaries of the ceramic, forming a physical bond with the electrode. In *reactively bonded* systems, a reactive oxide is added in order to facilitate a reaction, forming a product at the interface, and a chemical bond. *Mixed bonded* systems are probably the most commonly used electrode pastes over a wide range of applications. These contain a mixture of at least 2 (usually all) of the 3 different types of bonding agents described above.[3]

Most development of thick film electrodes has utilised glass frit bonded systems. Typical substrate materials used in these studies are ceramics for printed circuit board substrates, notably Al_2O_3 or AlN[4–12] along with pastes based on Ag, Ag/Pd, Cu and Au. Recent developments in electrode systems for multilayer capacitors are towards the use of cheaper, base metal systems such as Ni.[13]

Glass frit bonded thick films have been studied with regard to sintering behaviour, microstructural development, Ag/glass ratio, and Ag/glass powder characteristics.[6–8] The Ag/glass ratio was found to be especially important in terms of densification and electrical conductivity. Additions of glass can effectively increase the conductivity of the thick film,[6] as indicated for Ag based pastes containing different amounts of lead bismuth borosilicate glass on 96% alumina, using identical firing schedules (peak temperature 800°C). At low glass concentrations (5 vol.%), sintering is controlled by grain-boundary diffusion and shape accommodation, leaving the microstructure of the fired paste with many pores. At 10 vol.% glass additions, the liquid (glass) phase accelerates the re-packing of the metal particles by active rearrangement, resulting in rapid densification, high homogeneity, and a high contact area. At high glass concentrations (>30 vol.%) the densification rate increases still further, but the glass starts to interrupt the effective contact of the metal grains.

Recent research has examined electroceramic substrates such as $BaTiO_3$, $CaTiO_3$, and Al_2O_3[14–16] with binary glasses alone ($PbO–B_2O_3$, $PbO–SiO_2$, $Bi_2O_3–B_2O_3$). A grain boundary penetration bonding mechanism of Bi_2O_3 containing glass was seen in $BaTiO_3$, and it was suggested that this would be detrimental to the dielectric properties of the perovskite.[14] Investigations of the effect of oxide additions to electrode metal pastes are more scarce. In a study of a Cu electrode system on alumina, Bi_2O_3 was effective as a bond former leading to both penetrative and reactive bonds with the substrate. Cu_2O and PbO additives were found to be less effective.[14]

A system of classifying metal-ceramic interfaces[19] is shown in Fig. 1, although more than one such mechanism can be at work in any particular interface. The sessile drop technique is used extensively to investigate the surface energetics of a liquid drop resting on a horizontal solid surface and was developed in the late 19th century.[20] Young's equation describes the forces at work at the vapour-liquid-solid intersection in such a system at equilibrium:

$$\gamma_{sv} = \gamma_{lv} \cos \theta + \gamma_{sl} \tag{1}$$

where γ_{sv} is the solid/vapour interfacial energy, γ_{lv} is the liquid/vapour interfacial energy, γ_{sl} solid/liquid interfacial energy, and θ is the contact angle. If θ>90° then the drop is said to be non-wetting of the solid, if θ<90° then the liquid is said to be partially wetting. Only if θ=0° is the liquid drop said to completely wet the solid.

In this paper the need for bonding additives in Ag electrode pastes is illustrated, and an initial attempt made to correlate the mechanisms of bonding of these additives to the microstructures of the interfaces with PZT.

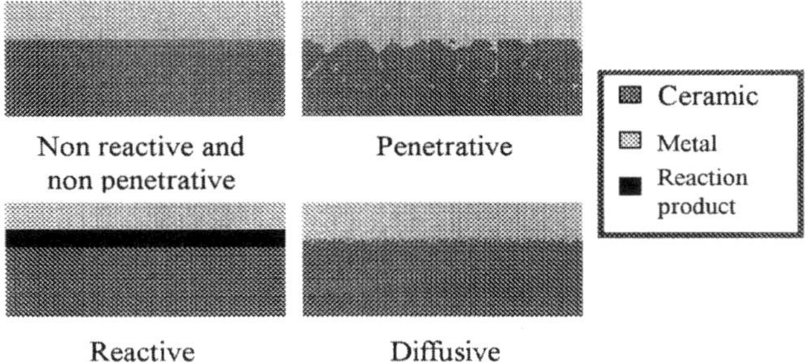

Figure 1 Classification of the metal–ceramic interface.[18]

2. EXPERIMENTAL TECHNIQUES

The scanning electron microscopes (SEM) used were a Camscan series 2A (operating at 20kV), and a JEOL–6400 (operating at 20kv), both with Link Analytical energy dispersive spectroscopy (EDS) systems, and capable of secondary electron imaging (SEI), and backscattered electron imaging (BEI). A probe size of >0.05μm diameter was used and it is likely that this would extend ~1μm in width and ~1μm in depth on interaction with the sample, the sampling volume for X-ray analysis is slightly less than this. Cross-sectional samples were sectioned using a diamond saw, ground with 800 and 1200 grade SiC papers, and finished with 6μm, 3μm, and 1μm diamond pastes. The specimens were then sputter coated with carbon to avoid charging. When PZT was prepared as a substrate, 800 and 1200 grade SiC papers were used, followed by 6μm, 3μm, 1μm and 0.25μm diamond pastes.

Cross sectional samples of commercial pre-metallised PZT, with a sprayed electrode, fired on at a peak temperature of ~750°C (no dwell) with a heating/cooling rate of ~10°C min^{-1}. (supplied by Morgan Matroc – Unilator division, Wrexham, Clywd, Wales) were prepared for SEM and EDS analysis.

The sessile drop technique was used to determine the contact angles of molten Ag (99.99% pure) on polished substrates of PZT. Both substrate and Ag were ultrasonically cleaned in methylated spirit prior to use. The sessile drop furnace configuration is shown in Fig. 2. Sessile drops of Ag on PZT, were photographed through a telephoto lens, at regular intervals up to 90 mins at temperatures of 996°C, 1020°C and 1045°C in air. A heating rate of ~15°C min^{-1} was used, and the furnace was allowed to cool to room temperature after the hold at temperature. The contact angles were determined by measuring the drop dimensions, and using the method described by Ellefson and Taylor[21] and the tables of Bashforth and Adams.[20] Cross sectional samples were prepared for SEM and EDS analysis using the method described above.

Electrode metal pastes 1–6 (Table 1) were applied to polished substrates of PZT. The samples were held at 150°C for 15 minutes to burn off the organic vehicle (binders) and dry the paste, before being raised to 700°C (for pastes 1–4) or 790°C (for pastes 5–6) at 10°C min^{-1} and holding for 1 hour, before cooling to room temperature at 10°C min^{-1}. A peak temperature of 790°C was used for pastes 5–6 because this was the recommended firing temperature for the glass frit. The frit used in the model pastes was a glass containing ~10wt.% SiO_2,

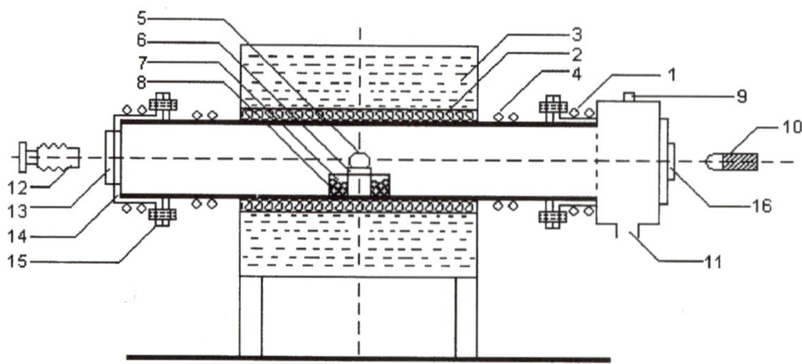

Figure 2 Sessile drop apparatus used for measurements: (1) furnace tube, (2) nichrome wire heating element, (3) insulation, (4) cooling pipes, (5) sessile drop, (6) PZT substrate, (7) alumina boat, (8) alumina support, (9) air admittance valve, (10) fibre optic illuminator (11) exit to diffusion and rotary pumps, (12) camera unit, (13) observation window, (14) water cooled end flange, (15) location of O-ring seal to furnace tube, (16) illumination window.

Table 1 Model electrode paste systems.

Paste number	Composition
1	Ag + organic binders
2	Ag + Bi_2O_3(5wt.%) + organic binders
3	Ag + CuO(1wt.%) + organic binders
4	Ag + Bi_2O_3(5wt.%) + CuO(1wt.%) + organic binders
5	Ag + glass frit(~8wt.%) + organic binders
6	Ag + Bi_2O_3(5wt.%) + CuO(1wt.%) + glass frit(~8wt.%) + organic binders

~13wt.% Al_2O_3, ~62wt.% PbO, and ~13wt.% B_2O_3. The organic vehicle (25–30 wt.%) used contained two hydrocarbon resins, one cellulose resin, and two different grades of glycol solvents. The electrode pastes/glass frit were supplied by Gwent Electronic Materials, Pontypool, Gwent, Wales.

3. RESULTS AND DISCUSSION

3.1 Commercial Electroded PZT

The sample of pre-metallised commercial PZT was examined to form a basis to which model pastes could be compared. In this mix bonded system the dominant mechanism is glass frit (lead borosilicate) bonding – a dense interlayer (~5–10μm thick) with bright Pb/Bi rich additions is present at the interface (shown in Fig. 3). EDS analysis of the interlayer shows a composition similar to that of the bulk PZT, but with low levels of Cu. The glass melts at

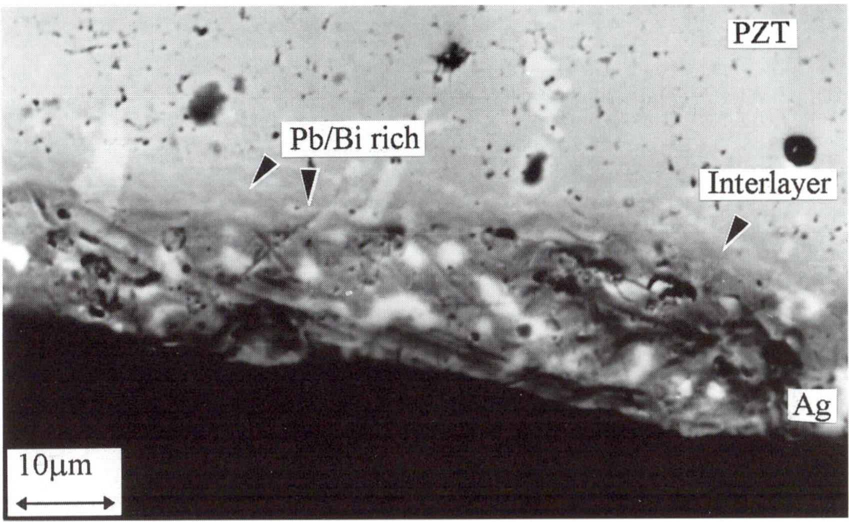

Figure 3 Backscattered electron image (BEI) of a commercial sample of PZT, the Ag electrode is frit/flux/reactive bonded.

temperature, and is drawn to the interface by gravity/capillary action. Reaction occurs at the interface, the resultant excess free energy providing the driving force for wetting. Pb rich phases with bright contrast occurred throughout the electrode with Ag, Bi, Zr, Ti, and Si often present in EDS spectra.

3.2 Sessile Drop Data

The sessile drop method was performed to confirm the non wetting behaviour of Ag on PZT. The experiments were performed at temperatures significantly higher than those used in the commercial firing of electrode paste systems. This was necessary in order to form a molten sessile drop (Ag melting temperature ~969°C). The contact angle data for pure Ag on PZT are shown in Fig. 4(a). Ag is seen to be non-wetting (0>90°) on PZT at temperatures of 996°C–1050°C in air. There was, however, a steady decrease in θ with increasing temperature. Such a trend arises from changes in metal and ceramic surface energies, and from faster interfacial reactions occurring at higher temperatures, rather than a fundamental difference in the mechanisms at work, such as the breakdown of an oxide film (as seen for example in ceramic–Al systems at ~900°C).[22–24] The early fluctuations in contact angle are possibly due to a lag in attaining equilibrium temperature in the furnace.

SEM was used to determine the possible adhesion mechanisms in the pure Ag/PZT system. EDS analysis of the interfacial volume revealed migration of ceramic species (Pb, Zr, and Ti) ~1μm into the metal, and migration of Ag ~1μm into the PZT. EDS traces of volumes 0.5μm either side of the interface are shown in Fig. 4(b). Similar migration is seen for the model electrode pastes, and the commercial premetallised PZT, fired at significantly lower temperatures (~700°C) than the sessile drop experiments. The penetration of silver by diffusion from electrode pastes into PZT during metallisation has been studied by Slinkina *et al.*[25]

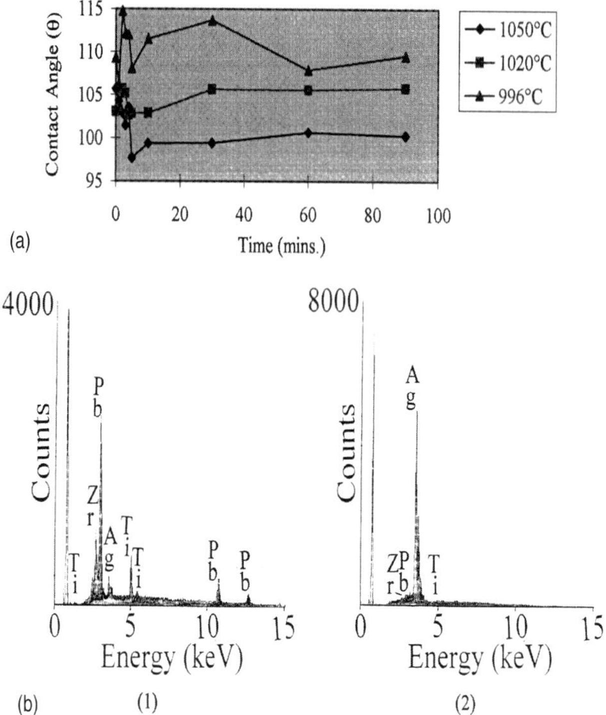

Figure 4 (a) contact angles of pure Ag on PZT with varying time and temperature, (b) EDS spectra of volumes 1/2μm either side of the interface between a Ag sessile drop and PZT, showing migration of Ag into the PZT(1), and Pb/Zr/Ti into the Ag(2).

between 650°C and 750°C by the radio tracer method. The method detected the diffusion of silver up to ~1000μm into the ceramic at 750°C, with the Ag concentration varying from ~$2x10^{19}$ at. cm^{-3} at a depth of ~1000μm, to ~$2x10^{18}$ at. cm^{-3} at the interface. X-ray microanalysis (detection limit > 0. 1 wt.% Ag) did not detect Ag in the ceramic. This may explain why in the present study Ag migration was only observed in volumes local to the interface.

The behaviour of Ag sessile drops could not be compared at 700°C since it is below the melting point of Ag (969°C), so no droplet would form. The bonding mechanism may well be different in the sessile drop samples where the pure metal is liquid, to the commercial systems where the Ag has many additives present, and at 700°C is not believed to melt.

The sessile drops (pure Ag) and paste 1 (pure Ag + organics) did not adhere well. Samples of these had to be mounted in resin to avoid destruction, a problem not encountered in the pastes with bonding additives. This illustrates the importance of paste additives on the bonding of electrode metals to PZT.

3.3 Model Electrode Pastes on PZT

Model electrode pastes (Table 1) were used in order to begin to correlate the bonding mechanisms to the additives present in the pastes. Results are summarised in Table 2 and Figures 5 and 6 show the interfaces between PZT and various model electrode pastes after 1 hour at

Table 2 Summary of microstructural observations in model electrode paste/PZT systems.

Electrode paste 1 (Ag) Fig. 5(a)	Electrode paste 2 (Ag + Bi_2O_3) Fig. 5(b)
• Extreme delamination and poor adhesion. • Ag porous.	• Limited penetration of the Bi phases into the ceramic • Bi_2O_3 abundant along the interface, and with Pb, Zr and Ti is also present in bright phases remote to it.
Electrode paste 3 (Ag + CuO) Fig. 5(c)	**Electrode paste 4 (Ag + Bi_2O_3 + CuO) Fig. 5(d)**
• Non continuous dark reaction product along the interface – Cu containing, but Pb based. • Significant quantities of unreacted CuO remain throughout the electrode microstructure.	• No evidence of penetration into the ceramic by Bi rich phases. • Bi_2O_3/CuO found at the interface, though not in any obvious reaction product. • Unreacted CuO abundant throughout the electrode, as were Pb/Zr/Ti/Cu containing Bi rich phases.
Electrode paste 5 (Ag + glass) Fig. 6(a)	**Electrode paste 6 (Ag + Bi_2O_3 + CuO + glass) Fig. 6(b)**
• Low electrical porosity, especially at interface. • Pb/Al rich phases concentrated towards the electrode–ceramic interface. • Al present in bulk Ag throughout the electrode microstructure, and to a depth of ~2μm in the ceramic.	• Porosity throughout the electrode volume. • Bi_2O_3 detectable only in low levels at the interface. Undetectable in bulk electrode/ceramic. • Low levels of Cu local to the interface, though not as an obvious reaction product. • Al present with Ag throughout the electrode, and to a depth of ~2μm in the ceramic.

700°C (pastes 1–4) or 790°C (pastes 5–6). There is evidence of grain boundary penetration of Bi_2O_3 (melting temperature ~825°C) into the PZT network to a depth of ~2μm (about 1 grain diameter) for paste 2. Bi_2O_3 is present at the interface intermittently (and non-penetratively) in paste 4 , but continuously in paste 2. The presence of these Bi rich liquid phases at temperatures lower than the melting point is probably due to the PbO–Bi_2O_3 eutectic that allows a liquid phase to form above ~600°C.[26]

The Bi_2O_3 phase appears to encourage transport of the elements originating from the ceramic (especially Pb) throughout the electrode microstructure. This increased mass transport is only seen in systems fluxed with Bi_2O_3.

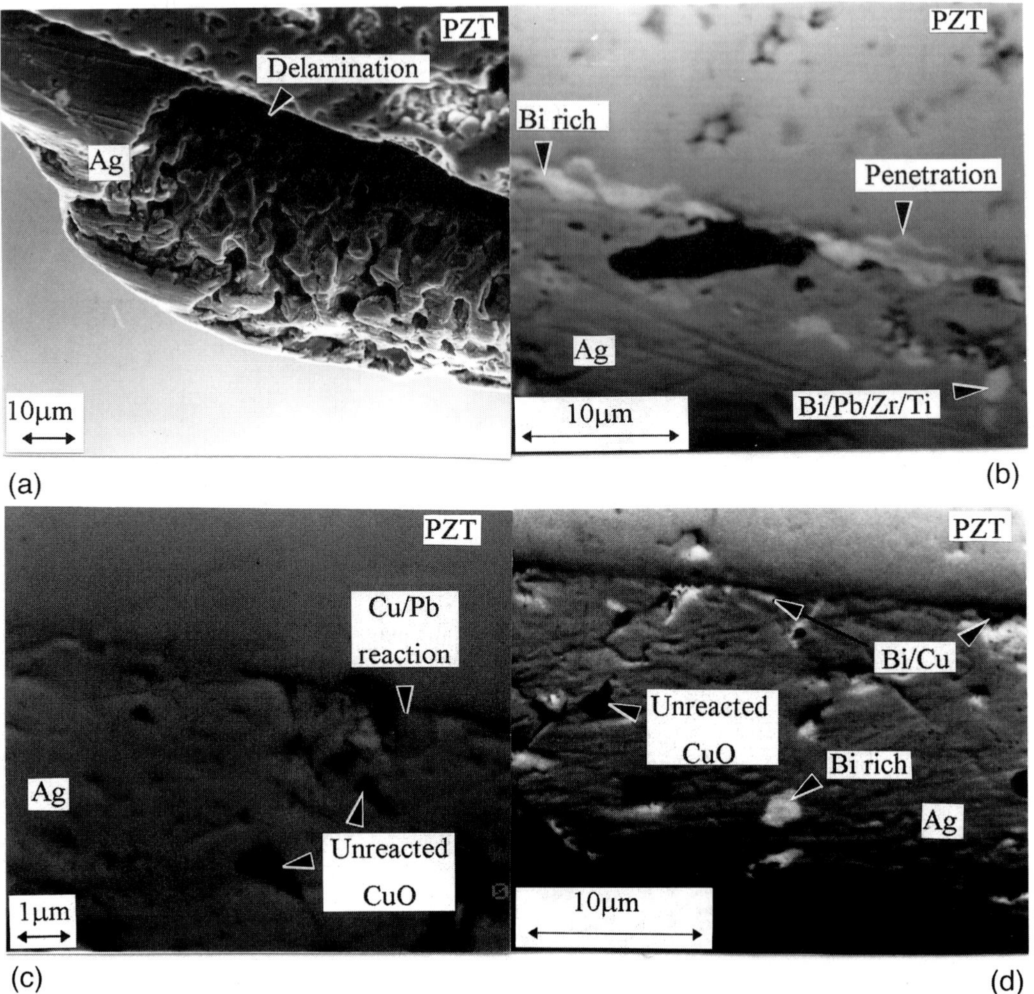

Figure 5 SEM micrographs of electrode pastes fired onto PZT at 700°C for 1 hour. (a) SEI of PZT/paste 1, (b) BEI of PZT/paste 2, (c) BEI of PZT/paste 3, (d) BEI of PZT/paste 4.

4. FURTHER DISCUSSION

The function of these additives and their response both individually and synergistically will now be discussed. The role of the CuO appears to be to partially reduce and react with the substrate. The function of the Bi_2O_3 appears to be to melt and penetrate into the grain boundaries or react with the ceramic substrate. The role of the frit also is to melt, wet and (chemically) bond with the substrate, and form a mechanically interlocked physical bond with the electrode metal network.

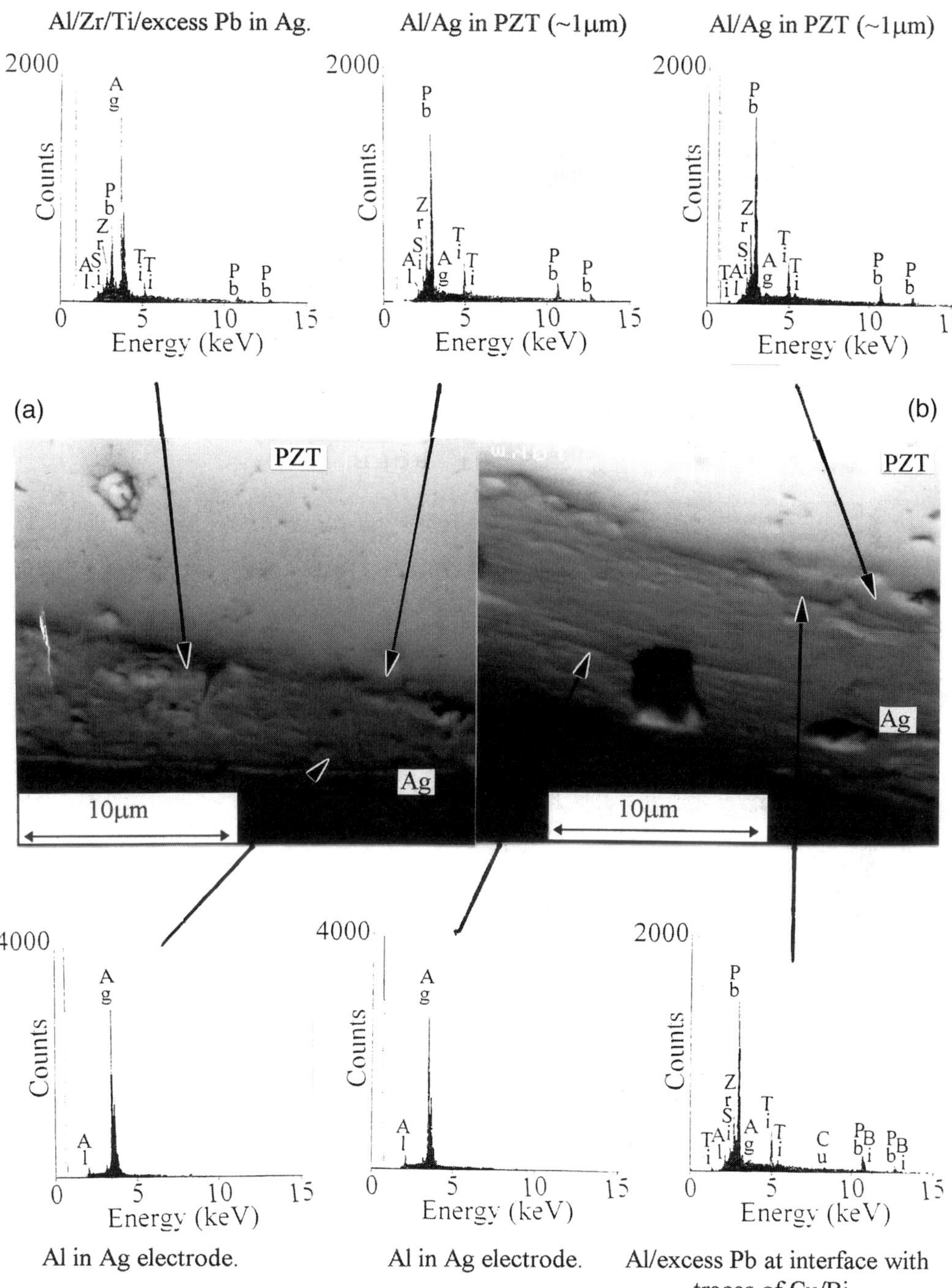

Figure 6 SEM micrographs of electrode pastes fired onto PZT at 790 °C for 1 hour. (a) BEI of paste5/PZT, (b) BEI of paste 6/PZT, showing EDS traces from the regions indicated.

Commercial PZT (Fig. 3) relies on glass frit (lead borosilicate) bonding, with a non-porous interlayer attributable to the glass/Bi rich phases. It is likely that on firing the glass melts, dissolves CuO and Bi_2O_3 additions and gets drawn towards the ceramic via the metal network (partially densified porous Ag particles), by a combination of gravity and capillary action through the pores, before wetting and bonding with the ceramic substrate.

Unfritted electrode pastes are often fired on to substrates at ~1000°C, a higher temperature than that required for fritted pastes (~700–800°C). On attempting to fire these model Ag-containing electrodes at ~1000°C it was found that melting of the Ag particles led to their balling up on the substrate. Consequently, such films need to be fired at lower temperatures than this (~850–950°C) to avoid melting of the Ag particles. To form a chemical bond between metal and ceramic requires a metal-metal oxide link. Such a link between a metal electrode and an electroceramic requires that some of the metal oxide additives in the electrode paste partially reduce to metal, and form an alloy type shared-electron bond with the electrode metal (Ag in this study). Metal-oxides may also bond with the ceramic by penetration or chemical reaction. The electrode paste of this study contains Bi_2O_3, CuO and a lead boro-aluminosilicate glass. Pure CuO is known to melt and decompose at 1026°C, and when exposed to temperatures close to this has a lower oxidation state than it has at lower temperatures. Oxide reduction occurs at the firing temperatures at which maximum adhesion occurs.[27] At lower temperatures, such as the 700°C/790°C range used for samples investigated to date, CuO will have a higher oxidation state, which may account for the apparent lack of continuity of the reaction product for paste 3 (Fig. 5(c)).

In unfritted systems (pastes 1–4) at 700°C, gravity cannot be relied upon to draw the CuO to the interface (Fig. 5(c)). Riemer,[27] in using Au films on alumina, suggests that the excellent compatibility between the bonding additives (CuO) and the metal at the temperature of decomposition results in gravity being ineffective in bringing the reactive additives to the interface. This can result in an oxide layer over the Au thick film attributable to the reactive bonding additives, which has a detrimental effect on the solderability of the electrode to wires connecting the component to the active system.

The present study shows Bi_2O_3 to be ineffective in the role of carrying the reactive bonding agent (CuO) to the interface (Fig. 5(d)). This leaves significant amounts of unreacted CuO in the bulk electrode. This was not the case for paste 6, where no CuO pockets were left within the electrode microstructure. Low levels of Cu (or CuO) were found by EDS local to the interface, together with Al, with no obvious reaction product. It is possible that the CuO had simply dissolved in the glass frit, and been transported to the interface in this melt.

Although the presence of glass frit enables Bi_2O_3 and CuO to be transported to the interface more easily, this does not necessarily mean that the three mechanisms of bonding (frit bonding, flux bonding, and reactive bonding) operate simultaneously. The CuO and Bi_2O_3 dissolving into the glass added for frit bonding, often means that the bonding mechanism is simply frit bonding but by a different glass composition to that originally added.[3] This may be a consequence of the empirical way in which the electrode paste technology has evolved – the theory that if one bonding mechanism works well, then all three should work better, does not necessarily hold.

Of the fritted systems, the fired electrode with paste 5 is notably denser than that with paste 6, and appears the densest of all of the 6 tested. The addition of glass frit in the ratios used is known to improve densification of the electrode.[5–7] Paste 6 has significantly lower levels of Bi_2O_3 within the electrode microstructure. Bi_2O_3 is noted for being volatile, and studies on Bi_2O_3 loss from another electroceramic, during firing of ZnO varistors have been performed from 900–1100°C.[28] Within 10 minutes of reaching the holding temperature there was weight loss due to the evaporation of Bi_2O_3. Considering that this loss was from a bulk sample, rather than a 10μm thick layer on a substrate, the probability of Bi_2O_3 loss is likely to be higher at similar temperatures. While the firing of electrode pastes and the production method of ZnO varistors are different, evaporation of Bi_2O_3 would provide a credible reason for the low Bi content in the interfacial microstructure of the fritted model electrode paste.

The concentration of Pb, Al, and Si rich phases towards the electrode/ceramic interface supports the idea of the glass melting at temperature and being drawn to the interface. The presence of Al in EDS spectra from ~2μm into the PZT suggests an interaction between the PZT and the glass frit. The glass frit is a possible source for the Al detected throughout the electrode microstructure for both pastes 5 and 6, in areas of little detectable Pb/Si content.

5. CONCLUSIONS

1. Pure Ag does not wet PZT in air at temperatures of 996°C–1050°C, although there is a steady decrease in contact angle with increasing temperature.
2. Ag alone (either pure or with organics) does not adhere well to PZT. The bonding additives are vital for reasonable adherence.
3. Penetrative bonding is not the major mechanism in the model electrode pastes/PZT systems studied. However, in Bi_2O_3 containing systems fired at 700°C, Bi rich phases are always present in interfacial regions. In the fritted systems, fired at 790°C, the Bi content is lower than Bi_2O_3 containing unfritted pastes fired at 700°C.
4. Bi rich phases are not restricted to interfacial areas, both in fritted and non-fritted systems.
5. Pb, Zr, and Ti dissolve readily into Bi_2O_3 and through it are transported to areas remote to the interface.
6. CuO is inefficient at reaching the interface in non-fritted systems. Reaction with the PZT can occur when it does reach the interface, even at 700 °C.
7 The glass frit is drawn towards the interface, with the frit content higher at the interface compared to the outer regions of the electrode.
8. In fritted systems, CuO is dissolved by the glass frit. The effectiveness of the CuO as a reactive agent is unclear.
9. Commercial pre-metallised PZT exhibited a dense interlayer (attributable to the glass frit), that also contained Bi rich additions, and low levels of Cu (dissolved by the glass and drawn to the interface).

REFERENCES

1. M. G. Nicholas, 'Ceramic-Metal Interfaces', *Science of Ceramic Interfaces*, J. Nowotny ed., Elsevier science publishers, Amsterdam, London, 1991, 411–437.

2. K. Suganama, 'Recent Advances in Joining Technology of Ceramics to Metals', *Iron and Steel Institute of Japan International*, 1990, **30** (12), 1046–1058.
3 R. W. Vest, 'Materials Science of Thick Film Technology', *Bull. Amer. Ceram. Soc.*, 1986, **65** (4), 631–636.
4. T. Yamaguchi, S. Takeoka, T. Iizuka, and T. Nakano, 'Glass-Substrate Interaction in Thick Film Technology', *Scripta Metallurgica et Materialia*, 1994, **31**(8), 1013–1018.
5. Y. Kurihara, S. Takahashi, K.Yamada, and T. Endoh, 'Ag-Pd Thick Film Conductor for AlN Ceramics', *IEEE Trans. Comp., Hybrids, Manuf. Technol.*, 1990, **13**(2), 306–312.
6. Y. S. Chung and H. Kim, 'Effect of Oxide Glass on the Sintering Behaviour and Electrical properties in Ag Thick Films', *IEEE Trans. Comp., Hybrids, Manuf Technol.*, 1988, **11**(2), 195–199.
7 K. Yajima and T. Yamaguchi, 'Sintering and Microstructure Development of Glass-Bonded Silver Thick Films', *J. Mater. Sci.*, 1984, **19,** 777–784.
8. K. Yajima and T. Yamaguchi, 'A Further Study on the Microstructure of Glass -Bonded Ag Thick-Film Conductors', *IEEE Trans. Comp., Hybrids, Manuf. Technol.*, 1984, **7**(3), 281–285.
9 C. E. Newsberg and S. H. Risbud, 'Binder Chemistry, Adhesion and Structure of Interaces in Thick-Film Metallized Aluminium Nitride Substrates', *J. Mater. Sci.*, 1992, **27**, 2670–2676.
10. M. G. Norton, 'Glass-Ceramic Interactions and Thick-Film Metallization of Aluminium Nitride', *J. Mater. Sci.*, 1991, **26**, 2322–2328.
11 T. Yamaguchi and M. Kageyama, 'Oxidation Behaviour of AlN in the Presence of Oxide and Glass for Thick Film Applications', *IEEE Trans. Comp., Hybrids, Manuf. Technol.*, 1989, **12** (3), 402–405.
12. N. Iwase, K. Anzm, K. Shinozaki, O. Hirao, T. D. Thanh, and Y. Sugiura, 'Thick Film and Direct Bond Copper Forming Technologies for Aluminium Nitride Substrate', *IEEE Trans. Comp., Hybrids, Manuf Technol.*, 1985, **8** (2), 253–258.
13. Y. Sakabe, 'Development of the Multilayer Ceramic Capacitor with Base-Metal Electrode', *Bull. Amer. Ceram. Soc.*, 1998, **77** (8), 57.
14. Y. Kuromitsu, S. F. Fang, S. Yoshikawa, and R. E. Newnham, 'Evolution of Interfacial Microstructure between Barium Titanate and Binary Glasses', *J. Am. Ceram. Soc.*, 1994, **77** (3), 852–856.
15. Y. Kuromitsu, S. F. Fang, S. Yoshikawa, and R. E. Newnham, 'Interaction between Barium Titanate and Binary Glasses', *J. Am. Ceram. Soc.*, 1994, **77**(2), 493–498.
16. Y. Kuromitsu, H. Yoshida, H. Takebe, and K. Morinaga, 'Interaction between Alumina and Binary Glasses', *J. Am. Ceram. Soc.*, 1997, **80**(6), 1583–1587.
17. T. Yamaguchi, H. Ayaki, and L. Asai, 'Reaction of $CaTiO_3$ with $PbO–B_2O_3$ and $PbO–SiO_3$ Glasses', *J. Am. Ceram. Soc.*, 1993, **76** (4), 993–997.
18. T Ogawa, M. Ootani, T. Asai, M. Hasegawa and O. Ito, 'Effect of Inorganic Binders on the Properties of Thick Film Copper Conductor', *IEEE Trans. Comp., Pack., Manuf. Technol. Part A*, Dec. 1994, **17** (4), 625–630.
19. T. Okamoto, 'Interfacial Structure of Metal-Ceramic Joints', *Iron and Steel Institute of Japan International*, 1990, **30** (12), 1033–1040.
20. F. Bashforth and J. C. Adams, *An Attempt to Test the Theory of Capillary Action*, Cambridge Univ. Press, London, 1883.
21. B. S. Ellefson and N. W. Taylor, 'Surface Properties of Fused Salts and Glasses: I Sessile-Drop Method for Determining Surface Tension and Density of Viscous Liquids at High Temperatures', *J. Am. Ceram. Soc.*, 1938, **21**, 193–205.
22. W. Kohler, 'Untersuchungen zur Benetzung von Al_2O_3-und SiC-Kristallen durch Al und Al-Legierungen', *Aluminium*, 1975, **51**, 433–447.
23. H. John and H. Hausner, 'Influence of Oxygen Partial Pressure on the Wetting Behaviour in the system Al/ Al_2O_3', *J. Mater. Sci. Lett.*, 1986, **5**, 549–551.
24. S. Y. Oh, J. A. Cornie and K. C. Russell, 'Wetting of Ceramic Particulates with Liquid Aluminium Alloys: Part II. Study of Wettability', *Met. Trans. A*, Mar. 1989, **20A,** 533–541.
25. M. V. Slinkina, G. I. Dontsov, and V. M. Zhukovsky, 'Diffusional Penetration of Silver from Electrodes into PZT Ceramics', *J Mater. Sci.*, 1993, **28**, 5189–5192.

26. E. M. Levin, H. F. McMurdie and F. P. Hall, *Phase diagrams for ceramists*, Figure 97, American Ceramic Society, Westerville, Ohio, USA, 1956.
27. D. Eriemer, 'High-Adhesion Thick-Film Gold Without Glass or Metal-Oxide Powder Additives', *IEEE Trans. Comp., Hybrids, Manuf Technol.*, 1985, **8** (4), 195–199.
28. M. Scastro, C. M. Aldao and J. M. Porto Lopez, 'Cobalt Oxide and Antimony Oxide Effects in the Zinc Oxide–Bismuth Oxide System', *Materials Research Bulletin*, 1994, **29** (12), 1287–1295.

REFRACTORIES

The Fracture Behaviour of Dual Phase Composite Refractories

R. J. HENDERSON and H. W. CHANDLER

Department of Engineering, University of Aberdeen, Aberdeen AB9 2UE, UK

ABSTRACT

Dual phase refractories, where residual stresses are present at the microstructural level, can exhibit non-linear stress-strain behaviour accompanied by permanent deformation. This non-linear deformation gives rise to increasing toughness as crack propagation occurs. In this paper a micromechanical model, combining a specifically developed constitutive model for ceramic composites and a model for transformation toughening is presented. This model will be shown to be capable of simulating toughness curves in dual phase refractories.

1. INTRODUCTION

Many ceramics, and in particular coarse grained refractories, exhibit non-linear stress-strain behaviour accompanied by permanent deformation upon unloading. In the materials we tested, where this behaviour occurred, the toughness increased with crack propagation. The shape of these toughness curves determines the onset of brittle fracture under mechanical and thermal loading. Therefore the optimisation of this toughening behaviour offers opportunities for improving the thermal shock resistance of refractory materials.

Many mechanisms, including crack bridging,[1,2] microcracking,[3] transformation toughening[4] and the release of residual stresses,[5] have been proposed to explain this increase in toughness. Recently many authors have favoured crack bridging as the dominant mechanism for this behaviour. For crack bridging to occur microcracks must form ahead of, and to the side of, the putative main crack path. In this paper, however, it will be shown that as well as encouraging this pattern of cracking, the release of residual stresses accompanied by microcracking is capable of explaining this behaviour in coarse grained refractories without the requirement of bridging.

In this paper the stress-strain response and toughness curves of two commercial refractories is compared. One of these materials has significant residual stresses present within the microstructure, whereas the other has a nominally stress free microstructure. An existing model for transformation toughening is modified to take into account the release of residual stresses and the toughness curves of the two materials are predicted using information obtained from a previously reported constitutive model.

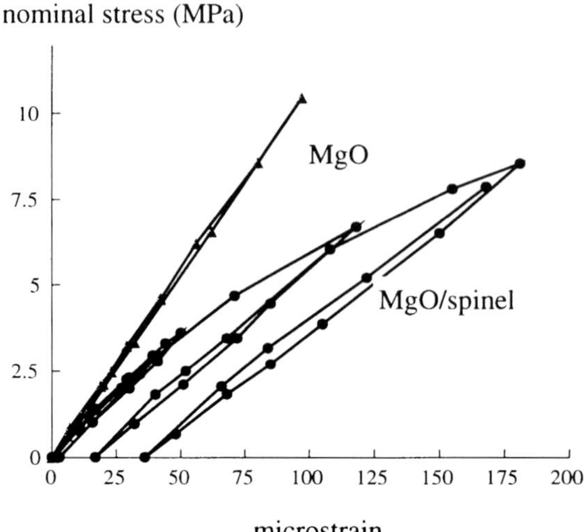

Figure 1 Stress–strain behaviour of the magnesia and magnesia spinel composites. The strain is measured on the component and the nominal stress is calculated from linear elasticity using beam theory.

2. EXPERIMENTAL

2.1 Materials

Two commercially available refractories were investigated; a magnesia composite consisting of large grains of magnesia in a fine grained magnesia matrix, and a very similar material with a small addition of spinel made to the matrix. The spinel modified matrix has a different thermal expansion coefficient to the magnesia grain and thus residual stresses are present within this material at room temperature.

2.2 Stress-Strain Behaviour

The stress-strain behaviour of the materials was obtained in four point flexure tests the strain being measured by strain gauges attached to the ceramic and the stress calculated from linear elasticity. The stress-strain response of the two refractories are shown in Fig. 1. It should be noted that the magnesia material exhibits a linear stress-strain response prior to fracture. However, the magnesia-spinel material has a non-linear response and significant permanent deformation upon unloading.

2.3 Toughness behaviour

A modified double cantilever beam geometry was used to obtain the toughness behaviour of these materials.[6] The toughness behaviour of the magnesia and magnesia spinel materials are shown in Fig. 2. As may be expected from the stress-strain curves of these materials the pure magnesia refractory has a constant value of toughness whilst the magnesia-spinel material exhibits rising toughness with crack propagation.

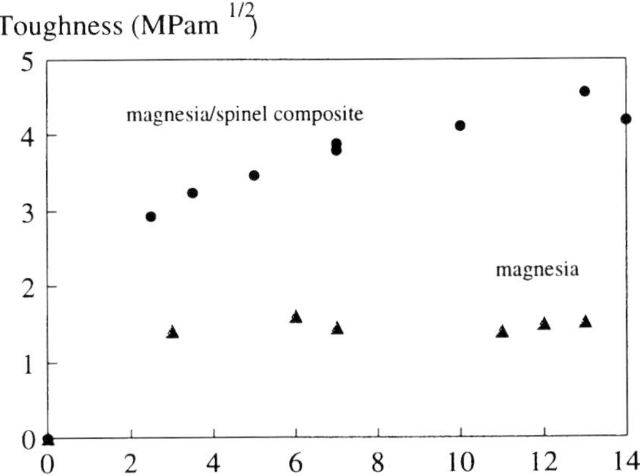

Figure 2 Toughness curves obtained from a modified cantilever beam geometry for the two refractories.

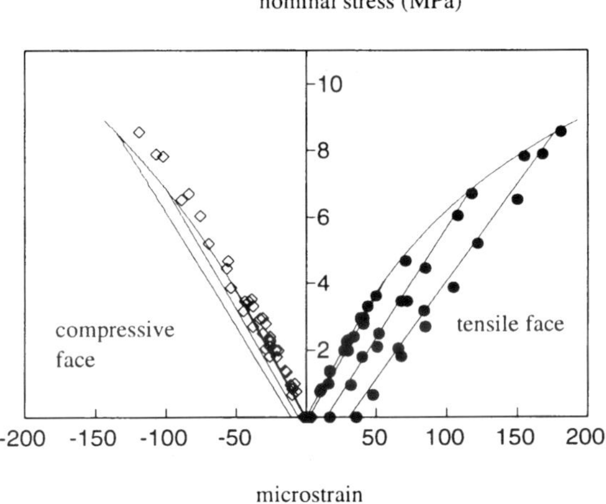

Figure 3 Comparison between the constitutive model (lines) and the experimental stress-strain results.

2.4 Modelling the mechanical response

Many mechanisms have been proposed which would be capable of explaining the toughness behaviour of these materials. However few of these mechanisms can explain both the constitutive and toughening behaviour observed in these dual phase refractories.

The simplest model to explain this behaviour is that the toughening behaviour is a result of crack bridging[7] and the permanent deformation upon unloading is a result of the crack faces being wedged open by debris. However as the crack closes one would expect the stiffness to increase as the crack faces indent more debris. This is not observed in experiments where the unloading curve is linear. One may also expect that as the two refractories are very similar they would exhibit similar mechanical behaviour if grain bridging was the dominant mechanism.

In this paper it is proposed that the mechanical behaviour can be explained by the combination of the release of residual stresses and microcracking. In materials where there is differential thermal expansion between the phases, then upon cooling from the firing temperature residual stresses will be present within the microstructure. In the magnesia-spinel system the particulate has a higher thermal expansion coefficient than the matrix[8] and therefore the particulate will be in tension and the matrix in compression. If sufficient tensile stress is then applied to the composite a crack will propagate at the interface between the matrix and particulate. This has the effect of relaxing the compressive stress within the matrix and an overall expansion of the composite occurs.

A constitutive model combining Hashins composite sphere and damage mechanics has been previously reported by the authors,[9,10] the results of which are shown in Figure 3. This model has only two adjustable parameters which control the tensile stress at which microcracking starts and finishes.

2.5 Toughness model

Various methods have been developed to model transformation toughening in zirconia ceramics.[11,12] Most of these models are based upon estimating the shielding stress intensity produced by an expansion in front of the crack tip. In these dual phase refractories a similar expansion (but of a lower magnitude) occurs. The model developed here is based upon that of Browne and Chandler[13] but is modified to include microcrack toughening.

Consider a crack propagating through a material exhibiting rising toughness. The applied stress intensity (K_{app}) required for further crack propagation depends not only upon the intrinsic toughness of the material (T_{tip}) but also upon the shielding stress intensity which results from the expansion ahead of the process zone (K_{micro}).

$$K_{app} = T_{tip} + K_{micro} \tag{1}$$

The shielding stress intensity, produced by an expansion (θ_T) ahead of the crack tip for a particulate embedded in a linear elastic matrix with Young's modulus E, has consistently been shown to be[14]

$$K_{micro} = \theta_T E \beta \sqrt{h} \tag{2}$$

where h is the width of the process zone and β is a geometrical parameter. In order to estimate time shielding stress intensity it is necessary to estimate the size of the process zone. If it is assumed that on the edge of the process zone the hydrostatic stress is equal to the critical stress required to induce microcracking σ_c , then the width of the process zone has been shown to be[14]

$$h = \Omega \left(\frac{K_{app}}{\sigma_c} \right)^2 \tag{3}$$

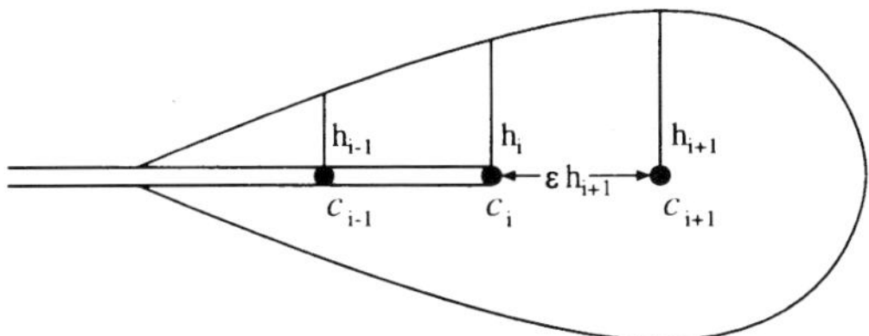

Figure 4 The sampling system used to estimate the zone width (h_i) at a set crack length (c_i). ε controls ratio of the crack length and process zone width.

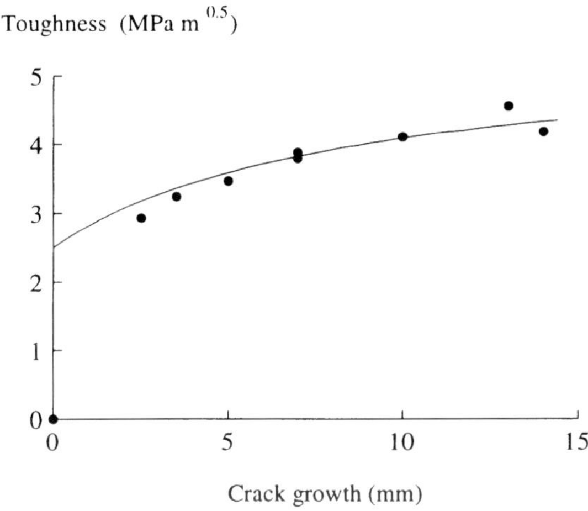

Figure 5 Comparison of the simulated toughness results and the experimental results (ε = 0.55).

where Ω is constant and σ_c is the critical stress to induce microcracking. This corresponds to the stress required to induce microcracking utilised in the constitutive model.

If the toughness increases as crack extension occurs, it is necessary to increase the applied stress intensity to obtain further crack propagation. This in turn increases the width of the process zone. Following Browne and Chandler a set of sampling points (Fig. 4) can be set up and the width of the process zone estimated at a number of crack extensions. With some modification[5] Browne and Chandler's result can be expressed in parametric form as

$$c = c_0 + \frac{b^2 \varepsilon}{(1-a)^2}\left(n + 2a^2\left(\frac{a^n - 1}{1-a}\right) + a^4\left(\frac{1-a^{2n}}{1-a^2}\right)\right) \tag{4}$$

$$\frac{K_{app}^2}{T_{tip}^2} = \left(\frac{1}{1-a}\right)^2\left(1 - 2a^{n+1} + a^{2n+2}\right) \tag{5}$$

where c is the crack length, $a = \theta_T E\beta\Omega^{1/2}/\sigma_c$, $b = T_{tip}\,\Omega^{1/2}/\sigma_c$ and n is a parameter that increases with the length of the crack. It should be noted that if $a > 1$ then the toughness will increase without bound. However if $a < 1$ an asymptotic value of toughness will be observed.

In order to model the experimental results obtained for the magnesia spinel refractory it is necessary to obtain the parameters a , b and ε. If an intrinsic value of toughness (T_{tip}) is estimated from the experimental results, both a and b can be found using information from the constitutive model. ε however is adjusted to obtain the best fit to experimental results. The results are shown in Fig. 5.

3. CONCLUSIONS

1. Coarse grained refractories where residual stresses are present at the microscale exhibit non-linear stress–strain behaviour and increasing toughness curves.
2. Both the constitutive and toughness behaviour can be explained by the release of residual stresses by microcracking.
3. Micromechanical models combining composite theory, damage mechanics and transformation toughening theory can simulate both the constitutive and toughening behaviour.
4. These micromechanical models can provide insight into the microstructural design of composite refractories.

REFERENCES

1. P. L. Swanson *et al.*, 'Crack-interface grain bridging as a fracture resistance mechanism in ceramics I, experimental study on alumina', *J. Am. Ceram. Soc.*, 1987, **70** (4), 279–89.
2. M. Gils, 'Quasi-Brittle Fracture of Ceramics', PhD thesis, Technische Universiteit Eindhoven, 1997.
3. A. G. Evans and K. T. Faber, 'Crack-growth resistance of microcracking brittle materials', *J. Am. Ceram. Soc.*, 1984, **67**, 255–60.
4. R. M. McMeeking and A. G. Evans, 'Mechanics of transformation-toughening in brittle materials', *J. Am. Ceram. Soc*, 1982, **65** (5), 242–246.
5. R. J. Henderson, 'The Thermomechanics of Dual Phase Refractories', PhD thesis, University of Aberdeen, 1997.
6. H. W. Chandler, R. J. Henderson, M. N. Al-Zubaidy, M. Saribiyk and A. Muhaidi, 'A fracture test for brittle materials', 1997, *J. European. Ceram. Soc.*, **17**, 759–763.
7. Y. Mai and B. R. Lawn, 'Crack-interface grain bridging as a fracture resistance mechanism in ceramics :II, theoretical fracture mechanics model', 1987, *J. Am. Ceram. Soc.*, **70** (4), 289–94.
8. Z. P. Chang and G. R. Barsch, 'Pressure dependence of single-crystal elastic constants and anharmonic properties of spinel', *J. Geophysical. Res.*, 1973, **78** (14), 2418–2433.
9. R. J. Henderson and H. W. Chandler, 'A model for the non-linear mechanical behaviour of refractories,' *Br. Ceram. Trans.*, 1997, **96**, 85–91.
10. R. J. Henderson and H. W. Chandler, 'The mechanical behaviour of high performance refractories', *Key Engineering Materials*, 1997, **132–136**, 504–507.
11. A. G. Evans, 'Perspective on the development of high-toughness ceramics', *J. Am. Ceram. Soc.*, **73** (2), 1990, 187–206.
12. D. M. Stump, 'The role of shear stress and shear strains in transformation-toughening', *Phil. Mag.*, 1991, **64** (4), 879–902.

13. D. J. Browne and H. W. Chandler, 'Computer simulations of R-curve behaviour in zirconia ceramics', *Br. Ceram. Proc.*, 1989, **42**, 159–165.
14. J. W. Hutchinson, 'Crack tip shielding by microcracking in brittle solids', *Acta Metall.*, 1987, **35** (7), 1605– 1619.

Dissolution of Dolomite and Doloma in a Model BOF Slag

Y. SATYOKO and W. E. LEE

Department of Engineering Materials, University of Sheffield, Sheffield S1 3JD, UK

ABSTRACT

Dissolution mechanisms of dolomite, $CaMg(CO_3)_2$, and doloma, CaO.MgO, in a model basic oxygen furnace (BOF) slag based on the CaO–MgO–SiO_2–FeO–MnO system have been characterised with regard to the reaction product phases at the dolomite/slag and doloma/slag interfaces by *post-mortem* microstructural analysis. It was found that after lh immersion at 1350°C in a stagnant molten slag, the CaO and MgO in doloma react with SiO_2 and FeO in the slag forming dicalcium silicate (2CaO.SiO_2 or C_2S), MgO-rich magnesiowüstite ((Fe,Mg)O) and dicalcium ferrite (2CaO.Fe_2O_3 or C_2F). The formation of (Fe,Mg)O and C_2F breaks up the C_2S layer, leaving it discontinuous. Similar reactions take place in dolomite, but at a later stage of dissolution since they are retarded by the intermediate decomposition of dolomite in which rapid CO_2 gas evolution quenches the resulting C_2S layer and forms a gap between the resulting doloma and slag. A more complete and continuous dicalcium silicate layer was observed in dolomite compared to doloma.

1. INTRODUCTION

Dolomite and doloma, in the form of pellets and bonded aggregates, are common basic slag additives in primary steelmaking processes, both basic oxygen furnace (BOF) and electric arc furnace (EAF). Dolomite is a naturally-occurring mineral made up of a double-carbonate $CaMg(CO_3)_2$, while doloma is the calcined product of dolomite in which the carbonates have been converted to the oxides, lime (CaO) and periclase (MgO), by the action of heat and the evolution of CO_2 gas, i.e. by a thermal decomposition reaction, in a shaft or rotary kiln at temperatures between 1400° and 1500°C. These two basic materials are used in steel refining as fluxing additions as well as lime to give a basic slag for the removal of oxidation products and sulphur. Dolomite and doloma enhance the dissolution of lime into the molten slag by forming an early basic slag at the start of refining. In general, the slag formation can be described as follows:[1]

$$[Si] + O_2 \rightarrow SiO_2$$

$$\left.\begin{array}{l} [P] + 5/4O_2 \rightarrow 1/2P_2O_5 \\ [Mn] + 1/2O_2 \rightarrow MnO \\ Fe + 1/2O_2 \rightarrow FeO \end{array}\right] \text{fluxed with CaO forming a molten slag}$$

Previous studies, for example, by Matsushima *et al.*[2] and Williams *et al.*[3] have shown that the formation of high melting point dicalcium silicate phase ($2CaO.SiO_2$, m.pt. 2130°C) at the lime/slag interface retards the dissolution of lime. The slow dissolution of lime means less efficient fluxing of impurity oxides and prolonged contact between basic refractory lining and SiO_2-rich (acid) slag resulting in faster lining corrosion. However, by making the slag basic ($CaO/SiO_2 \approx 2$–3) early in the refining process, formation of the dicalcium silicate (C_2S) phase can be suppressed. This is made possible by adding MgO (a basic oxide) in the form of dolomite and/or doloma.

The presence of MgO crystals in the CaO matrix of burnt dolomite or doloma is believed to cause the C_2S layer from the CaO–SiO_2 reaction to be discontinuous.[4] In addition, MgO has a high potential to form low melting solid solutions on reaction with iron oxide in the slag, such as magnesiowüstite, (Fe,Mg)O (melting temperatures, T_m, 1369°C for pure FeO and 2800°C for pure MgO). Thus, magnesia is readily fluxed by iron oxide forming a magnesiowüstite solid solution and liquid. This liquid then penetrates and breaks up the C_2S layer, allowing the lime to dissolve completely and serve its purpose of fluxing the impurities from the oxidation of molten steel, forming a liquid solution. Nevertheless, Green[5] reported that the level of MgO addition to slag must be controlled, since high MgO content leads to a viscous slag which means inefficient refining. The use of dolomite and/or doloma in modern steelmaking processes has been an important breakthrough in the development of lime-fluxing materials, substituting for fluorite, CaF_2, which is highly corrosive to refractories.

The mechanisms by which doloma and dolomite react with molten steelmaking slag during initial dissolution of the materials are different. However, in the later stage of dissolution when dolomite has decomposed into lime and periclase, dolomite dissolves in the same manner as does doloma. In the investigations carried out by Umakoshi *et al.*[4] on the dissolution of burnt dolomite in molten Fe_tO–CaO–SiO_2 slags, it was found that lime and periclase in the burnt dolomite react individually with the slag, forming dicalcium silicate ($2CaO.SiO_2$) and magnesiowüstite solid solution, (Fe,Mg)O, as the reaction products at the interface. In dolomite, this reaction mechanism does not occur immediately since dolomite initially undergoes thermal decomposition on exposure to heat during contact with molten slag. The reaction is endothermic and results in rapid CO_2 gas evolution. Dolomite dissolution is believed to be retarded since the evolution of CO_2 gas initially prevents direct contact and reaction between the slag and lime/periclase in the resulting doloma and both CO_2 gas evolution and the endothermic decomposition produce a cooling effect which slightly decreases the slag temperature at the interface.

The current fundamental study on the dissolution of dolomite and doloma in a BOF slag has two principal objectives, namely to characterise the mechanisms by which dolomite and doloma dissolve into a model BOF slag in the system CaO–MgO–SiO_2–FeO–MnO and to compare their rates of dissolution by means of a qualitative method employing *post-mortem* microstructural analysis.

2. EXPERIMENTAL

Samples of as-mined dolomite and as-manufactured doloma (calcined dolomite) were sup-

Table 1 Typical chemical compositions (wt.%) of Lafarge-Lime's Thrislington Dolomite and Doloma

Dolomite		Doloma	
$CaCO_3$	58.3	CaO	61.5
$MgCO_3$	39.7	MgO	36.9
SiO_2	0.70	SiO_2	0.84
Al_2O_3	0.28	Al_2O_3	0.09
Fe_2O_3	1.22	Fe_2O_3	0.65
		S	0.02

Table 2 Batch compostition of the BOF model slag and its basicity.

Oxides	CaO	MgO	SiO_2	FeO	MnO
wt.%	30	4	30	30	6
Basicity (wt.% ratio)	$CaO/SiO_2 = 1$				

plied by Lafarge-Lime., Worksop, Notts. Table 1 gives the composition of dolomite and doloma samples from Redland's Thrislington works, County Durham. The chemical analyses were done by X-ray fluorescence (XRF) at the work's quality control laboratory. A BOF model slag composition (typical of an early BOF slag) which was designed for a static dissolution test is given in Table 2. The slag batch was made by mixing the appropriate amount of the starting powders comprising BDH reagent grade calcium carbonate ($CaCO_3$ precipitated, 99% purity) and magnesium carbonate (basic, light, $xMgCO_3.yMg(OH)_2.2H_2O$, 40–45% MgO), Aldrich reagent grade ferrous oxalate ($FeC_2O_4.2H_2O$, 99% purity) and manganese (II) oxide (MnO, 99% purity) and Tilcon Loch Aline silica sand (99.8% SiO_2). Dolomite and doloma pellets with dimensions of approximately 20 mm × 10 mm diameter were ground to shape using dry SiC grinding paper (120 grit). The model slag preparation and the static dissolution test of dolomite and doloma are shown schematically in Figs 1 and 2.

Characterisation of dolomite, doloma and model slag before and after the dissolution test was carried out by means of X-ray powder diffractometry (XRD) and scanning electron microscopy (SEM) with the associated energy dispersive spectroscopy (EDS) facility. A Philips 1050 diffractometer was used for XRD analysis and a Camscan Series 2 electron microscope was used for SEM. A Link AN 10 000 system with a Be window detector was used for EDS. All SEM samples were prepared using standard ceramographic grinding and polishing procedures. In addition, a Stanton Redcroft DTA 673–4 differential thermal analyser was used to characterise the decomposition reaction of as-mined dolomite (particle size 50–75 μm) upon heating at $10°C min^{-1}$ in nitrogen atmosphere.

The initial dissolution test reported here was performed at 1350 °C with 1h immersion time. The slag and test specimens were cooled slowly in the furnace to room temperature before they were prepared for *post-mortem* examination. Furnace cooling was necessary to

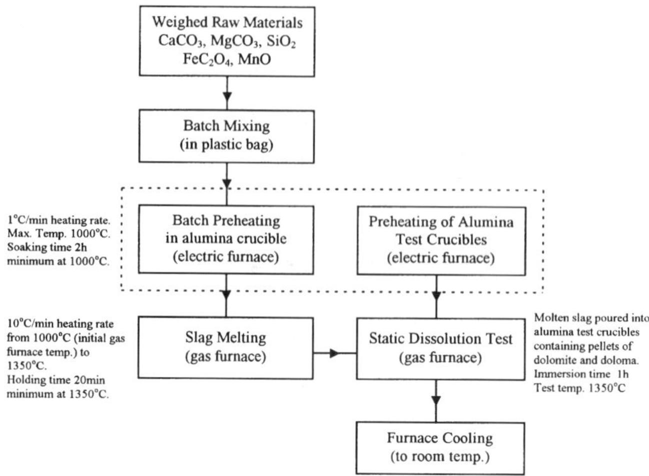

Figure 1 Outline of slag preparation and the static dissolution test at 1350°C and 1h immersion.

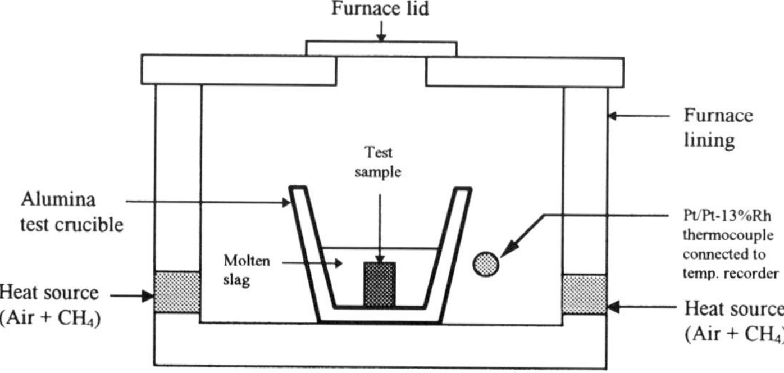

Figure 2 Schematic illustration of the crucible test arrangement for the static dissolution experiment in a gas furnace fired with an atmosphere consisting of a mixture of air and methane (CH_4).

prevent cracking of the alumina crucibles due to thermal shock and to prevent sample failure arising from the volume expansion (~10%) associated with the transition of β–Ca_2SiO_4 to γ–Ca_2SiO_4 present at the slag/additive interface on cooling over the 735°–670°C temperature range.[6]

3. RESULTS AND DISCUSSION

The microstructure of as-mined dolomite (Fig. 3(a)) reveals single-phase dolomite grains with porosity, while the doloma microstructure after calcining for 3.5–4 hrs at 1450–1550°C (Fig. 3(b)) contains two phases, light lime (CaO) matrix and darker magnesia (MgO) grains. XRD reveals the characteristic peaks of these phases along with some calcite impurity in the dolomite (Figs 4(a) and 4(b)). DTA of dolomite decomposition (Fig. 5) shows typical endotherms associated with the two stage decomposition occurring via the following reactions;[7]

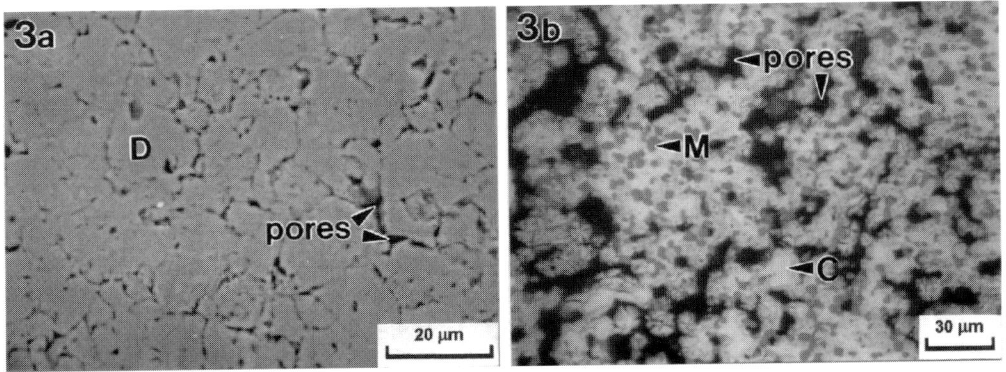

Figure 3 Backscattered electron SEM images (BEI) of the microstructure of (a) as-mined dolomite etched in 5% HF for 30 seconds and (b) as-manufactured doloma (calcined dolomite). D = dolomite, $CaMg(CO_3)_2$; C = lime, CaO; M = magnesia (periclase), MgO (samples courtesy of Lafarge-Lime, Thrislington Works, Durham, UK).

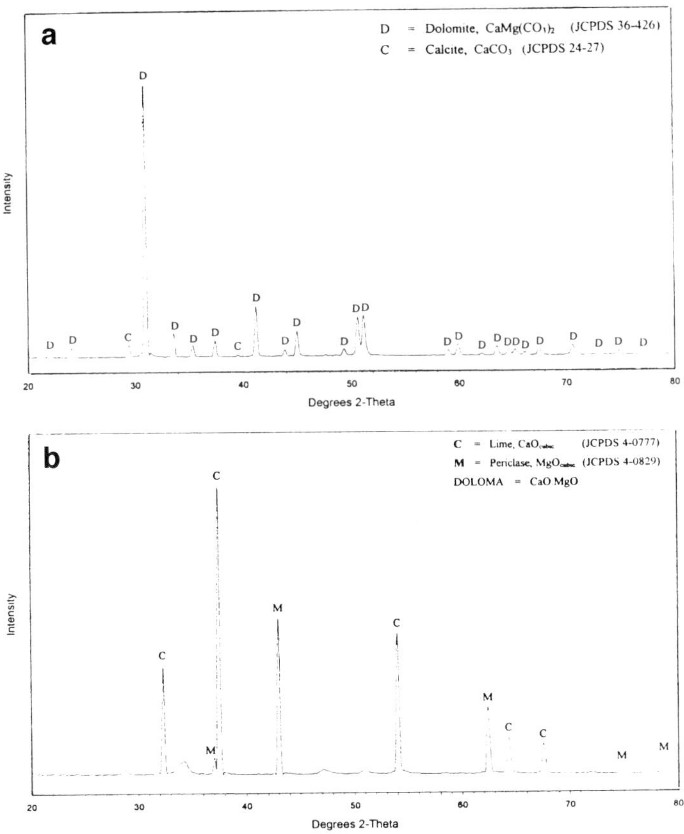

Figure 4 XRD patterns showing the characteristic peaks of (a) as-mined dolomite consisting of dolomite [D] and calcite [C] crystalline phases and (b) doloma (calcined dolomite) consisting of lime[C] and periclase [M] phases.

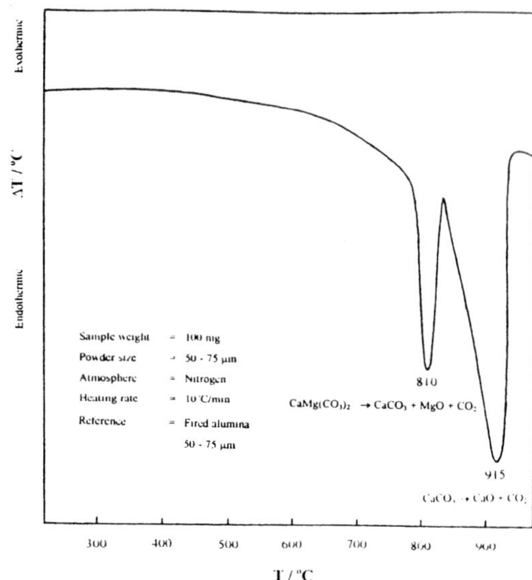

Figure 5 Typical DTA curve of the thermal decomposition of dolomite mineral quarried nearby Lafarge-Lime's Thrislington works, Durham.

at 810°C: $CaMg(CO_3)_2\ (s) \rightarrow CaCO_3\ (s) + MgO\ (s) + CO_3\ (g)$ (Stage 1)
at 915°C: $CaCO_3\ (s) \rightarrow CaO\ (s) + CO_2\ (g)$ (Stage 2)

The phases and microstructures in the BOF model slag melted at 1300°C and held for 20 minutes in the molten state before being cast onto a steel plate and cooled in air to room temperature are shown in Figs 6 and 7. The microstructure of the BOF model slag consists of angular and dendritic magnesioferrite spinel ($MgO.Fe_2O_3$ or MF) in a SiO_2–rich matrix which, since no other peaks occur in the XRD pattern and because of the presence of a broad hump at 30° 2θ (Fig. 6), is believed to be glass.

Figures 8(a)–(b) and 9(a)–(b) show the reaction interfaces at the doloma/slag and dolomite/slag boundaries, respectively, after lh immersion at 1350°C. The slag microstructure on the right hand side of Fig. 8(a) comprises dicalcium silicate (C_2S), tricalcium silicate (C_3S, $3CaO.SiO_2$) and manganowüstite ((Fe,Mn)O). This is quite unlike the model slag (Fig. 7) since the basic addition has changed the overall composition, making it far more basic. The microstructure of doloma/slag reaction interface viewed at higher magnification (Fig. 8(b) reveals the formation of a relatively incomplete C_2S layer at the doloma surface being broken up by dicalcium ferrite (C_2F, $2CaO.Fe_2O_3$) and a MgO-rich phase. EDS reveals Mg and Fe in this phase, suggesting it is magnesiowustite, (Fe, Mg)O. These reaction products form from individual reactions between lime in the doloma and silica in slag (C_2S and possibly C_3S), between lime in the doloma and iron oxide in slag (C_2F) and between periclase in the doloma and iron oxide in slag ((Fe, Mg)O).

The C_2S phase at the surface of decomposing dolomite appears to form a continuous layer along the dolomite/slag interface (Fig. 9). The formation of this continuous layer may be

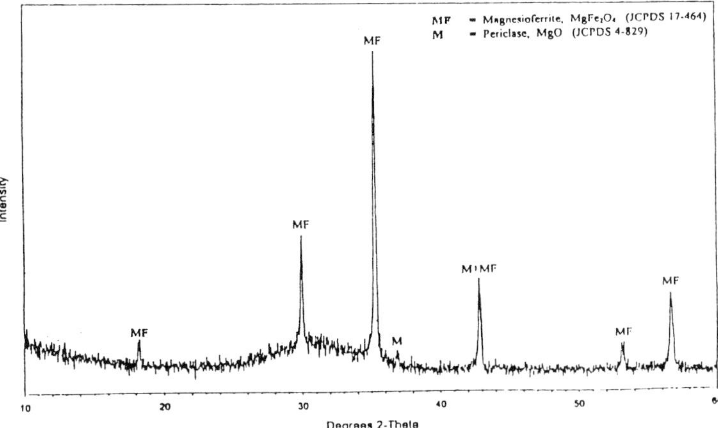

Figure 6 XRD pattern of the BOF model slag showing crystalline peaks of magnesioferrite (MF) and periclase (M) phase and a broad peak indicating the presence of a glassy phase in the slag.

Figure 7 SEM micrograph (BEI) of the BOF model slag prior to dolomite/doloma dissolution test showing angular and dendrite morphology of magnesioferrite spinel phase (MF) in a silica-rich matrix believed to be glassy. MF = $MgO.Fe_2O_3$.

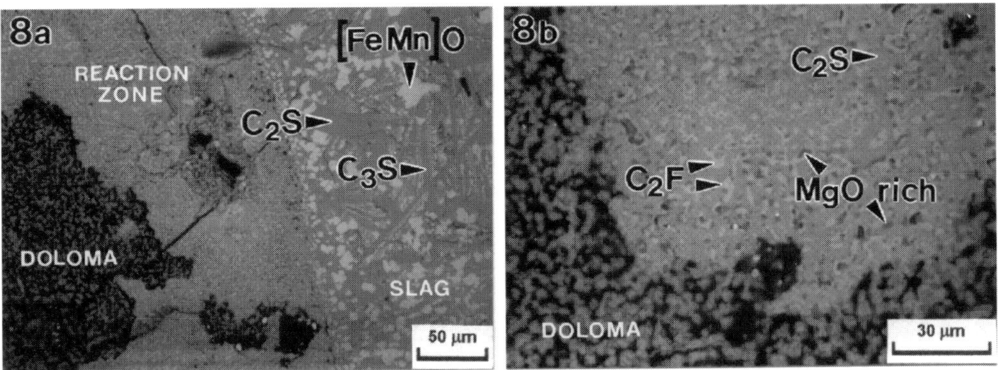

Figure 8 Backscattered electron micrographs of doloma/slag interfaces after 1h dissolution at 1350°C showing reaction products consisting of C_2S, dicalcium ferrite ($C_2F = 2CaO.Fe_2O_3$) and a MgO-rich phase.

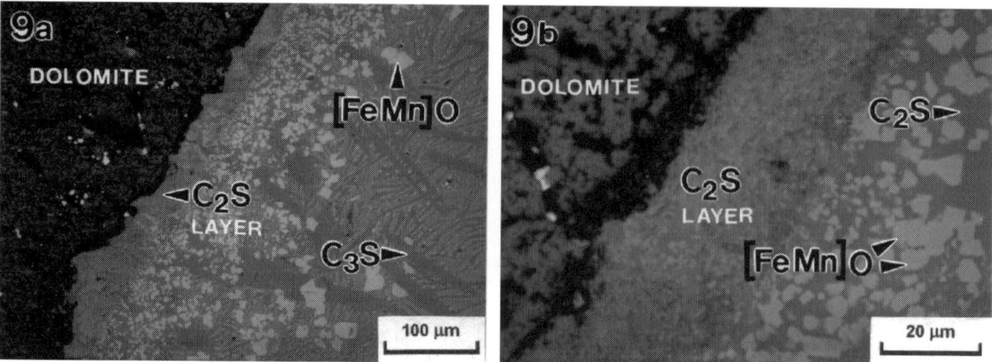

Figure 9 Backscattered electron micrographs of dolomite/slag interfaces after 1h dissolution at 1350°C showing C_2S layer at the dolomite surface. Also shown is the change in dolomite microstructure to doloma as a result of exposure to heat during contact with the molten slag.

associated with the instantaneous cooling of the C_2S resulting from the CO_2 gas evolution and the endothermic reaction during dolomite decomposition. This layer may then act as an intermediate barrier between the dolomite (now doloma) and slag, possibly retarding the dissolution process. Also shown is the gap separating the C_2S layer and the dolomite surface, which suggests that the evolution of CO_2 gas prevents direct contact between slag and dolomite at the initial stage of dissolution.

In both doloma and dolomite dissolution, a manganowüstite phase, (Fe,Mn)O, is observed to precipitate at the boundary between the slag and the reaction interfaces (Figs 8(a) and 9(a)). This is unlike the situation at the doloma/slag or dolomite/slag interface where MgO preferentially reacts with the slag to give magnesiowüstite, (Fe,Mg)O. The formation of magnesiowüstite solid solution which breaks up the C_2S layer shows the faster dissolution of doloma relative to lime, in which the C_2S phase forms a continuous and dense layer, retarding its dissolution into the slag. This characteristic of doloma makes it a suitable fluxing addition for forming an early basic slag and so enhancing the dissolution of lime, the primary slag-forming material. As mentioned earlier, magnesiowüstite has a lower liquidus temperature than C_2S, thus acting as a flux so that the C_2S build-up can be suppressed.

Dissolution Mechanisms in Static Molten Slag

A model of dissolution of burnt dolomite (i.e. doloma) in a typical BOF slag has been reported by Umakoshi *et al.*[4] as shown in Fig. 10. Initially, the liquid slag penetrates the pores and/or cracks in the doloma structure and due to the concentration gradient of CaO and MgO between doloma and liquid slag, the CaO and MgO diffuse out of doloma and react with SiO_2 and FeO in the slag. As the concentration of these oxides equilibrates at the doloma/slag boundary layer, they start forming a mixture of phases comprising C_2S, (Fe, Mg)O, and C_2F, with high melting point C_2S being the predominant phase. The formation of lower melting points (Fe, Mg)O and C_2F (T_m 1438°C) breaks up the C_2S layer and dissolution of doloma then proceeds with further penetration of the slag into the discontinuous C_2S layer.

A similar mechanism may be observed in the case of unburnt dolomite, since the carbon-

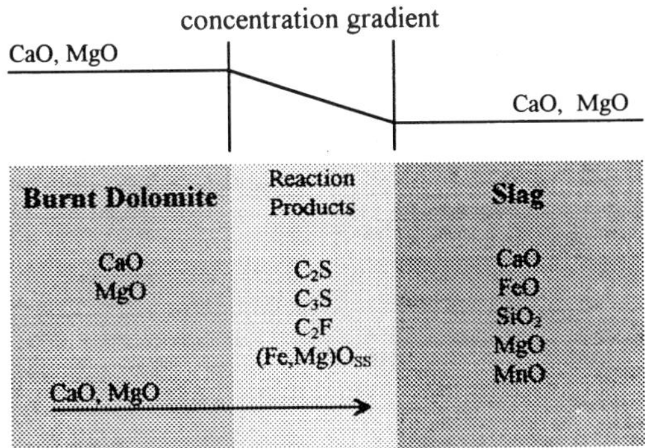

Figure 10 Dissolution of burnt dolomite into molten slag.[4]

ates in dolomite would eventually convert to their oxides upon exposure to heat, i.e. on contact with molten slag. However, due to the presence of CO_2 gas resulting from dolomite decomposition, then the mechanism is expected to be more complicated. The CO_2 gas quenches the C_2S layer formed from the reaction between CaO in decomposing dolomite and SiO_2 in the slag. In addition, the CO_2 gas forms a gap which acts as a barrier preventing direct interaction between dolomite and slag. Due to these effects, the dissolution of dolomite is expected to be retarded compared with that of doloma.

Furthermore, this work has shown that the C_2S layer at the dolomite/slag interface was relatively more complete (continuous) than that at the doloma/slag interface when both dolomite and doloma were subjected to the same conditions during dissolution in a stagnant molten slag after lh immersion at 1350°C. The direct reaction between doloma and slag as opposed to the indirect reaction between dolomite and slag suggests that at the initial stage of dissolution doloma dissolves faster than dolomite.

4. CONCLUSIONS

A lh static dissolution test at 1350°C suggests that doloma initially dissolves faster than dolomite in a stagnant melt of CaO–MgO–SiO_2–FeO–MnO slag. The molten slag penetrates into the doloma through porosity and/or cracks, then reacting with MgO and CaO, forming low melting phases such as magnesiowustite ((Fe, Mg)O) and dicalcium ferrite (2CaO.Fe_2O_3). The formation of these phases breaks up the dicalcium silicate (C_2S) layer at the doloma/slag interface, making it easy for further penetration by the slag.

The dissolution of dolomite requires initial thermal decomposition in which the carbonates transform into oxides with CO_2 gas being liberated. This gas phase acts as a barrier between the resulting doloma and slag, thus retarding the reaction. Also, the formation of the C_2S layer due to reaction between SiO_2 in slag and CaO in the decomposed dolomite is likely to be accelerated by localised cooling due to CO_2 gas evolution. The presence of this layer will retard the dissolution of dolomite.

ACKNOWLEDGEMENT

The authors wish to thank Lafarge-Lime UK, Worksop, Notts, for funding the research and providing the samples for experimental work.

REFERENCES

1. E.T. Turkdogan, *Fundamentals of Steelmaking,* Institute of Materials, London, UK, 1996.
2. M. Matsushima, S. Yadoomaru, K. Mori and Y. Kawai, 'A Fundamental Study on the Dissolution Rate of Solid Lime into Liquid Slag', *Iron and Steel Inst. Japan Transactions*, 1997, **17**(8), 442–449.
3. P. Williams, M. Sunderland and G. Briggs, 'Interactions Between Dolomitic Lime and Iron Silicate Melts', *Ironmaking & Steelmaking*, 1982, **9**(4), 150–162.
4 M. Umakoshi, K. Mori and Y. Kawai, 'Dissolution Rate of Burnt Dolomite in Molten Fe_tO–CaO–SiO_2 Slags', *Iron and Steel Inst. Japan Transactions*, 1984, **24**(7), 532–539.
5. J. Green, Unpublished report, Steetley Refractories Ltd., Worksop, Notts, UK, 1976.
6. A. Muan and E. F. Osborn, *Phase Equilibria Among Oxides in Steelmaking,* Addison-Wesley, Reading, MA, USA, 1965.
7. R. M. McIntosh, J. H. Sharp and F. W. Wilburn, 'The Thermal Decomposition of Dolomite', *Thermochimica Acta*, 1990, **165**(2), 281–296.

Reactions of Silicon and Carbon in Reducing Atmospheres

S. A. FRANKLIN*

Kvaerner Metals, Ashmore House, Richardson Road, Stockton-on-Tees, TSI8 3RE

B. RAND*

School of Materials, University of Leeds, Leeds, L52 9JT

ABSTRACT

Silicon is commonly added to carbon containing refractories to enhance their performance. The silicon is shown to react to form various ceramic phases (SiO_2, SiC, Si_2N_2O and Si_3N_4) with reactions via the gaseous species SiO being important. Metastable phases, particularly SiC form whilst elemental silicon remains and these phases may persist in a refractory material. The reaction

$$2Si + CO = SiC + SiO$$

is shown to be important in the generation of silicon monoxide gas and thus in the overall reaction of silicon. The equilibrium partial pressure of SiO for this reaction is very high (> 1 atm), much higher than the SiO partial pressure when no silicon remains ($< 3 \times 10^{-3}$ atm). The SiO formed then reacts to form ceramic phases. This mechanism is shown to explain morphological features in carbon-containing refractories. A method of producing coatings of silicon containing ceramic phases (SiO_2, SiC, Si_2N_2O and Si_3N_4 based on the above reaction is given.

1. INTRODUCTION

The reactions in the Si–C–N–O system are technologically important in such diverse areas as refractories, ceramic matrix composites and reaction bonded engineering ceramics with ceramic phases including silica, silicon carbide, silicon nitride, silicon oxynitride and carbon, all of which are important electronic, engineering or refractory materials. Additionally these phases are used both as matrices and reinforcements of ceramic matrix composites.

This paper describes part of a systematic study of the fabrication – property – microstructure relationships in alumina-carbon refractories. These materials may contain many bond types including partial sintering between oxide particles, carbon bonding and a (possibly dominant) contribution from the reaction products of a silicon 'anti-oxidant' addition to the material. A method for quantifying the contribution to the properties of the oxidic sintering

*Both authors were formerly at the School of Materials, University of Sheffield, Sheffield

has already been described.[1] This paper describes the conditions required to form the different silicon based compounds which may contribute to the bonding and further papers are in the course of preparation describing the properties of these complex refractories.

The results discussed in this paper are of relevance to applications where ceramic phases are formed in situ by the reaction of elemental silicon; such as in oxide-carbon refractories that contain silicon as an additive to enhance the strength, abrasion and oxidation resistance, reaction bonding of engineering ceramics such as silicon carbide. This paper concentrates on the reactions occurring within a carbon containing refractory. Refractory materials based on composites of carbon and a refractory oxide are widely applied within the steel industry; magnesia-carbon is used in BOS converters, electric arc furnaces and in the high wear areas of steel ladles, alumina carbon is applied to continuous casting refractories and zirconia-carbon is used in specialised areas. The alumina and zirconia do not usually react with other components of the refractory, whereas the magnesia is more reactive and has been observed to form forsterite (Mg_2SiO_4)[2] or can be reduced to magnesium vapour by the reaction

$$C(s) + MgO(s) = CO(g) + Mg(g) \tag{1}$$

Microstructural examination of used carbon containing refractories containing a silicon antioxidant addition often reveals a microstructual feature consisting of a hollow polycystalline aggregate formed by the reaction of silicon with other constituents of the material and with pore gases. This paper uses thermodynamic considerations to predict the reactions by which this feature forms. The results of an experimental test of these proposed reactions is then described.

The thermodynamics concentrate on reactions in the Si–C–N–O system that prevail within the body of a refractory in service. The effect of reactions with a refractory oxide (alumina, magnesia or zirconia) have not been considered hence this work is of more direct relevance to alumina- and zirconia-based materials than to magnesia carbon materials. It was known that non-equilibrium phases may still be present after a period in service at elevated temperatures: to account for this phenomena, this work emphasises non-equilibrium conditions that are significant, especially while elemental silicon remains.

The thermodynamics of the Si–C–N–O system have been extensively studied owing to their industrial importance, however studies relevant to the conditions existing within a carbon containing refractory have received less attention; those published to date consider the equilibrium phases present and do not shed light on the behaviour of non equilibrium phases. These considerations add to the understanding of the properties of carbon containing refractories.

2. MICROSTRUCTURAL OBSERVATIONS

To develop an understanding of the development of microstructure in alumina–carbon refractories, model materials were produced from polycrystalline commercial alumina particles having a wide range of sizes. These were mixed with 10wt% silicon powder (Dunstan & Wragg, >99.3% Si, <75μm). These materials were fired for 3h at 1600°C in carbon filled saggars, such that the atmosphere represents that found within the pores of a refractory material. Full details of the raw materials, fabrication and test methods are given elsewhere.[3] The silicon level chosen is much higher than that used commercially, to facilitate the study of its effects on the materials. It should also be noted that the unfired materials do not contain carbon (either as graphite or as an organic binder). Any reaction of the silicon would be with the N_2–CO mixture formed by the reaction of carbon in the saggar with air.

The fired materials were found to contain no unreacted silicon but contained silicon carbide and silicon oxynitride. Optical examination showed the silicon oxynitride to be present as hollow polycrystalline particles (Fig. 1). SEM examination showed that on occasion these particles were regular in shape in contrast to the angular alumina and silicon carbide particles (Fig. 2). These features are often found in used alumina carbon refractories although they are generally smaller and less regular in shape in a commercially produced material that has less silicon and a lower porosity than the model materials produced for this study. The same microstructural features are however common to both the model and commercial materials. To explain the formation of these features, the thermodynamics of the Si–C–N–O system were investigated.

3. THERMODYNAMIC CONSIDERATIONS

The thermodynamic data used in this work were taken from published tables[4] with the exception of Fegley's data[5] for silicon oxynitride.

3.1 Gas Phase

The composition of gases within the pores of a carbon-containing refractory material may be considered to be in equilibrium with the carbon of the refractory. At temperatures encountered in service, the reaction of carbon with oxygen is rapid and the small pore volume (which is predominantly open porosity) may be assumed to reach equilibrium very quickly. It is therefore assumed that the gas composition in the pores will be air in equilibrium with carbon, this will comprise a mixture of nitrogen, carbon monoxide and carbon dioxide. The carbon monoxide:carbon dioxide ratio defines the equilibrium oxygen partial pressure under these conditions. Figure 3 shows the gas compositions. At service temperatures greater than 1100K the pore space will be filled with a mixture comprising nitrogen and carbon monoxide with only a very low partial pressure of carbon dioxide. As long as carbon is present in the system, any diffusion or pressure differences that cause air to be drawn into the pores of the material will not result in changes in composition as the oxygen in the air will react with carbon at the surface of the material, resulting in the gas composition shown in Fig. 3. It is considered that any reactive gaseous species, such as silicon vapour, will react rapidly with the CO–N_2 mixture and that they will be present in such small proportions that the gas composition may be considered to be that already described, with the addition of the Si(g) or SiO(g). The total pressure is considered to be 1.0 atm at all stages of reaction, as the permeability of the refractories is such that any pressure differences through the sample will rapidly disappear as gas flows through the pores; the apparent porosity of a carbon containing refractory is typically 5–25% with 10–12% being more usual after the carbonaceous bond has been coked.

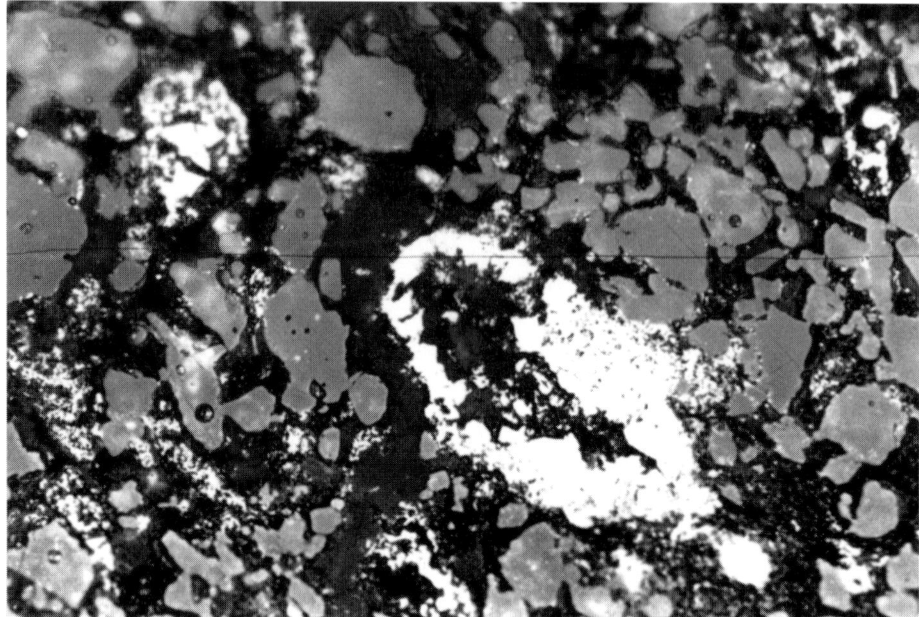

Figure 1 Optical micrograph of alumina and silicon fired in a coke filled saggar at 1600°C. The hollow agglomerate consists of silicon oxynitride. 720x

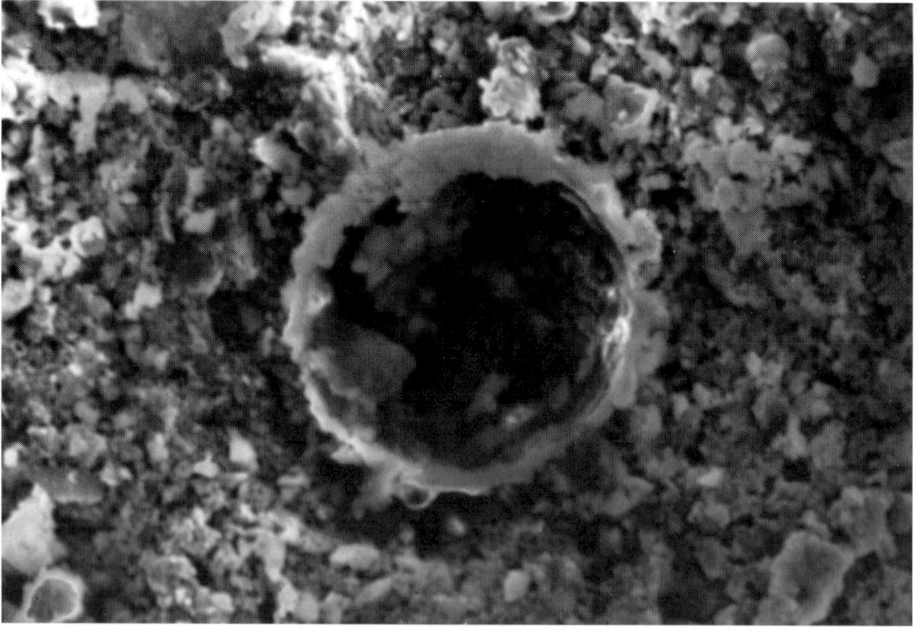

Figure 2 Scanning electron micrograph of alumina and silicon fired in a coke filled saggar at 1600°C. The hollow agglomerate consists of silicon oxynitride. 850x

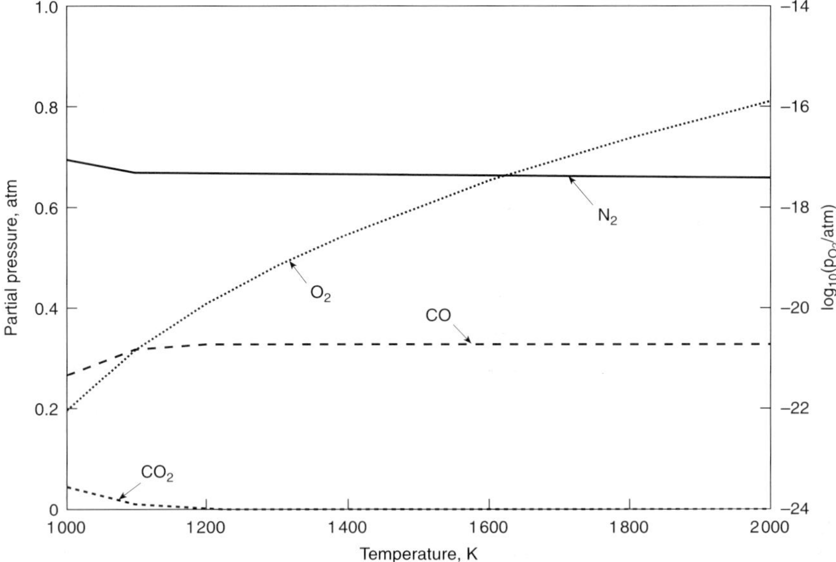

Figure 3 Equilibrium gas composition for air in a porous carbon bed.

3.2 Equilibrium Position

Figure 4 shows the free energy of reaction for the formation of each of the ceramic phases in the Si–C–N–O system, in equilibrium with the nitrogen and oxygen partial pressures that are themselves in equilibrium with carbon at each temperature, as shown in Fig. 3. Each reaction is expressed in terms of the reaction of 1 mol of silicon. All the phases are stable with respect to their elements at the temperatures considered with the exception of silica, which is unstable above 1850K. The stable phase below 1490K is silica, with silicon oxynitride being stable between 1490K and 1900K, SiC is the stable phase above this temperature.

3.3 Reaction Mechanism

Through a careful consideration of the relative stability of the various metastable phases and the free energy of reaction for different reactions in the Si–C–N–O system, the likely reaction mechanisms occurring may be deduced. Although silica is the stable phase at low temperatures, it is unlikely to form through the reaction

$$Si(s/l) + O_2(g) = SiO_2(s/l) \tag{2}$$

as the partial oxygen pressure is so low, instead the reaction

$$Si(s/l) + 2CO(g) = SiO_2(s/l) + 2C(s) \tag{3}$$

is likely to be important. Yamaguchi[6] showed that this reaction is important in protecting the carbon from oxidation, because the silicon is oxidised preferentially, resulting in carbon deposition. This reaction will involve the deposition of carbon onto the silicon particles,

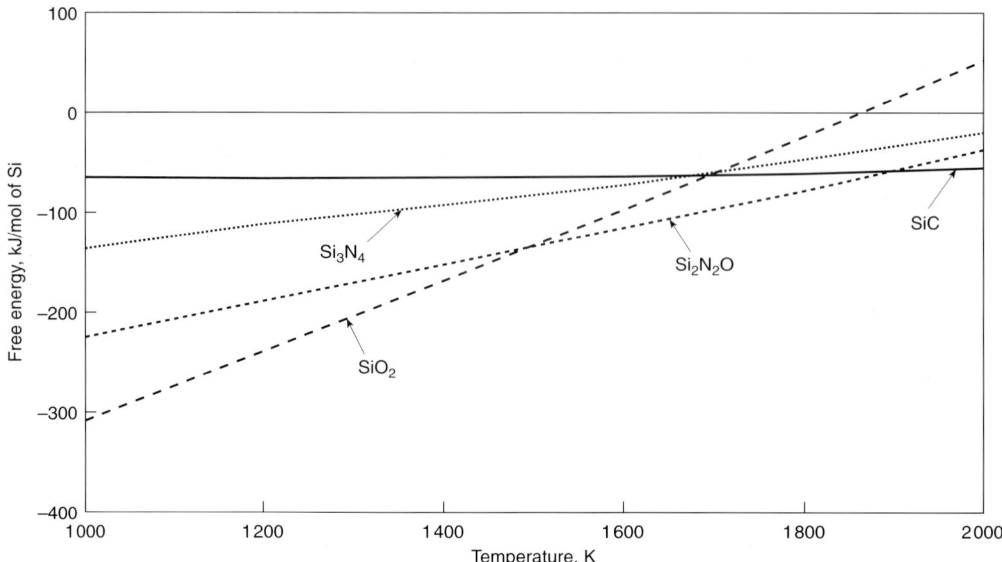

Figure 4 Free energy of reaction of 1 mol of silicon to form various compounds in the atmosphere shown in Fig. 3.

such that the local conditions may be sufficiently far from equilibrium for the reaction

$$Si(s/l) + C(s) = SiC(s) \tag{4}$$

also to occur at a significant rate; this reaction is likely to occur where the silicon and carbon are in contact, where gases are prevented from coming into contact and reacting with the silicon. The reaction, overall

$$3Si(s/l) + 2CO(g) = 2SiC(s) + SiO_2(s/l) \tag{5}$$

may therefore be expected to be important below 1850K as long as free silicon is available. Since silica is stable with respect to the carbide, Fig. 4, the silicon carbide may be oxidised when little elemental silicon remains so it is essentially an intermediate product in the formation of SiO_2. Furthermore, the oxidation of the carbide may be expected to be sluggish due to the protective action of a layer of silica which would be expected to form on the carbide particles.

Silicon has a low vapour pressure, the vapour pressure of silicon in equilibrium with the metal is less than 10^{-5} atm at all the temperatures considered in this work, Fig. 5. The silicon vapour pressure in equilibrium with stable silicon compounds will be many orders of magnitude lower than this; the vapour is therefore not considered to be important in the reactions within this system. The gas silicon monoxide, SiO, may however exist at much higher partial pressures than silicon and will be an important intermediate product in the reaction of silicon. Figure 5 also shows the equilibrium vapour pressure of silicon monoxide for

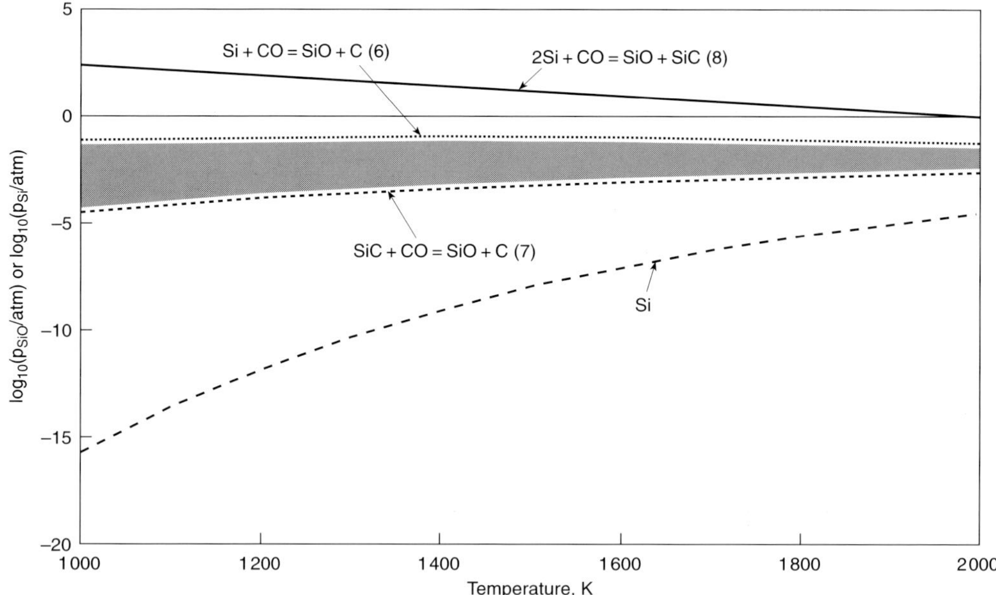

Figure 5 Equilibrium vapour pressure for various reactions under the gas pressures shown in Fig. 3.

various reactions, in equilibrium with the CO, CO_2 and O_2 partial pressures shown in Fig. 3. Considering the reaction of silicon with carbon monoxide

$$\mathrm{Si(s/l) + CO(g) = SiO(g) + C(s)} \tag{6}$$

the equilibrium partial pressure of silicon monoxide is of the order of 0.1 atm whilst silicon remains. This indicates that the reaction of silicon to form ceramic phases can occur via the gas phase with silicon monoxide being present in significant quantities. The silicon monoxide may react with carbon to form silicon carbide,

$$\mathrm{SiO(g) + 2C(s) = SiC(s) + CO(g)} \tag{7}$$

Reactions (6) and (7) together are equivalent to reaction (4) overall. This process can occur if the kinetics are such that the partial pressure of silicon monoxide is below that for the equilibrium of reaction (6) but above that for reaction (7). This is shown as the shaded region of Fig. 5. In graphical terms, when the partial pressure of SiO is above the equilibrium value for a reaction (represented by the curves), the reaction will proceed in a direction to reduce the SiO partial pressure (i.e. from right to left as the reactions are written on the legend). Conversely, when the pressure is below the equilibrium value for a given reaction, it will proceed in a direction to create more SiO, provided the other reactants are present. When the partial pressure lies between two curves (as in the shade region of Fig. 5), the

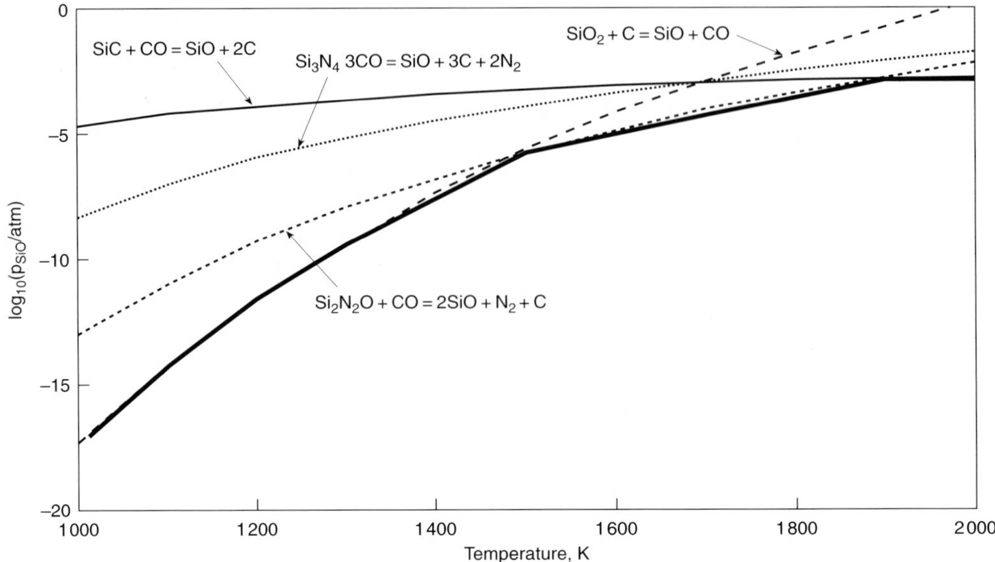

Figure 6 Equilibrium vapour pressure for various reactions under the gas pressures shown in Fig. 3.

reaction represented by the upper curve will occur in the forward direction whilst the reaction forming the lower boundary will occur in reverse. Overall, one condensed phase will be converted into another via SiO gas.

The importance of the reaction

$$2Si(s/l) + CO(g) = SiO(g) + SiC(s) \tag{8}$$

was recognised by Van Konijnenberg,[7] but the high equilibrium pressure of silicon monoxide was not discussed, the paper explains that through reactions (8) and (7) silicon may react to form the carbide at temperatures below its melting point and that the oxygen acts as a 'catalyst' in the reaction of silicon and carbon. Figure 5 shows that the equilibrium partial pressure of silicon monoxide for reaction (8) is in excess of 1 atm, so that as long as there is unreacted silicon metal the forward reaction will be expected to occur, since this reaction will be far from equilibrium. However, as soon as the partial pressure of silicon monoxide exceeds that for reaction (7), further silicon carbide formation will occur. It may be inferred that reaction (8) may be important in the reaction and transport of silicon, and that silicon carbide may form from the metal although it is only metastable below 1900K; once all the silicon has reacted, the carbide may then react further to form other phases.

Figure 6 shows the equilibrium partial pressure of silicon monoxide for other reactions in the Si–C–N–O system, in equilibrium with the gases indicated in Fig. 3. This shows that the partial pressure of silicon monoxide in equilibrium with each phase reflects their relative stability as shown in Fig. 4; the equilibrium partial pressure is shown by the bold line. However, reactions other than those shown in Fig. 6 are thermodynamically favoured when silicon and carbon coexist in a material.

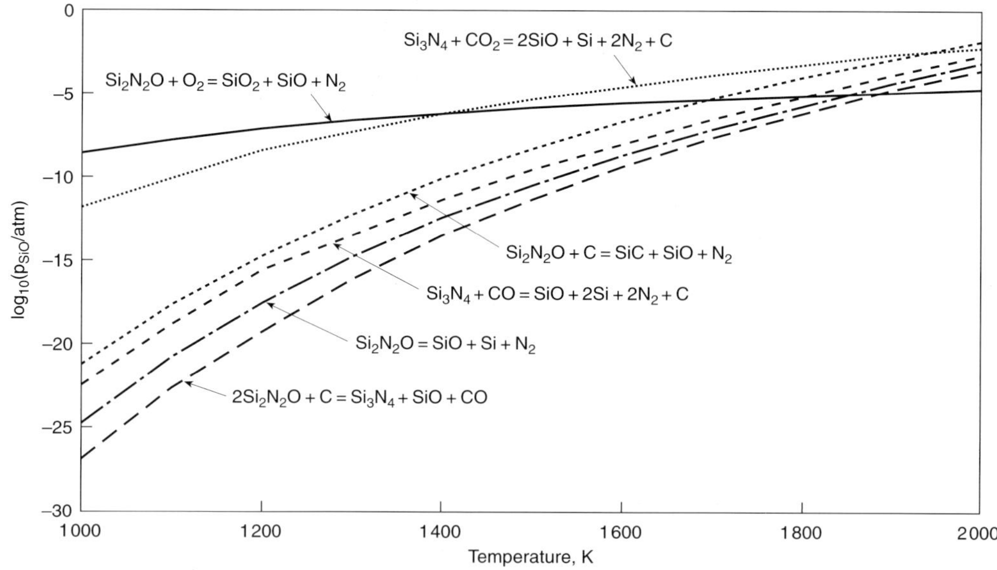

Figure 7 Equilibrium vapour pressure for various reactions under the gas pressures shown in Fig. 3.

Figure 7 shows that at partial pressures of silicon monoxide below the equilibrium partial pressure, the gas can react with the metal to form silicon nitride or oxynitride whilst silicon remains. Whilst elemental silicon remains, SiO gas can react with silicon, nitrogen and carbon (and other species) to form silicon nitride or silicon oxynitride at partial SiO pressures many orders of magnitude lower than the equilibrium partial SiO pressure. However, under these conditions, reaction (8) is likely to occur in the forward direction since the equilibrium partial pressure of SiO for this reaction is so high. In this way, the SiO will be continuously formed by reaction (8) only to react to form other ceramic phases.

After all the silicon has reacted, the equilibrium partial pressure of silicon monoxide will be that in equilibrium with the stable phase, silica, silicon oxynitride or silicon carbide, as shown on Fig. 6.

4. EXPERIMENTAL STUDY OF REACTIONS OF SILICON

To confirm the above findings, some samples of a commercially produced silicon powder were fired in saggars filled with carbon but not in contact with the carbon; the atmosphere may be assumed to be that in Fig. 3. The samples of silicon were fired in covered mullite crucibles for 5h or 10h at 1873K. The samples were analysed by qualitative X-ray diffraction. Full details of the methods used are given elsewhere.[3]

Figure 8 shows the X-ray diffraction traces of the fired samples. In both the samples all the silicon had reacted, with silicon carbide and oxynitride being the main reaction products. The proportion of silicon oxynitride was greater in the sample that had been fired for 10h than in the sample that had been fired for 5h.

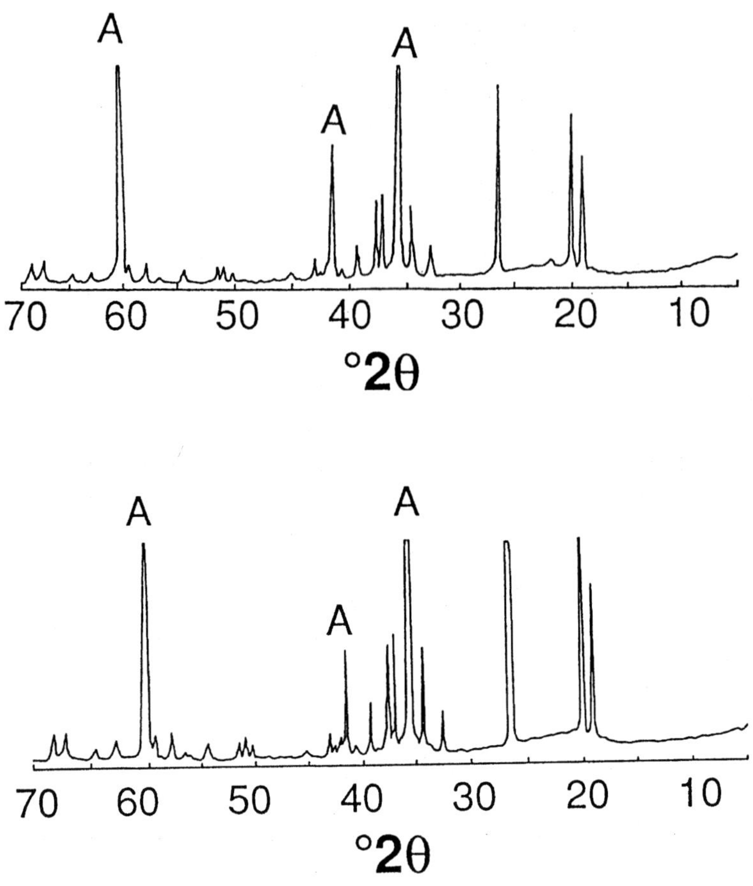

Figure 8 X-ray diffraction traces of silicon samples fired in coke filled saggars at 1600°C for 5h (top) and 10h (bottom). A=SiC peak, other peaks are all identified as Si_2N_2O. Note the relative sizes of the SiC peak at 41.6°2θ and the Si_2N_2O peaks at 19.0°2θ, 20.0°2θ and 26.5°2θ.

5. DISCUSSION

It has been shown that the silicon is likely to react to form the carbide and silicon monoxide. The reaction of the silicon has been proposed[6] to prevent the oxidation of the carbon by 'gettering' the oxygen and resulting in the deposition of carbon, through the reaction

$$Si(s/l) + 2CO(g) = SiO_2(s/l) + 2C(s) \quad (3)$$

however, this reaction or the formation of silicon nitride or oxynitride will result in a pressure difference developing through the sample as the gases are 'condensed' through the formation of ceramic phases. This pressure difference will result in the air being drawn into the material and will cause oxidation at the surface of the sample as the oxygen in the air reacts to form carbon monoxide. However, the reaction to form silicon carbide and monoxide (reaction (8)) does not result a decrease in the amount of gas and so does not result in accelerated surface oxidation. When the silicon monoxide reacts to form phases other than SiC the amount of gas is decreased and surface oxidation may be accelerated. The silicon monoxide will diffuse through the material before it reacts, and will oxidise to form silica at the surface of a material, where a higher oxygen partial pressure exists. This will stabilise silica formation and will have two effects.

1. the deposition of the silica will serve to reduce the permeability of a refractory and will help to limit the extent to which oxidation occurs, as is observed during the preheat of magnesia–carbon ladle bricks.

2. the deposition of silicon as silica at the sample surface will result in a gradient in the partial pressure of silicon monoxide which will result in a tendency for this gas to diffuse towards the sample surface; this has the net effect of transporting carbon from the sample surface to the body of the material (as CO), where it reacts with silicon to form SiC and SiO, the SiO the diffuses towards the surface removing silicon and oxygen from the interior of the sample.

It is proposed that these two effects explain the enhanced resistance to oxidation of carbon containing refractories with silicon additions.

The formation of silicon compounds by the reaction of silicon monoxide has the potential to be used to deposit layers of SiC, Si_3N_4, Si_2N_2O or SiO_2; these could be used, for example, to coat fibres to enable control of the adhesion between fibre and matrix, to deposit wear resistant or corrosion resistant layers onto critical components, without the need for compounds such as $SiCl_4$. Furthermore, since the layers are deposited at elevated temperatures there will be a reduced thermal stress in components used at high temperatures. The deposition of ceramic phases might be achieved by a schedule such as

1. Oxygen, air or argon-oxygen would be passed through a heated carbon bed so that the oxygen reacts to form carbon monoxide,

2. The carbon monoxide containing gas is passed over silicon, or ferrosilicon that has been heated to above its melting point, some reaction to SiO and SiC will occur. The silicon carbide would be a by-product or could be reduced to the metal for recycling.

3. By control of temperature, pressure and atmosphere the silicon monoxide could be deposited as SiC, Si_3N_4, Si_2N_2O or SiO_2 or even combinations of these phases further down the 'line' onto any suitable substrate.

6. CONCLUSION

When heated in an atmosphere of carbon monoxide and nitrogen, silicon will react to form SiC and SiO gas through the reaction

$$2Si(s/l) + CO(g) = SiC(s) + SiO(g) \quad (8)$$

although silicon carbide is not the thermodynamically stable phase at temperatures below 1900K; once all the silicon has reacted the silicon carbide may then react to the stable phase, possibly via the gas SiO. From the morphology of Si_2N_2O, it may be inferred that the reaction of silicon carbide to form Si_2N_2O occurs with silicon monoxide gas as an intermediate product.

These reactions have several implications for the behaviours of carbon containing refractories, in that they will result in a pressure difference becoming established between the centre and outside of the sample as the silicon reacts with gases to produce condensed phases; the pressure difference will only be small but will result in a net flow of air into the sample which may oxidise the carbon at the materials surface. The microstructure of the silicon oxynitride found in used alumina carbon refractories is a result of its formation via silicon monoxide gas.

REFERENCES

1. S. A. Franklin and B. Rand, *Brit. Ceram. Trans.*, 1996, **95**(3), 93.
2. P. O. R. C. Brant, *PhD Thesis*, University of Sheffield, Sheffield, 1988.
3. S. A. Franklin, *Ph.D. Thesis*, University of Sheffield, Sheffield, 1989.
4. JANAF thermochemical tables, 2nd Ed, NBS, Washington USA, 1971.
5. M. B. Fegley, *Comm. Amer. Ceram. Soc.*, C124, 1981, **64**(9).
6. A. Yamaguchi, *Taikabutsu Overseas*, 1984, **4**(3), 14.
7. J. T. Van Konijnenberg, *Science of Ceramics 9*, K J de Vries, Nederlandse Keramische Vereniging, 1977, 339.

Barium Hexaaluminate Refractories

T. J. DAVIES, H. G. EMBLEM†, M. BIEDERMANN, Q.-G. CHEN
and W. A. AL-DOURI

Manchester Materials Science Centre, University of Manchester/UMIST, Grosvenor Street, Manchester M1 7HS, UK

ABSTRACT

Barium hexaaluminate is a well-defined refractory compound. Barium hexaaluminate refactorics are strong (MOR ca 600 MPa), possible industrial uses are in the glass and metallurgical industries, also as moulds and cores for metal casting. A gel precursor of barium hexaaluminate can be prepared from a solution of a barium salt mixed with a solution of an aluminium salt, to give the oxide stoichiometry $Ba0.6A1_20_3$, by heating to remove volatiles or by adding a separate gelling agent such as ammonium acetate. The gelling agent can also be a salt already present, for instance barium acetate, which gives better control of gelation. The gel is dried and fired to form barium hexaaluminate. Filaments which convert on firing to a barium hexaaluminate ceramic fibre can be drawn as the gel is forming or extruded from a semi-rigid gel. Refractory components can be prepared from a suitably graded refractory grain mix bonded with the gel. The aluminium salt may be replaced by an aluminium alkoxide. Alkaline gelation is desirable to obtain a rigid coherent gel or a ceramic fibre. FT–IR, XRD and ^{27}Al NMR studies showed that barium was not incorporated directly into the gel structure, no barium aluminium oxides being formed before the barium species liquify. A powder mixture suitable for firing to form barium hexaaluminate was prepared from alumina and barium carbonate or barium hydroxide. Strong ceramic shapes have been made from the sintered powder mix. XRD studies showed that rapid firing to 1600–1700°C gives barium hexaaluminate directly, slow firing to 1600–1700°C giving barium monoaluminate as one intermediate phase. Chromium (III) addition reduces the temperature required to form an aluminium–barium oxide, and ultimately, the temperature of formation of barium hexaaluminate.

INTRODUCTION

Oxides containing aluminium and barium are refractory, with several possible industrial uses.[1–4] From phase diagram studies,[3] composition limits for a possible engineering ceramic material or refractory composition could be $BaO.4.6Al_2O_3$ to $BaO.6.6Al_2O_3$, with the well defined compound $BaO.6Al_2O_3$, barium hexaaluminate, being preferred. Barium hexaaluminate, like calcium hexaaluminate and strontium hexaaluminate, has the space group D^4_{6h}-hexagonal crystal system with two molecules per unit cell.[5] It may be prepared by a sol-gel route (route (i)) or by a dry powder route (route (ii)).

Route (i)

Barium hexaaluminate may be obtained[1] from a gel precursor prepared by mixing a solution of a barium salt with a solution of an aluminium salt to give the oxide stoichiometry

† Sadly, Harold Emblem died before this volume was ready to go to press.

$BaO.6Al_2O_3$ then forming a gel which is fired to obtain the oxide $BaO.6Al_2O_3$. The preferred aluminium salts are the aluminium chlorhydrates $Al_2(OH)_{6-n}Cl_n$ with n in the range 1–5, e.g. $Al_2(OH)_5Cl$, available in aqueous solution as 'Chlorhydrol' solution. Polymerisation and hence gelation may be induced by adding a separate gelling agent, for instance a solution of ammonium acetate, which is basic, or by heating the solution to remove volatiles. A barium or aluminium salt already present can produce polymerisation and gelation, barium acetate being one such salt. The gels may be used to bind a refractory grain mix.[6] Under suitable conditions,[1,3,7] gel filaments, which convert to a ceramic fibre when fired, may be obtained from a gel precursor corresponding to the oxide $BaO.6Al_2O_3$.

Route (ii)

Barium hexaaluminate may be prepared[1–4] from the source materials barium oxide, barium hydroxide or barium carbonate and alumina or a hydrous alumina such as boehmite. The dry powder mix, of oxide stoichiometry $BaO.6Al_2O_3$ is milled to break down particle aggregates, then sintered. The resulting sinter is shaped by uniaxial and/or isostatic compression, followed by sintering at 1600–1700°C to complete conversion to barium hexaaluminate. The small grain average particle size and small flaw size (ca 1 μm) observed in the sintered products are probably responsible for the good modulus of rupture (ca 600 MPa at ambient temperature) obtained. It is likely that this strength would be maintained at high temperature, leading to possible use as an engineering ceramic, or for preparing refractory shapes.

The present paper summarises results from previous work and presents some new results. The use of barium acetate to obtain gels from precursor solutions of oxide stoichiometry $BaO.6Al_2O_3$ is described. Results of varying the weight ratio barium chloride (or bromide): barium acetate are given and the results obtained[8] when 0.1 M HCl, CH_3COOH or NH_4OH solutions are added, or the solution diluted, are summarised. The barium source for gels of oxide stoichiometry $BaO.6Al_2O_3$ can also include an aqueous suspension of barium carbonate. The use of ammonium acetate to form a gel of oxide stoichiometry $BaO.6Al_2O_3$ is also described. The aluminium salt may be replaced by an aluminium alkoxide. FT–IR, XRD and ^{27}Al NMR studies showed that barium was not incorporated directly into the gel structure obtained from a gel precursor solution prepared from 'Chlorhydrol' solution and having the oxide stoichiometry $BaO.6Al_2O_3$. The preparation and properties of sintered compacts of oxide stoichiometry $BaO.6Al_2O_3$ from a mix of barium carbonate and alumina grain are described (route (ii)). The sintering behaviour of gel filaments and components prepared from a dried and powdered gel, each of oxide stoichiometry $BaO.6Al_2O_3$ obtained from 'Chlorhydrol' solution (route (i)) is compared with the sintering behaviour of compacts having oxide stoichiometry $BaO.6Al_2O_3$ prepared according to route (ii).

EXPERIMENTAL

The aluminium salt was the chlorhydrate $Al_2(OH)_5Cl$ available in solution as 'Chlorhydrol' solution [$Al_2(OH)_5Cl$, nominal 22–23wt% Al as the Al_2O_3 equivalent, Wilfrid Smith Ltd, Edgeware, Middlesex, UK]. Soluble barium salts, barium carbonate and ammonium acetate were all of laboratory reagent grade barium chloride dihydrate or barium bromide dihydrate were dissolved in Chlorhydrol solution. Barium acetate was dissolved separately in distilled water. The barium acetate:water ratio was constant at 1:25 (w/w) to make certain that the barium acetate was completely dissolved. This solution was added to the barium chloride (or barium bromide) – 'Chlorhydrol' solution in the quantity required to give the oxide stoichiometry $BaO.6Al_2O_3$. Barium bromide dihydrate is more soluble in water than barium chloride dihydrate,[9] and is therefore preferable for the preparation of gel precursor solutions.

Unless otherwise stated, gel times for the various solutions were determined at ambient temperature. Filament formation was assessed by inserting then pulling out of the solution a 10 mm diameter glass rod during gelation. The apparatus and procedure used for continuous filament formation are described in Ref. 3. Cylindrical compacts, diameter 10 mm, height 3 mm were prepared from dried and powdered gels by compaction at 270 MPa in a steel die. The compacts were sintered in air at given temperatures for given times. FT–IR and XRD observations[3,4] together with ^{27}AINMIR observations[4,7] were used to characterise phases present before and after sintering.

Barium carbonate was suspended in water to give weight ratio $BaCO_3:H_2O$ of 0.185:1. Barium chloride dihydrate was dissolved in 'Chlorhydrol' solution to give weight ratio $BaCl_2.2H_2O$: 'Chlorhydrol' solution of 0.685:10. On mixing in quantities to give the oxide stoichiometry $BaO.6Al_2O_3$ a clear gel was formed in about 2 minutes.

The aluminium alkoxide was 'Aliso B' (an aluminium (isopropoxide)(sec-butoxide) Rhone-Poulenc Chemicals Ltd, Manchester, UK). Triethanolamine, ethylene glycol and acetylacetone were all of laboratory reagent grade.

A mixture of 'Aliso B' and barium acetate, having the oxide stoichiometry $BaO.6Al_3O_3$ was dissolved in the minimum amount of triethanolamine. The solution was warmed to 50–60°C and on cooling gave a solution from which filaments could be drawn. When the filaments were sintered at 1600–1700°C, XRD observatiuns showed that a barium hexaaluminate ceramic fibre was formed. Also, a mixture of Aliso B' and barium acetate, having oxide stoichiometry $BaO.6Al_2O_3$ was dissolved in the minimum amount of a mixture of ethylene glycol and acetylacetone (equal parts by volume); the solution was warmed to 50–60°C and on cooling gave a solution from which filaments could be drawn. XRD observations showed that when the filaments were sintered at 1600–1700°C, barium hexaaluminate ceramic fibres were formed. A mixture of 'Aliso B' and barium acetate, having the oxide stoichiometry $BaO.6Al_2O_3$ was dissolved in a mixture of water and acetylacetone, the amount of water being no more than one mole per mole of 'Aliso B', the amount of acetylacetone just sufficient to form a homogeneous solution. The solution formed a gel on standing for several hours. Gel formation was accelerated by warming the solution to 50–60°C.

Alumina grain ('low soda') was either RA107LS (average particle size 0.5 μm; surface area $6.7 m^2 g^{-1}$, Na_2O 0.05 wt%) or RA207LS (average particle size 0.5 μm; surface area 7.0 $m^2 g^{-1}$, Na_2O 0.08 wt%), both supplied by Alcan Chemicals Ltd, Gerrards Cross, Bucks., UK. Sintered components of oxide stoichiometry $BaO.6Al_2O_3$ were prepared from a mixture of alumina grain powder and barium carbonate that had been high-energy milled to break down particle aggregates. In the first instance, procedure A, the milled powder mix was compacted uniaxially at 100 MPa, then cold isostatically pressedat 1723 MPa, followed by sintering for 2h at 1700°C. In the second instance, procedure B, the milled powder was sintered at 1200°C, then re-milled, followed by compaction and sintering as for procedure

A. Modulus of rupture determinations were by four-point bend testing at ambient temperature. Phase developments during rapid or slow heating were followed by XRD observations as for filaments, fibres or compacts prepared from gels of oxide stoichiometry $BaO.6Al_2O_3$. The identity of crystalline phases was confirmed by XRD patterns with the aid of JCPDS International Center for Diffraction Data, Inorganic Materials Alphabetical Index, 1984.

A mixture of barium hydroxide dihydrate and alumina grain with the oxide stoichiometry $BaO.6Al_2O_3$ when sintered and re-milled, gave a powder suitable for compaction and sintering (1750°C for at least 30 min) to prepare simple shapes, (longer for large, more complex shapes).

RESULTS AND DISCUSSION

Route (i)

The results obtained using an ammonium acetate solution to gel a solution of oxide stoichiometry $BaO.6Al_2O_3$, prepared from 'Chlorhydrol' solution and barium chloride, are given in Table 1. The influence of the barium chloride:barium acetate weight ratio on the gel time, for solutions of oxide stoichometrv $BaO.6Al_2O_3$ is shown in Table 2. Table 3 summarises the effect of either 0.1 M HCl solution, 0.1 M $CH_3.COOH$ solution or 0.1 M NH_4OH solution, or dilution, on the gelation behaviour of solutions with oxide stoichiometry $BaO.6Al_2O_3$ and weight ratio barium chloride:barium acetate of 2.8 or 7.3. Details are given in Tables 4–7. In each case, increasing the temperature accelerates gelation. The effect of the barium bromide:barium acetate weight ratio on the gelation of solutions having the oxide stoichiometry $BaO.6Al_2O_3$ is shown in Table 8. All gels were clear. The results given in Tables 1–8 show that barium acetate, rather than ammonium acetate, gives better control of gel and filament formation from precursor solutions having the oxide stoichiometry $BaO.6Al_2O_3$.

XRD studies of gel particles obtained from a solution of oxide stoichiometry $BaO.6Al_2O_3$, prepared from 'Chlorhydrol' solution, dried in air, showed[4] that barium was present as barium chloride species. After heating to 650°C, barium was present as hydrated barium chloride. The FT–IR spectra of gels derived from aluminium chlorhydrate solution ('Chlorhydrol') using ammonium acetate solution or a barium chloride-barium acetate solution, weight ratio 7:3 or 2:8 giving the oxide stoichiometrv $BaO.6Al_2O_3$, closely resemble each other.[4] This together with the XNRD observations,[3,4] shows that barium is not part of the gel network. This is confirmed[4,7] by ^{27}Al NMR spectra of Al–O groups, in the solution at the start of gelation, during gelation, also in the gel formed – oxide stoichiometry $BaO.6Al_2O_3$ with barium acetate as gel-inducing agent. The NMR spectra are similar, suggesting that Al–O groups in the solution probably have a structure similar to that of Al–O groups in hydrous alumina gels. Barium is not incorporated directly in the polymeric gel structure, which comprises only (=Al–O–Al=) units; (–Ba–O–Al=) units were not formed at ambient temperature. The hydrolysis and condensation of aluminium ions has been reviewed by Singhal and Keefer[10] who have also followed these reactions by ^{27}Al NMR observations.

An aluminium alkoxide, such as 'Aliso B' will form a rigid coherent gel when treated with an aminoalcohol and water. When the aminoalcohol is triethanolamine, a solution having the oxide stoichiometry $BaO.6Al_2O_3$ is produced, from which filaments can be drawn. The

Table 1 Preparation and properties of gels derived from a precursor solution of oxide stoichiometry $BaO.6Al_2O_3$ using ammonium acetate as gel inducing agent.

Strength of ammonium acetate solution	Volume of ammonium acetate solution (cm^3)	Quantity of gel pecursor solution	Gel time (min)	Gel characteristics
1g in 1.5 cm^3 H_2O	1.0	5.0 cm^3	2.5	Elastic, gaining strength rapidly
1g in 2.5 cm^3 H_2O	0.5	5.0 cm^3	10	Elastic, good strength development
1g in 2.5 cm^3 H_2O	25.0	250 cm^3	12	Elastic, good strength development
1g in 1.0 cm^3 H_2O	1.0	5.3 cm^3	0.5	Hard gel, transparent

Table 2 Influence of barium chloride:barium aceteate ratio on gel time for solutions with oxide stoichiometry $BaO.6Al_2O_3$

Weight ratio barium chloride:barium acetate	Time to form a clear rigid gel
1:9	23 min
2:8	44 min
3:7	40 min
4:6	27 min
5:5	166 min
7:3	175 min
8:2	24 h
9:1	<2d

Table 3 Effect of additives on gelation of $BaO.6Al_2O_3$ gel precursor solutions

Additive	Effect on gelation behaviour
0.1 M HCl solution	Accelerates gelation
0.1 M CH_3COOH solution	Accelerates gelation
0.1 M NH_4OH solution	Accelerates gelation
Water	Slows down gelation

Table 4 Influence of 0.1 M HCl on gel time of solutions with oxide stoichiometry $BaO.6Al_2O_3$

Weight ratio barium chloride: barium acetate	Molar ratio Al_2O_3/HCl	pH	Gel time	Time available for drawing filaments
2:8	-	4.9	44 min	1 min
2:8	1:0.00025	4.9	5 min	1 min
2:8	1:0.0005	4.9	6 min	1 min
2:8	1:0.001	4.8	9 min	1 min
7:3	-	4.4	175 min	50 min
7:3	1:0.00025	4.3	52 min	20 min
7:3	1:0.0005	4.3	78 min	40 min
7:3	1:0.001	4.3	113 min	60 min

Table 5 Influence of 0.1 M CH_3COOH on gel time of solutions with oxide stoichiometry $BaO.6Al_2O_3$

Weight ratio barium chloride: barium acetate	Molar ratio Al_2O_3/CH3COOH	pH	Gel time	Time available for drawing filaments
2:8	-	4.9	44 min	1 min
2:8	1:0.00025	4.9	10 min	1 min
2:8	1:0.0005	4.9	9 min	2 min
2:8	1:0.001	4.8	4 min	1 min
7:3	-	4.4	175 min	50 min
7:3	1:0.00025	4.35	36 min	25 min
7:3	1:0.0005	4.3	56 min	25 min
7:3	1:0.001	4.3	83 min	40 min

Table 6 Influence of 0.1 M NH_4OH on gel time of solutions with oxide stoichiometry $BaO.6Al_2O_3$

Weight ratio barium chloride: barium acetate	Molar ratio Al_2O_3/NH_4OH	pH	Gel time	Time available for drawing filaments
2:8	-	4.8	45 min	1 min
2:8	1:0.0005	-	20 min	2 min
2:8	1:0.001	-	17 min	no
7:3	-	4.3	175 min	50 min
7:3	1:0.00025	4.3	57 min	15 min
7:3	1:0.0005	4.3	55 min	10 min
7:3	1:0.001	4.3	60 min	25 min

Table 7 Influence of water on gelation of solutions having oxide stoichiometry $BaO.6Al_2O_3$

Weight ratio barium chloride: barium acetate	Molar ratio Barium acetate: water	Gel time	Time available for drawing filaments
2:8	1:21	7 min	1 min
2:8	1:29	15 min	1 min
2:8	1:36	43 min	1 min
2:8	1:43	45 min	3 min
2:8	1:57	185 min	35 min

Table 8 Influence of barium bromide:barium acetate ratio on gelation of solutions having oxide stoichiometry $BaO.6Al_2O_3$

Weight ratio barium bromide: barium acetate	Gel time	Gel properties
9:1	3 to 4 days	Elastic, clear
7:3	10-24 hours	Elastic, clear
6:4	5-8 hours	Rigid, clear
5:5	5 hours	Rigid, clear
4:6	3 hours	Rigid, clear
3:7	45 min	Hard, clear
1:9	15 min	Hard, clear

solvent may also be a mixture of ethylene glycol and acetylacetone. When the solvent is a mixture of water and acetylacetone, the solution formed a gel on standing. The filaments converted to barium hexaaluminate ceramic fibres when sintered. The best filaments were formed under alkaline conditions, using triethanolamine.

Route (ii)

The preparation and properties of sintered compacts having the preferred oxide stoichiometry $BaO.6Al_2O_3$ are given in Table 9. The lower modulus of rupture values obtained for sintered compacts prepared by procedure A could be due to CO_2 evolution from barium carbonate producing microcracking during sintering and inhibiting densification. Alumina grain RA107LS and barium carbonate mixed and sintered according to procedure B gave compacts with an interacting network of small crystals.[4] The observed average particle size (and average flaw size) of ca 1 μm may explain the good strength of ca 600 MPa: if grain growth is suppressed, it is possible that this strength would be maintained at high temperature.[4] Sintered components prepared from the preferred composition $BaO.6Al_2O_3$; using RA107LS alumina and following procedure B are stronger than the alumina–chromia,[11] and the magnesia–alumina–chromia[12] sintered compacts prepared previously, they are also stronger than the sintered compacts prepared[13] from alumina–chromia and mullite grain mixes.

Table 9 Properties of sintered compacts, oxide stoichiometry $BaO.6Al_2O_3$ prepared from a barium carbonate and alumina grain mix. Fabrication and sintering procedure for compacts: high energy milling to mix components, then route (A) compact uniaxially at 100 MPa, then cold isostatically press at 1723 MPa followed by sintering at 1700°C for 2h, or route (B) sinter at 1200°C after mixing then re-milling, followed by compaction and sintering as for (A). Modulus of rupture determined by four-point bend test at ambient temperature.

Alumina grade	Fabrication and sintering procedure	Modulus of rupture (MPa ± 10 MPa)	Density when sintered (kg m^{-3})
RA107LS	A	392	2550
RA207LS	A	276	1760
RA107LS	B	632	3420
RA207LS	B	370	1940

Sintering of filaments and powder compacts

The behaviour of gel filaments (oxide stoichiometry $BaO.6Al_2O_3$), prepared from 'Chlorhydrol' solution, during sintering is given in Refs 3 and 7. No barium–aluminium oxides are formed before the barium species liquify. Tribarium monoaluminate is formed first. This could be formed by reaction between the barium-rich surface phase and the aluminium-rich core phase. As the temperature of sintering is increased, tribarium monoaluminate converts to barium monoaluminate and ultimately with increasing temperature of sintering, barium hexaaluminate is formed. When the filaments are heated rapidly to high temperatures, (ca 1400°C), barium hexaaluminate is formed directly. The grain structure of a barium hexaaluminate ceramic fibre[3,7] resembles the microstructure found[4] in a sintered compact, prepared according to procedure B, route (ii).

The behaviour of compacts prepared from a powder mix of alumina and barium carbonate, oxide stoichiometry $BaO.6Al_2O_3$ when sintered, is given in Table 10. Compacts were also prepared from a precursor gel of oxide stoichiometry $BaO.6Al_2O_3$ obtained from 'Chlorhydrol' solution, after the gel had been air-dried then milled. The behaviour of these compacts during sintering is also given in Table 10. In each case, barium monoaluminate is formed as an intermediate phase during slow sintering to 1600–1700°C. The formation of barium monoaluminate during the slow sintering of mixtures of barium carbonate and alumina (oxide stoichiometry $BaO.6Al_2O_3$) is consistent with accounts[14] of the sintering behaviour of mixes of alkaline earth oxides and alumina. Refractory shapes can also be prepared[1,6] from a refractory grain mix bonded with a barium hexaaluminate precursor gel. Examples of suitable refractory grain mixes include an alumina grain mix (fused and/or tabular), also chromia and alumina–chromia grain mixes, as well as an alumina–chromia solid solution grain mix. After drying to remove volatiles, refractory components with good bonding between grains were obtained by firing to ca 1700°C. The barium monoaluminate intermediate may form the basis of cement or grouting compositions.

Table 10 Formation of barium hexaluminate during sintering

Barium hexaluminate source	Sintering conditions	Observations
Mix of alumina and barium carbonate powders having the oxide stoichiometry $BaO\text{-}6Al_2O_3$	Slow sintering (more than 4h) to 1600-1700° C	Barium monoaluminate formed at 800-1100° C as an intermediate phase, with little densification, converts to barium hexaluminate at 1600-1700° C, with densification
	Fast sintering to 1600-1700° C	Barium hexaluminate formed directly, with desification
Precursor gel, having the oxide stoichiometry $BaO\text{-}6Al_2O_3$	Slow sintering (more than 4h) to 1600-1700° C	Tribarium monoaluminate $3BaO\text{-}6Al_2O_3$ formed intially, then barium monoaluminate and alumina are formed; barium hexaaluminate first obtained at 1200-1300° C, complete conversion to barium hexaaluminate at 1600-1700° C
	Fast sintering to 1600-1700° C	Barium hexaaluminate formed directly

Chromium (III) addition can reduce the temperature required to form an aluminium–barium oxide on firing.[3] When the alumina source is 'Chlorhydrol' solution, the chromitim (Ill) source[1] can be chromium(III) acetate. When the alumina source is an aluminium alkoxide, the chromium (III) source may be an aluminium–chromium double alkoxide; these double alkoxides form a rigid coherent gel when treated[15] with an aminoalcohol and water (alkaline conditions). With a mix of alumina grain and barium carbonate the chromium (III) source may be[1] chromium (III) oxide. The system Al_2O_3–Cr_2O_3 is a simple binary, forming a complete range of solid solutions, which show[16] (except for a narrow composition range) expansion of the alumina lattice with increasing chromium content. This lattice expansion can allow barium to react more readily during firing and probably explains why chromium reduces the temperature of formation of an aluminium–barium oxide and ultimately the temperature of formation of barium hexaaluminate.

CONCLUSIONS

- Barium hexaaluminate may be prepared from precursor solutions containing the re quired amount of barium and aluminium.
- The aluminium source may be an aluminium chlorhydrate solution ('Chlorhydrol') or an aluminium alkoxide ('Aliso B').

- Ammonium acetate and barium acetate both form a gel from precursor solutions prepared from 'Chlorhydrol' and having the oxide stoichiometry $BaO.6Al_2O_3$, with barium acetate giving better control of gel formation; barium is not incorporated directly into the gel structure.
- The barium source may include barium chloride, barium bromide or barium carbonate; barium bromide can be preferable to barium chloride because of the higher solubility in water.
- For 'Aliso B', the barium source may be barium acetate.
- Gel filaments which convert to a ceramic fibre on firing can be obtained from a gel having oxide stoichiometry $BaO.6Al_2O_3$.
- Under suitable conditions, strong (MOR ca 600 MIPa) barium hexaaluminate components can be prepared from an alumina and barium carbonate grain mix; a probable retention of strength at high temperature would indicate a use as an engineering ceramic.
- Compacts prepared from micronised gel powders of oxide stoichiometry $BaO.6Al_2O_3$ obtained from 'Chlorhydrol' solution (route (i)), or from an alumina/barium carbonate mix of oxide stoichiometry $BaO.6Al_2O_3$ (route (ii)), form barium hexaaluminate when sintered, with barium monoaluminate being an intermediate product. Chromium (III) addition can redude the temperature required to form barium hexaaluminate.

ACKNOWLEDGEMENT

Alcan International Limited is thanked for supporting the work described in this paper.

REFERENCES

1. H. G. Emblem and T. J. Davies, PCT Published Patent Application No. 94/29220 (1994) to Alcan International Limited.
2. T. J. Davies, W. A. Al-Douri, M. Biedermann. Q.-G. Chen and H. G. Emblem, *J. Mater. Sci. Lett.*, 1996, **15**, 482–484.
3. W. A. Al-Douri, M. Biedermann, Q.-G. Chen, T. J. Davies and H. G. Emblem, *Eur. J. Solid State Inorg. Chem.*, 1996, **33**, 507–518.
4. W. A. Al-Douri, M. Biedermann, Q.-G. Chen, T. J. Davies and H. G. Emblem, *Brit. Ceram. Proc.*, 1997, **57**, 77–86.
5. G. Maczura, K. Goodboy and J. J. Koenig, *Kirk-Othmer Encyclopedia of Chemical Technology, 3rd Edn.*, John Wiley and Sons, New York, NY, 1978, Vol. 2, 239.
6. T. J. Davies, W. A. Al-Douri, Q.-G. Chen and H. G. Emblem, *J. Mater. Sci. Lett.*, 1997, **6**, 1673–1674.
7. Q.-G. Chen and T. J. Davies, *Brit. Ceram. Trans.*, 1997, **96**, 170–174.
8. M. Biederrnarin, *M.Sc. Thesis*, UMIST, 1993.
9. H. G. Emblem and K. Hargreaves, *Reviews in Inorganic Chemistry*, 1995, **15**, 109–144.
10. A. Singhal and K. D. Keefer, *J. Mater. Res.*, 1994, **8**, 1973.
11. T. J. Davies, H. G. Emblem, C. S. Nwobodo, A. A. Ogwu and V. Tsantzalou, *J. Mater. Sci.*, 1991, **26**, 1061–1068.
12. W. A. Al-Douri, T. J. Davies, A. A. Ogwu and H. G. Emblem, *J. Mater. Sci. Lett.*, 1994, **13**, 1543–1545.
13. H. G. Emblem, T. J. Davies, A. Harabi and G. K. Sargeant, *Silicates Industriels*, 1993, **LVIII**, 17–22.
14. I. Teoreanu and N. Ciocea, *Mater. Constr.* (Bucharest), 1976, **6**(1), 17–19, 26, and *Chem. Abstr.*, 1977, **86**, 126049 s.

15. Zirconal Process Limited, British Patent Application 2, 143, 809A (1985), Inventors: T. J. Davies, H. G. Emblem, K. Jones and P. Parkes.
16. H. G. Emblem and T. J. Davies, *Reviews in Inorganic Chemistry*, 1993, **13**, 103–124.

Author Index

Subject Index

BOOKS ON CERAMICS, GLASSES, CERMETS, BIOMATERIALS AND ELECTRONIC MATERIALS *from* IOM Communications

March 1999

IOM Communications
IOM Communications Ltd is a wholly-owned subsidiary of the Institute of Materials
Reg. Charity No. 1059475

Ceramic Interfaces: Properties and Applications

Edited by R. St C. Smart and J. Nowotny

Rapid development of research activity has been observed in the area of ceramic interfaces. It becomes increasingly important to update progress in this area in an accessible form. That is the purpose of the present volume.

The science of ceramic interfaces is multi-disciplinary overlapping several existing, well established disciplines, such as solid-state chemistry, high temperature chemistry, solid-state electrochemistry, surface science, catalysis and metallurgy. Therefore, scientists representing all these disciplines were invited to contribute to the present volume.

For each chapter, the authors were asked to provide a brief review of their area of research or the particular topic of their chapter. The book is divided into six sections, namely: I Oxide Surfaces: Properties, Structure, Modelling; II Grain Boundaries: Structure, Modelling; III Grain Boundaries: Electrical and Transport Properties; IV Ceramic Forming: Densification, Phase Change; V Metal–Ceramic Interfaces; VI Solution Processing. The leading chapter for each section gives a more developed review of the research area by a leading international expert in the field.

1998 247mm x 174mm 536pp ISBN 1 86125 064 9 (H)
Order Code: B699 European Union: £85.00 All other countries: US$170.00

Growth and Processing of Electronic Materials

Edited by Neil McN. Alford

This volume comprises papers and posters given at the Materials Congress '98 in the workshop 'Growth and Processing of Materials for Electronics'. The area of materials for electronics is strong in the UK. It is far stronger than its profile within the materials community suggests. The impetus for the Growth and Processing session came from the Electronic Applications Divisional Board of The Institute of Materials. The organising committee were Professor Arthur Willoughby, Dr Eric Yeatman, Dr Caroline Millar and Professor Neil Alford. Dr Sue Dunkerton and Professors Colin Humphreys, Peter Goodhew, Roger Whatmore and Nihal Sinnadurie made up the advisory committee. However, the success of the workshop arose from the enthusiasm of the contributors.

The range of topics is both broad and of high quality. The processing includes thin film, thick film (mixed oxide and sol-gel) and bulk mixed oxide methods. Professor Don Pashley opened the session with an invited talk on epitaxy growth mechanisms. Achieving epitaxy in electronics applications is sometimes absolutely essential and this lecture demonstrated the science of the processes.

The workshop covered an enormous range of materials. The materials include relaxor ferroelectrics, piezoelectrics, dielectrics, superconductors and semiconductors. The applications include infra-red detectors, dielectric resonators, micro-actuators and humidity sensors. There were interesting new structures for piezoelectrics from Birmingham University, some beautiful microscopy from Liverpool and Durham, interesting applications of electron energy loss spectroscopy from Cambridge.

Dielectric resonator materials and superconductors were covered by South Bank and by Leeds who have a strong tradition in electroceramics. There were papers on preparation and characterisation of both piezoelectrics and relaxor ferroelectrics. Piezoelectric thin films made by combining liquid metal precursors were described by Imperial College.

Solar cell research was described by both Liverpool and Cranfield. Industrial activity on Cd–

Hg–Te infra-red detectors was described by GEC Marconi Infra-Red Ltd and there was a down-to-earth paper on packaging from TWI.

1998 247mm x 174 mm 144pp ISBN 1 86125 072 X (H)
Order Code: B703 European Union: £45.00 All other countries: US$90.00

Fundamentals of Functionally Graded Materials

Subra Suresh and Andreas Mortensen

Functionally graded materials feature gradual transitions in microstructure and composition which are engineered so as to meet functional performance requirements that vary with location within a single component and to optimise the overall performance of the component. This monograph constitutes a revised, expanded, updated and integrated collection of two review articles by the authors which appeared in *International Materials Reviews*. An emphasis is placed on the fundamentals of processing and thermomechanical response of graded metals and metal/ceramic composites, in an attempt to maintain an adequate depth of coverage of a reasonably broad range of topics including process metallurgy, composite synthesis, the mechanics and micromechanics of composites, and fracture mechanics. The applications of fundamental principles to practical situations are also pointed out wherever appropriate.

Contents: INTRODUCTION: Functionally graded materials: rationale and definition; Historical; Structure and scope of this monograph; PROCESSING: General considerations; Constructive FGM processes; Transport-based FGM processes; THERMOMECHANICAL RESPONSE: Determination of the effective properties of metal–ceramic composites; Thermoelastic deformation of graded multilayers; Onset and progression of plastic flow in graded multilayers; Experimental measurement of stresses and deformation in graded multilayers; Large deformation of graded multilayers; Processing induced stresses in graded composites; Edge effects and singular fields; Fracture and fatigue of graded materials; Engineering surfaces with graded materials to resist indentation and tribological damage; Summary. Index

1998 247mm x 174mm 170pp ISBN 1 86125 063 0 (H)
Order code: B698 European Union £20.00 All other countries US$40.00

Better Ceramics Through Processing

British Ceramic Proceedings Series No. 58

Edited by J. Binner and J. Yeomans

Contents: Development and evaluation of alumina calcination; The preparation of Ce a-sialons and related unstable phases; Crystallisation morphologies of refractory aluminosilicate glass ceramics; Crystallisation of $CaO–P_2O_5–SiO_2–Al_2O_3–TiO_2$ glass-ceramics; Controlled heat treatment – the way to prepare new oxynitride glass ceramics; Crystallisation of calcium phosphate glass ceramics for biomedical applications; The deformation characteristics of clay pastes; Comparison of microstructural evolution in kaolinite powders and dense clay bodies; Characterisation of nano-sized colloidal suspensions for the preparation of multilayer alumina fibre-reinforced mullite CMCs using electrophoretic filtration deposition (EFD); Processing properties of ceramic paste in radial flow; Co-extrusion of multilayered tubes; Solid freeform fabrication methods for engineering ceramics; Developments in microwave processing of ceramics at Staffordshire University; Silicon carbide coated carbon fibre tows by chemical vapour deposition; Hydrothermal synthesis and characterisation of strontium doped lanthanum manganite perovskite powders for use as a cathode material in SOFCs; Low-temperature, aqueous processing of lead zirconate titanate (PZT) ceramics; The effect of processing conditions on the dielectric properties of SBN ceramics; The effect of atmosphere controlled sintering on the piezoelectric shear mode activity of PZT based ceramics; The effect of grain size on dielectric response in the PMNT–PFNW system; Author index; Subject index.

1998 247mm x 174mm 216pp ISBN 1 86125 033 9 (H)
Order Code: B674 European Union: £75.00 All other countries: US$150.00

Advances in the Characterisation of Ceramics

British Ceramic Proceedings Series No. 57

Edited by R. Freer

Covers development in scanning probe microscopies, electron, ion and photon based spectroscopies, structure imaging, analysis and atomic co-ordination, mechanical evaluation of surfaces, characterisation of polishing damage and creep, and measurement of dielectric, piezoelectric and thermal properties. Materials featured include refractories, engineering ceramics, glasses, composites and electronic ceramics.

1997 247 x 174mm 272pp ISBN 1 86125 025 8 (H)
Order Code: B646 European Union: £75.00 All other countries: US$150.00

Whitewares: Production, Testing and Quality Control

W. Ryan and C. Radford

Provides a comprehensive introduction to ceramic whitewares, with particular emphasis on the testing of materials used in whitewares production, the control of manufacturing processes and quality control of the finished product. The text is highly readable and will appeal to anyone with an interest in ceramics.

1997 235 x 150mm 352pp ISBN 1 86125 040 1 (P)
Order Code: CT006 European Union: £27.00 All other countries: US$54.00

Ceramic Oxygen Ion Conductors and Their Technological Applications

British Ceramic Proceedings Series No. 56

Edited by B. C. H. Steele

15 papers ranging from Surfaces and Interface Properties of Oxides Exhibiting Fast Oxygen Transport, through Coulometric Titration of Ceria Solid Solutions and Single and Multiple Grain Boundary Diffusion through Membranes, to Nature of Electronic Defects in Yttria-stabilised Zirconia and their Influence on Oxygen Diffusion.

1996 247 x 174mm 192pp ISBN 0 901716 98 7 (H)
Order Code: B645 European Union: £60.00 All other countries: US$120.00

21st Century Ceramics

British Ceramic Proceedings Series No. 55

Edited by D. P. Thompson and H. Mandal

An excellent summary of current developments in, and the likely evolution of, the following ceramic areas: bioceramics; nanoceramics; novel processes; developments in processing technology; electroceramics; nitrogen ceramics and composite ceramics. Provides a good overview for industrialists considering investment in this area.

1996 247 x 174mm 328pp ISBN 0 901716 85 5 (H)
Order Code: B616 European Union: £75.00 All other countries: US$150.00

Ceramic Films and Coatings

British Ceramic Proceedings Series No. 54

Edited by W. E. Lee

Divided into two parts, this volume covers the uses of ceramic films for improving wear, oxidation and corrosion resistance, including their use for protecting Ti in prosthetics within the human body. The second part examines the electronic and optical properties of ceramic films including surface acoustic wave devices.

1995 247 x 174 mm 272pp ISBN 0 901716 79 0 (H)
Order Code: B612 European Union: £75.00 All other countries: US$150.00

Novel Synthesis and Processing of Ceramics

British Ceramic Proceedings Series No. 53

Edited by F. R. Sale

Proceedings of the Conference held in 1994 as part of the Annual Convention of the Institute's Ceramic Industry Division. Contents include novel production processes using electrochemical routes and aqueous and alkoxide precursors; sol-gel technology; manufacture of sialon membrane filters; and the production of nano-sized materials from aerogels.

1995 247 x 174mm 352pp ISBN 0 901716 70 7 (H)
Order Code: B608 European Union: £75.00 All other countries: US$150.00

Electroceramics: Production, Properties and Microstructures

British Ceramic Proceedings Series No. 52

Edited by W. E. Lee and A. Bell

Proceedings of the Symposium held as part of the Condensed Matter and Materials Conference in 1993. Contents include: ferroelectric thin films for integrated device applications; effect of thermal processing conditions on the structure and properties of sol-gel derived PZT tine layers; citrate gel-route processing of ZnO varistors; development and evaluation of oxide cathodes for ceramic fuel cell operation at intermediate temperatures; monitoring the integrity of MOS gate oxides.

1995 257x175mm 342pp ISBN 0 901716 42 1 (H)
Order Code: B565 European Union: £75.00 All other countries: US$150.00

Nanoceramics

British Ceramic Proceedings Series No. 51

Edited by R. Freer

Proceedings of a conference held in Cambridge, UK in April 1992. Papers cover the synthesis, properties and fabrication of nanoceramics.

1993 210x148mm 224pp ISBN 0 901716 41 3 (H)

Order Code: B564 European Union: £57.50 All other countries: US$115.00

Engineering Ceramics: Fabrication Science and Technology

British Ceramic Proceedings Series No. 50

Edited by D. P. Thompson

This 50th volume in the British Ceramics Proceedings series includes papers presented at a meeting of the Basic Science Section held in Manchester, UK in December 1991.

1993 215x152mm 256pp ISBN 0 901716 40 5 (H)

Order Code: B563 European Union: £57.50 All other countries: US$115.00

Chemical Bonding in Transition Metal Carbides

A. H. Cottrell

This book presents a mainly qualitative understanding of this family of materials which ranges from simple NaCl type compounds to complex chromium carbides. Contents: close packed sphere models; the affinity of transition metals for carbon; bonds; the cohesive energy; titanium carbide; the early transition metals; tungsten and molybdenum; chromium and the later transition metals.

1995 216x138mm 128pp ISBN 0 901716 68 5 (H)

Order Code: B613 European Union: £18.00 All other countries: US$36.00

Dictionary of Ceramics (Third Edition)

A. Dodd and D. Murfin

This updated edition of Dr A E Dodd's classic ceramics dictionary contains over 2000 new terms. The new edition includes terminology covering new developments in engineering ceramics, electroceramics, whiteware processes, environmental legislation. The coverage of glass, vitreous enamel and the cement industries has been widened and relevant areas of basic science i.e. crystal structure, fracture mechanics and sintering, included. Longer entries have been introduced, especially for the advanced ceramics section.

1994 246x172mm 384pp ISBN 0 901716 56 1 (H)

Order Code: B588 European Union: £45.00 All other countries: US$90.00

Health and Safety in Ceramics (Third Edition)

Edited by R. C. P. Cubbon

Identifies the numerous potential hazards encountered in the commercial, educational and craft production of ceramic-ware, and provides authoritative and uncomplicated guidance on safe practice.

1991 210x148mm 48pp ISBN 0 901092 42 8 (P)
Order Code CT004 European Union: £6.00 All other countries: US$12.00

Fundamental Principles of Sol-Gel Technology

R. W. Jones

Covers categories of colloids, sol-gel transition, formation of sols, sol-gel transition in alkoxides, gel to glass/ceramic transition, applications and recipes.

1989 209x148mm 128pp ISBN 0 901462 69 1
Order Code B475 European Union: £10.00 All other countries: US$20.00

Glasses and Their Applications

H. Rawson

A wide-ranging survey of the applications of glasses from containers and glazing to developing uses such as fibre optic communications.

1991 297 x 210mm 256pp ISBN 0 901462 89 6 (H)
Order code B499 European Union: £10.00 All other countries: US$20.00

Functional Materials in New Millennium Systems

The Cyril Hilsum Symposium

Edited by M. J. Kelly

Contents: The Versatility of III–V Compounds: What Hopes and Expectations for the Future; The Exploitation of III–V Technology in Electronic Systems; New Applications of Narrow Gap Semiconductor Materials; Liquid Crystal Materials – The Beginning and Where We Are Today; The Impact of Materials on LCD Technologies; Design of Anisotropically Conducting Adhesives with Uniform Conductor Arrays; Light Emitting Materials; Bulk and Thick Film High-Temperature Superconductors; Field-Emitter Array Developments for Vacuum Microelectronics; Sensor Materials for an Electronic Nose; The Linking Thread.

1997 247mm x 174mm 214pp ISBN 1 86125 014 2 (H)
Order code: B664 European Union: £45.00 All other countries: US$90.00

Overseas Mission on Biomaterials to Japan

Edited by W. Bonfield

This is the report of a DTI-supported a UK overseas mission to Japan between 26 January and 7 February 1998 sponsored by the Institute of Materials. Essential reading for all in the biomaterials industry and research.

1998 297mm x 210mm 88pp ISBN 1 86125 67 3
Order code: B702 European Union: £35.00 All other countries: US$70.00

ORDER FORM

Qty	Order code	Title	Price	Subtotal
	B699	Ceramic Interfaces: Properties & Applications	£85/$70	
	B703	Growth & Processing of Electronic Materials	£45/$90	
	B698	Fundamentals of Functionally Graded Materials	£20/$40	
	B674	Better Ceramics Through Processing	£75/$150	
	B646	Advances in the Characterisation of Ceramics	£75/$150	
	CT006	Whitewares	£27/$54	
	B645	Ceramic Oxygen Ion Conductors	£60/$120	
	B616	21st Century Ceramics	£75/$150	
	B612	Ceramic Films and Coatings	£75/$150	
	B608	Novel Synthesis and Processing of Ceramics	£75/$150	
	B565	Electroceramics: Production Properties etc.	£75/$150	
	B564	Nanoceramics	£57.50/$115	
	B563	Engineering Ceramics	£57.50/$115	
	B613	Chemical Bonding in Transition Met. Carbides	£18/$36	
	B588	Dictionary of Ceramics	£45/$90	
	CT004	Health and Safety in Ceramics	£6/$12	
	B475	Fundamental Principles of Sol-Gel Tech.	£10/$20	
	B499	Glasses and Their Applications	£10/$20	
	B664	Functional Materials in New Millennium Sys	£45/$90	
	B702	Overseas Mission on Biomaterials to Japan	£35/$70	
		Post & Packing European Union: £5 All other countries: $10		
			TOTAL	

NB: Members of the Institute of Materials are entitled to a 20% discount

I enclose a cheque/ Please debit my credit card for the above amount

Type of credit Card: ❑ Visa ❑ MasterCard ❑ American Express

Credit card number ______________________________

Name ______________________________

Address ______________________________

____

Signature ______________________________ Date ______________

Send orders to: IOM Communications, Shelton House, Stoke Road, Shelton, Stoke-on-Trent, ST4 2DR, UK
Tel: +44 (0) 1782 202 116 Fax: +44 (0) 1782 202 421
Email: Orders@materials.org.uk VAT Reg. No. GB 649 1646 11